Solutions Manual to Accompany Introduction to Abstract Algebra

Fourth Edition

W. Keith Nicholson
University of Calgary
Calgary, Alberta, Canada

WILEY

A JOHN WILEY & SONS, INC., PUBLICATION

Published by John Wiley & Sons, Inc., Hoboken, New Jersey
Published simultaneously in Canada

For general information on our other products and services or for technical support, please contact our Customer Care Department within the United States at (800) 762-2974, outside the United States at (317) 572-3993 or fax (317) 572-4002.

Wiley also publishes its books in a variety of electronic formats. Some content that appears in print may not be available in electronic formats. For more information about Wiley products, visit our web site at www.wiley.com.

Library of Congress Cataloging-in-Publication Data:

Nicholson, W. Keith.
    Introduction to abstract algebra / W. Keith Nicholson. – 4th ed.
        p. cm.
    Includes bibliographical references and index.
    ISBN 978-1-118-28815-3
  1.   Algebra, Abstract.   I. Title.
    QA162.N53 2012
    512'.02–dc23   2011031416

10 9 8 7 6 5 4 3 2 1

Contents

Chapter 0

Preliminaries

0.1 PROOFS

1. (a) (1) If $n = 2k$, k an integer, then $n^2 = (2k)^2 = 4k^2$ is a multiple of 4.

   (2) The converse is true: If n^2 is a multiple of 4 then n must be even because n^2 is odd when n is odd (Example 1).

   (c) (1) Verify: $2^3 - 6 \cdot 2^2 + 11 \cdot 2 - 6 = 0$ and $3^3 - 6 \cdot 3^2 + 11 \cdot 3 - 6 = 0$.

   (2) The converse is false: $x = 1$ is a counterexample. because
   $$1^3 - 6\,1^2 + 11 \cdot 1 - 6 \neq 0.$$

2. (a) Either $n = 2k$ or $n = 2k + 1$, for some integer k. In the first case $n^2 = 4k^2$; in the second $n^2 = 4(k^2 + k) + 1$.

   (c) If $n = 3k$, then $n^3 - n = 3(9k^3 - k)$; if $n = 3k + 1$, then
   $$n^3 - n = 3(9k^3 + 9k^2 + 2k);$$
   if $n = 3k + 2$, then $n^3 - n = 3(9k^3 + 18k^2 + 11k + 2)$.

3. (a) (1) If n is not odd, then $n = 2k$, k an integer, $k \geq 1$, so n is not a prime.

   (2) The converse is false: $n = 9$ is a counterexample; it is odd but is not a prime.

   (c) (1) If $\sqrt{a} > \sqrt{b}$ then $(\sqrt{a})^2 > (\sqrt{b})^2$, that is $a > b$, contrary to the assumption.

   (2) The converse is true: If $\sqrt{a} \leq \sqrt{b}$ then $(\sqrt{a})^2 \leq (\sqrt{b})^2$, that is $a \leq b$.

4. (a) If $x > 0$ and $y > 0$ assume $\sqrt{x + y} = \sqrt{x} + \sqrt{y}$. Squaring gives $x + y = x + 2\sqrt{xy} + y$, whence $2\sqrt{xy} = 0$. This means $xy = 0$ so $x = 0$ or $y = 0$, contradicting our assumption.

Student Solution Manual to Accompany Introduction to Abstract Algebra, Fourth Edition.
W. Keith Nicholson.
© 2012 John Wiley & Sons, Inc. Published 2012 by John Wiley & Sons, Inc.

(c) Assume all have birthdays in different months. Then there can be at most 12 people, one for each month, contrary to hypothesis.

5. (a) $n = 11$ is a counterexample because then $n^2 + n + 11 = 11 \cdot 13$ is not prime. Note that $n^2 + n + 11$ *is* prime if $1 \leq n \leq 9$ as is readily verified, but $n = 10$ is also a counterexample as $10^2 + 10 + 11 = 11^2$.

(c) $n = 6$ is a counterexample because there are then 31 regions. Note that the result holds if $2 \leq n \leq 5$.

0.2 SETS

1. (a) $A = \{x \mid x = 5k, \ k \in \mathbb{Z}, \ k \geq 1\}$

2. (a) $\{1, 3, 5, 7, \ldots\}$

(c) $\{-1, 1, 3\}$

(e) $\{ \ \} = \varnothing$ is the empty set by Example 3.

3. (a) Not equal: $-1 \in A$ but $-1 \notin B$.

(c) Equal to $\{a, l, o, y\}$.

(e) Not equal: $0 \in A$ but $0 \notin B$.

(g) Equal to $\{-1, 0, 1\}$.

4. (a) $\varnothing$, $\{2\}$

(c) $\{1\}$, $\{3\}$, $\{1, 2\}$, $\{1, 3\}$, $\{2, 3\}$, $\{1, 2, 3\}$

5. (a) True. $B \subseteq C$ means each element of B (in particular A) is an element of C.

(c) False. For example, $A = \{1\}$, $B = C = \{\{1\}, 2\}$.

6. (a) Clearly $A \cap B \subseteq A$ and $A \cap B \subseteq B$; If $X \subseteq A$ and $X \subseteq B$, then $x \in X$ implies $x \in A$ and $x \in B$, that is $x \in A \cap B$. Thus $X \subseteq A \cap B$.

7. If $x \in A \cup (B_1 \cap B_2 \cap \ldots \cap B_n)$, then $x \in A$ or $x \in B_i$ for all i. Thus $x \in A \cup B_i$ for all i, that is $x \in (A \cup B_1) \cap (A \cup B_2) \cap \ldots \cap (A \cup B_n)$. Thus

$$A \cup (B_1 \cap B_2 \cap \ldots \cap B_n) \subseteq (A \cup B_1) \cap (A \cup B_2) \cap \ldots \cap (A \cup B_n),$$

and the reverse argument proves equality. The other formula is proved similarly.

9. $A = \{1, 2\}$, $B = \{1, 3\}$, $C = \{2, 3\}$.

10. (a) Let $A \times B = B \times A$, and fix $a \in A$ and $b \in B$ (since these sets are nonempty). If $x \in A$, then $(x, b) \in A \times B = B \times A$. This implies $x \in B$; so $A \subseteq B$. Similarly $B \subseteq A$.

(c) If $x \in A \cap B$, then $x \in A$ and $x \in B$, so $(x, x) \in A \times B$. If $(x, x) \in A \times B$, then $x \in A$ and $x \in B$, so $x \in A \cap B$.

11. (a) $(x, y) \in A \times (B \cap C)$

if and only if $x \in A$ and $y \in (B \cap C)$

if and only if $(x, y) \in A \times B$ and $(x, y) \in A \times C$

if and only if $(x, y) \in (A \times B) \cap (A \times C)$.

(c) $(x, y) \in (A \cap B) \times (A' \cap B')$
  if and only if $x \in A \cap B$ and $y \in A' \cap B'$
  if and only if $(x, y) \in A \times A'$ and $(x, y) \in (B \times B')$
  if and only if $(x, y) \in (A \times A') \cap (B \times B')$.

0.3 MAPPINGS

1. (a) Not a mapping: $\alpha(1) = -1$ is not in $\mathbb{N}$.

   (c) Not a mapping: $\alpha(-1) = \sqrt{-1}$ is not in $\mathbb{R}$.

   (e) Not a mapping: $\alpha(6) = \alpha(2 \cdot 3) = (2, 3)$ and $\alpha(6) = \alpha(1 \cdot 6) = (1, 6)$.

   (g) Not a mapping: $\alpha(2)$ is not defined.

2. (a) Bijective. $\alpha(x) = \alpha(x_1)$ implies $3 - 4x = 3 - 4x_1$, so $x = x_1$, and α is one-to-one. Given $y \in \mathbb{R}$, $y = \alpha\left[\frac{1}{4}(3 - y)\right]$, so α is onto.

   (c) Onto: If $m \in N$, then $m = \alpha(2m - 1) = \alpha(2m)$. Not one-to-one: In fact we have $\alpha(1) = 1 = \alpha(2)$.

   (e) One-to-one: $\alpha(x) = \alpha(x_1)$ implies $(x + 1, x - 1) = (x_1 + 1, x_1 - 1)$, whence $x = x_1$. Not onto: $(0, 0) \neq \alpha(x)$ for any x because $(0, 0) = (x + 1, x - 1)$ would give $x = 1$ and $x = -1$.

   (g) One-to-one: $\alpha(a) = \alpha(a_1)$ implies $(a, b_0) = (a_1, b_0)$ implies $a = a_1$. Not onto if $|B| \geq 2$ since no element (a, b) is in $\alpha(A)$ for $b \neq b_0$.

3. (a) Given $c \in C$, let $c = \beta\alpha(a)$ with $a \in A$ (because $\beta\alpha$ is onto). Hence $c = \beta(\alpha(a))$, where $\alpha(a) \in B$, so β is onto.

   (c) Let $\beta(b) = \beta(b_1)$. Write $b = \alpha(a)$ and $b_1 = \alpha(a_1)$ (since α is onto). Then
   $$\beta\alpha(a) = \beta(\alpha(a)) = \beta(b) = \beta(b_1) = \beta(\alpha(a_1)) = \beta\alpha(a_1),$$
   so $a = a_1$ (because $\beta\alpha$ is one-to-one), and hence $b = b_1$ as required.

   (e) Let $b \in B$. As α is onto, let $b = \alpha(a)$, $a \in A$. Hence
   $$\beta(b) = \beta(\alpha(a)) = \beta\alpha(a) = \beta_1\alpha(a) = \beta_1(\alpha(a)) = \beta_1(b).$$
   Since $b \in B$ was arbitrary, this shows that $\beta = \beta_1$.

5. (a) If $\alpha^2 = \alpha$, let $x \in \alpha(A)$, say $x = \alpha(a)$. Then $\alpha(x) = \alpha^2(a) = \alpha(a) = x$. Conversely, let $\alpha(x) = x$ for all $x \in \alpha(A)$. If $a \in A$, write $\alpha(a) = x$. Then $\alpha^2(a) = \alpha(\alpha(a)) = \alpha(x) = x = \alpha(a)$, so $\alpha^2 = \alpha$.

   (c) $\alpha^2 = (\beta\gamma)(\beta\gamma) = \beta(\gamma\beta)\gamma = \beta(1_A)\gamma = \beta\gamma = \alpha$.

7. (a) If $y \in \mathbb{R}$, write $\alpha^{-1}(y) = x$. Hence $y = \alpha(x)$, that is $y = ax + b$. Solving for x gives $\alpha^{-1}(y) = x = \frac{1}{a}(y - b)$. As this is possible for all $y \in \mathbb{R}$, this shows that $\alpha^{-1}(y) = \frac{1}{a}(y - b)$ for all $y \in \mathbb{R}$.

   (c) First verify that $\alpha^2 = 1_\mathbb{N}$, that is $\alpha\alpha = 1_\mathbb{N}$. Hence $\alpha^{-1} = \alpha$ by the definition of the inverse of a function.

9. Let $\beta\alpha = 1_A$. Then α is one-to-one because $\alpha(a) = \alpha(a_1)$ implies that $a = \beta\alpha(a) = \beta\alpha(a_1) = a_1$; and β is onto because if $a \in A$ then $a = \beta\alpha(a) = \beta(\alpha(a))$ and $\alpha(a) \in B$. Hence both are bijections as $|A| = |B|$ (Theorem 2), and hence α^{-1} and β^{-1} exist. But then $\beta^{-1} = \beta^{-1}1_A = \beta^{-1}(\beta\alpha) = \alpha$. Similarly $\alpha^{-1} = \beta$.

11. Let $\varphi(\alpha) = \varphi(\alpha_1)$ where α and α_1 are in M. Then $(\alpha(1), \alpha(2)) = (\alpha_1(1), \alpha_1(2))$, so $\alpha(1) = \alpha_1(1)$ and $\alpha(2) = \alpha_1(2)$. Thus $\alpha = \alpha_1$ (by Theorem 1), so φ is one-to-one. Conversely, let $(x, y) \in B \times B$, and define $\alpha_2 : \{1, 2\} \to B$ by $\alpha_2(1) = x$ and $\alpha_2(2) = y$. Then $\alpha_2 \in$ M, and $\varphi(\beta) = (\alpha_2(1), \alpha_2(2)) = (x, y)$. Thus φ is onto. Then $\varphi^{-1} : B \times B \to$ M has action $\varphi^{-1}(x, y) = \alpha_2$ where $\alpha_2(1) = x$ and $\alpha_2(2) = y$.

13. For each $a \in A$ there are m choices for $\alpha(a) \in B$. Since $|A| = n$, there are m^n choices in all, and they all lead to different functions α because α is determined by these choices.

15. (a) $\Rightarrow$ (b)   Given $b \in B$, write $A_b = \{a \in A \mid \alpha(a) = b\}$. Then $A_b \neq \emptyset$ for each b (α is onto), so choose $a_b \in A_b$ for each $b \in B$. Then define $\beta : B \to A$ by $\beta(b) = a_b$. Then $\alpha\beta(a) = \alpha(\beta(b)) = \alpha(a_b) = b$ for each b; that is $\alpha\beta = 1_B$.

    (c) $\Rightarrow$ (a)   If $b_0 \in B - \alpha(A)$, we deduce a contradiction. Choose $a_0 \in A$, and define $\beta : B \to B$ by:

    $$\beta(b) = \begin{cases} b & \text{if } b \neq b_0 \\ \alpha(a_0) & \text{if } b = b_0. \end{cases}$$

    Then $\alpha(a) \neq b_0$ for all $a \in A$, so

    $$\beta\alpha(a) = \beta(\alpha(a)) = \alpha(a) = 1_B(\alpha(a)) = 1_B\alpha(a)$$

    for all $a \in A$. Hence, $\beta\alpha = 1_B\alpha$, so $\beta = 1_B$ by (c). Finally then $b_0 = \beta(b_0) = \alpha(a_0)$, a contradiction.

0.4   EQUIVALENCES

1. (a) It is an equivalence by Example 4.
    $$[-1] = [0] = [1] = \{-1, 0, 1\}, \quad [2] = \{2\}, \quad [-2] = \{-2\}.$$
    (c) Not an equivalence. $x \equiv x$ only if $x = 1$, so the reflexive property fails.

    (e) Not an equivalence. $1 \equiv 2$ but $2 \not\equiv 1$, so the symmetric property fails.

    (g) Not an equivalence. $x \equiv x$ is *never* true. Note that the transitive property also fails.

    (i) It is an equivalence by Example 4. $[(a, b)] = \{(x, y) \mid y - 3x = b - 3a\}$ is the line with slope 3 through (a, b).

2. In every case $(a, b) \equiv (a_1, b_1)$ if $\alpha(a, b) = \alpha(a_1, b_1)$ for an appropriate function $\alpha : A \to \mathbb{R}$. Hence $\equiv$ is the kernel equivalence of α.

    (a) The classes are indexed by the possible sums of elements of U.

    Sum is 2: $[(1, 1)] = \{(1, 1)\}$
    Sum is 3: $[(1, 2)] = [(2, 1)] = \{(1, 2), (2, 1)\}$
    Sum is 4: $[(1, 3)] = [(2, 2)] = [(3, 1)] = \{(1, 3), (2, 2), (3, 1)\}$
    Sum is 5: $[(2, 3)] = [(3, 2)] = \{(2, 3), (3, 2)\}$
    Sum is 6: $[(3, 3)] = \{(3, 3)\}$.

(c) The classes are indexed by the first components.

First component is 1: $[(1,1)] = [(1,2)] = [(1,3)] = \{(1,1),(1,2),(1,3)\}$

First component is 2: $[(2,1)] = [(2,2)] = [(3,2)] = \{(2,1),(2,2),(2,3)\}$

First component is 3: $[(3,1)] = [(3,2)] = [(3,3)] = \{(3,1),(3,2),(3,3)\}$.

3. (a) It is the kernel equivalence of $\alpha : \mathbb{Z} \to \mathbb{Z}$ where $\alpha(n) = n^2$. Here $[n] = \{-n, n\}$ for each n. Define $\sigma : \mathbb{Z}_\equiv \to B$ by $\sigma[n] = |n|$, where $|n|$ is the absolute value. Then $[m] = [n] \Leftrightarrow m \equiv n \Leftrightarrow |m| = |n|$. Thus σ is well-defined and one-to-one. It is clearly onto.

(c) It is the kernel equivalence of $\alpha : \mathbb{R} \to \mathbb{R}$ where $\alpha(x, y) = y$. Define $\sigma : (\mathbb{R} \times \mathbb{R})_\equiv \to B$ by $\sigma[(x, y)] = y$. Then

$$[(x,y)] = [(x_1,y_1)] \Leftrightarrow (x,y) \equiv (x_1,y_1) \Leftrightarrow y = y_1,$$

so σ is well-defined and one-to-one. It is clearly onto.

(e) Reflexive: $x \equiv x \in \mathbb{Z}$;

Symmetric: $x \equiv y \Rightarrow x - y \in \mathbb{Z} \Rightarrow y - x \in \mathbb{Z} \Rightarrow y \equiv x$;

Transitive: $x \equiv y$ and $y \equiv z$ gives $x - y \in \mathbb{Z}$ and $y - z \in \mathbb{Z}$. Hence

$$x - z = (x - y) + (y - z) \in \mathbb{Z}, \text{ that is } x \equiv z.$$

Now define $\sigma : \mathbb{R}_\equiv \to B$ by $\sigma[x] = x - \lfloor x \rfloor$ where $\lfloor x \rfloor$ denotes the greatest integer $\leq x$. Then $[x] = [y] \Rightarrow x \equiv y \Rightarrow x - y = n,\ n \in \mathbb{Z}$. Thus $x = y + n$, so $\lfloor x \rfloor = \lfloor y \rfloor + n$. Hence,

$$x - \lfloor x \rfloor = (y + n) - (\lfloor y \rfloor + n) = y - \lfloor y \rfloor,$$

and σ is well-defined. To see that σ is one-to-one, let $\sigma[x] = \sigma[y]$, that is $x - \lfloor x \rfloor = y - \lfloor y \rfloor$. Then $x - y = \lfloor y \rfloor - \lfloor x \rfloor \in \mathbb{Z}$, so $x \equiv y$, that is $\lfloor x \rfloor = \lfloor y \rfloor$. Finally, σ is onto because, if $0 \leq x < 1$, $\lfloor x \rfloor = 0$, so $x = \sigma[x]$.

5. (a) If $a \in A$, then $a \in C_i$ and $a \in D_j$ for some i and j, so $a \in C_i \cap D_j$. If $C_i \cap D_j \neq C_{i'} \cap D_{j'}$, then either $i \neq i'$ or $j \neq j'$. Thus

$$(C_i \cap D_j) \cap (C_{i'} \cap D_{j'}) = \emptyset$$

in either case.

7. (a) Not well defined: $\alpha(2) = \alpha\left(\frac{2}{1}\right) = 2$ and $\alpha(2) = \alpha\left(\frac{4}{2}\right) = 4$.

(c) Not well defined: $\alpha\left(\frac{1}{2}\right) = 3$ and $\alpha\left(\frac{1}{2}\right) = \alpha\left(\frac{2}{4}\right) = 6$.

9. (a) $[a] = [a_1] \Leftrightarrow a \equiv a_1 \Leftrightarrow \alpha(a) = \alpha(a_1)$. The implication $\Rightarrow$ proves σ is well defined; the implication $\Leftarrow$ shows it is one-to-one. If α is onto, so is σ.

(c) If we regard $\sigma : A_\equiv \to a(A)$, then σ is a bijection.

Chapter 1

Integers and Permutations

1.1 INDUCTION

1. In each case we give the equation that makes p_k imply p_{k+1}.

   (a) $k(2k-1) + (4k+1) = 2k^2 + 3k + 1 = (k+1)(2k+1)$

   (c) $\frac{1}{4}k^2(k+1)^2 + (k+1)^3 = \frac{1}{4}(k+1)^2(k^2 + 4k + 4) = \frac{1}{4}(k+1)^2(k+2)^2$

   (e) $\frac{1}{12}k(k+1)(k+2)(3k+5) + (k+1)(k+2)^2$
   $= \frac{1}{12}(k+1)(k+2)(3k^2 + 17k + 24) = \frac{1}{12}(k+1)(k+2)(k+3)(3k+8)$

   (g) $\frac{k}{3}(4k^2 - 1) + (2k+1)^2 = \frac{k}{3}(2k-1)(2k+1) + (2k+1)^2$
   $= \frac{1}{3}(2k+1)[2k^2 + 5k + 3] = \frac{1}{3}(2k+1)(k+1)(2k+3)$
   $= \frac{1}{3}(k+1)[4(k+1)^2 - 1]$

   (i) $1 - \frac{1}{(k+1)!} + \frac{k+1}{(k+2)!} = 1 - \frac{1}{(k+2)!}[(k+2) - (k+1)] = 1 - \frac{1}{(k+2)!}$

2. In each case we give the inequality that makes p_k imply p_{k+1}.

   (a) $2^{k+1} = 2 \cdot 2^k > 2 \cdot k \geq k + 1$.

   (c) If $k! \leq 2^{k^2}$, then $(k+1)! = (k+1)k! \leq (k+1)2^{k^2} \leq 2^{(k+1)^2}$ provided $k + 1 \leq 2^{2k+1}$. This latter inequality follows, again by induction on $k \geq 1$, because $2^{2k+3} = 4 \cdot 2^{2k+1} \geq 4(k+1) \geq k + 2$.

   (e) $\frac{1}{\sqrt{1}} + \cdots + \frac{1}{\sqrt{k}} + \frac{1}{\sqrt{k+1}} \geq \sqrt{k} + \frac{1}{\sqrt{k+1}} = \frac{\sqrt{k^2+k}+1}{\sqrt{k+1}} \geq \frac{k+1}{\sqrt{k+1}} = \sqrt{k+1}$.

3. In each case we give the calculation that makes p_k imply p_{k+1}.

   (a) If $k^3 + (k+1)^3 + (k+2)^3 = 9m$, then
   $(k+1)^3 + (k+2)^3 + (k+3)^3 = 9m - k^3 + (k+3)^3 = 9m + 9k^2 + 27k + 27$.

   (c) If $3^{2k+1} + 2^{k+2} = 7m$, then
   $$3^{2k+3} + 2^{k+3} = 9(7m - 2^{k+2}) + 2^{k+3} = 9 \cdot 7m - 2^{k+2}(9 - 2).$$

Student Solution Manual to Accompany Introduction to Abstract Algebra, Fourth Edition.
W. Keith Nicholson.
© 2012 John Wiley & Sons, Inc. Published 2012 by John Wiley & Sons, Inc.

5. If $3^{3k} + 1 = 7m$ where k is odd, then passing to $k + 2$,
$$3^{3(k+2)} + 1 = 3^6(7m - 1) + 1 = 3^6 \cdot 7m - (3^6 - 1)$$
$$= 3^6 \cdot 7m - 728 = 7(3^6 \cdot m - 104).$$

7. It is clear if $n = 1$. In general, such a $(k+1)$ digit number must end in 4, 5 or 6, and there are 3^k of each by induction. We are done since $3 \cdot 3^k = 3^{k+1}$.

9. It is clear if $n = 1$. Given $k + 1$ secants, remove one and color the result unambiguously by induction. Now reinsert the removed secant. On one side of this secant, leave all regions the original color (including the new regions of that side created by the new secant). On the other side, interchange colors everywhere (including those regions newly created). This is an unambiguous coloring.

10. (a) If $k \geq 2$ cents can be made up, there must be a 2-cent or a 3-cent stamp. In the first case, replace a 2-cent stamp by a 3-cent stamp; in the second case, replace a 3-cent stamp by two 2-cent stamps.

    (c) If $k \geq 18$ can be made up, either one 7-cent stamp is used (replace with two 4-cent stamps) or five 4-cent stamps are used (replace with three 7-cent stamps).

11. $a_0 = 0$, $a_1 = 7$, $a_2 = 63 = 7.9$, $a_3 = 511 = 7 \cdot 73$. The conjecture is that $2^{3n} - 1$ is a multiple of 7 for all $n \geq 0$. If $2^{3k} - 1 = 7x$ for some $n \geq 0$, then we have $2^{3(k+1)} - 1 = 2^3(7x + 1) - 1 = 7(2^3 + 1)$.

12. (a) If S_n is the statement "$1^3 + 2^3 + 3^3 + \cdots + n^3$ is a perfect square", then S_1 is true. If $k \geq 1$, assume that $1^3 + 2^3 + \cdots + k^3 = x^2$ for some integer x. Then $1^3 + 2^3 + \cdots + (k+1)^3 = x^2 + (k+1)^3$ and it is not clear how to deduce that this is a perfect square without some knowledge about how x is dependent upon k. Thus induction fails for S_n. However, if we strengthen the statement to $1^3 + 2^3 + \cdots + n^3 = \left[\frac{1}{2}n(n+1)\right]^2$, induction *does* go through (see Exercise 1(c)). The reason is that now the inductive hypothesis brings more information to the inductive step and so allows the (stronger) conclusion to be deduced.

13. $\binom{n}{r-1} + \binom{n}{r} = \frac{n!}{(r-1)!(n-r+1)!} + \frac{n!}{r!(n-r)!} = \frac{n!}{r!(n+1-r)!}[r + (n+1-r)] = \binom{n+1}{r}$.

14. (a) $\binom{n}{0} + \binom{n}{1} + \cdots + \binom{n}{n} = (1+1)^n = 2^n$ by the binomial theorem (Example 6 with $x = 1$).

15. We use the well-ordering principle to prove the principle of induction. Let $p_1, p_2, p_3, \cdots$ be statements such that p_1 is true and $p_k \Rightarrow p_{k+1}$ for every $k \geq 1$. We must show that p_n is true for every $n \geq 1$. To this end consider the set $X = \{n \geq 1 \mid p_n \text{ is false}\}$; we must show that X is empty. But if X is nonempty it has a smallest member m by the well-ordering principle. Hence $m \neq 1$ (because p_1 is true), so $m - 1$ is a positive integer. But then p_{m-1} is true (because m is the *smallest* member of X) and so p_m is true (because $p_{m-1} \Rightarrow p_m$). This contradiction shows that X must be empty, as required.

17. If p_n is "n has a prime factor", then p_2 is true. Assume $p_2, \ldots, p_k$ are all true. If $k + 1$ is a prime, we are done. If $k + 1 = ab$ write $2 \leq a \leq k$ and $2 \leq b \leq k$, then a (and b) has a prime factor by strong induction. Thus $k + 1$ has a prime factor.

18. (a) $a_n = 2(-1)^n$ $a_{n+1} = -a_n = -2(-1)^n = 2(-1)^{n+1}$

    (c) $a_n = \frac{1}{2}[1 + (-1)^n]$
    $a_{n+1} = 1 - a_n = 1 - \frac{1}{2}[1 + (-1)^n] = \frac{1}{2}[2 - 1 - (-1)^n] = \frac{1}{2}[1 + (-1)^{n+1}]$

19. Given n lines, another line intersects all existing lines (because no two are parallel) at new intersection points (none of these are concurrent) and so enters $n + 1$ regions. Hence it creates $n + 1$ new regions; so $a_{n+1} = a_n + (n + 1)$. Then $a_0 = 1$, $a_1 = 1 + 1$, $a_2 = 1 + 1 + 2$, $a_3 = 1 + 1 + 2 + 3$; and this suggests $a_n = 1 + (1 + 2 + \cdots + n)$. Hence Gauss' formula (Example 1) gives

$$a_n = 1 + \tfrac{1}{2}n(n + 1) = \tfrac{1}{2}(n^2 + n + 2).$$

This is valid for $n = 0$; if it holds for $n = k \geq 1$ then

$$a_{k+1} = a_k + (k + 1) = \tfrac{1}{2}[(k^2 + k + 2) + 2(k + 1)] = \tfrac{1}{2}[(k + 1)^2 + (k + 1) + 2].$$

21. (a) Let p_n denote the statement $a_n = (-1)^n$. Then p_0 and p_1 are true by hypothesis. If p_k and p_{k+1} are true for some $k \geq 0$, then $a_k = (-1)^k$, $a_{k+1} = (-1)^{k+1}$ and so

$$a_{k+2} = a_{k+1} + 2a_k = (-1)^{k+1} + 2(-1)^k = (-1)^k[-1 + 2] = (-1)^k = (-1)^{k+2}.$$

    Thus p_{k+2} is true and the principle applies.

23. $p_1 \Rightarrow p_2$ fails.

24. (a) Prove p_1 and p_2 are true.

25. If p_k is true for some k, then $p_{k-1}, p_{k-2}, \ldots, p_1$ are all true by induction using the first condition. Given m, the second condition implies that p_k is true for some $k \geq m$, so p_m is true.

27. (a) Apply the recursion theorem with $s_0 = a_0$ and $s_n = s_{n-1} + a_n$.

1.2 DIVISORS AND PRIME FACTORIZATION

1. (a) $391 = 23 \cdot 17 + 0$ (c) $-116 = (-9) \cdot 13 + 1$

2. (a) $n/d = 51837/386 = 134.293$, so $q = 134$. Thus $r = n - qd = 113$.

3. If $d > 0$, then $|d| = d$ and this is the division algorithm. If $d < 0$, then $|d| = -d > 0$ so $n = q(-d) + r = (-q)d + r$, $0 \leq r \leq |d|$.

5. Write $m = 2k + 1$, $n = 2j + 1$. Then $m^2 - n^2 = 4[k(k + 1) - j(j + 1)]$. But each of $k(k + 1)$ and $j(j + 1)$ is even, so $8 \mid (m^2 - n^2)$.

7. (a) $10(11k + 4) - 11(10k + 3) = 7$, so $d \mid 7$. Thus $d = 1$ or $d = 7$.

9. (a) $72 = 42 + 30$ (c) $327 = 6 \cdot 54 + 3$

    $42 = 30 + 12$ $54 = 3 \cdot 18$

    $30 = 2 \cdot 12 + 6$ Thus $\gcd(327 \cdot 54) = 3$ and

    $12 = 2 \cdot 6$ $3 = 1 \cdot 327 - 6 \cdot 54$

    Thus, $\gcd(72, 42) = 6$ and

    $6 = 30 - 2(42 - 30)$

    $\quad = 3 \cdot 30 - 2 \cdot 42$

    $\quad = 3(72 - 42) - 2 \cdot 42$

    $\quad = 3 \cdot 72 - 5 \cdot 42$

(e) $377 = 13 \cdot 29$

   Hence $29 \mid 377$, so

   $\gcd(29, 377) = 29$. Thus

   $29 = 0 \cdot 377 + 1 \cdot 29$

(g) $72 = 0 \cdot (-176) + 72$

   $-175 = (-3) \cdot 72 + 41$

   $72 = 41 + 31$

   $41 = 31 + 10$

   $31 = 3 \cdot 10 + 1$

   Hence $\gcd(72, -175) = 1$ and

   $1 = 31 - 3(41 - 31)$

   $ = 4(72 - 41) - 3 \cdot 41$

   $ = 4 \cdot 72 - 7(-175 + 3 \cdot 72)$

   $ = (-17) \cdot 72 - 7 \cdot (-175)$

11. If $m = qd$, then $\frac{m}{k} = q\frac{d}{k}$, so $\frac{d}{k} \mid \frac{m}{k}$. Similarly, $\frac{d}{k} \mid \frac{n}{k}$. If $d = xm + yn$, then $\frac{d}{k} = x\frac{m}{k} + y\frac{n}{k}$, so any common divisor of $\frac{m}{k}$ and $\frac{n}{k}$ is a divisor of $\frac{d}{k}$.

13. It is prime for $n = 1, 2, \ldots, 9$; but $10^2 + 10 + 11 = 121 = 11^2$.

15. If $d = \gcd(m, n)$ and $d_1 = \gcd(m_1, n_1)$, then $d \mid m$ and $d \mid n$, so $d \mid m_1$ and $d \mid n_1$ by hypothesis. Thus $d \mid d_1$.

17. If $1 = xm + yn$ and $1 = x_1 k + y_1 n$, then

$$1 = (xm + yn)(x_1 k + y_1 n) = (xx_1)mk + (xmy_1 + yx_1 k + yny_1)n.$$

Thus $\gcd(mk, n) = 1$ by Theorem 4.

   Alternatively, if $d = \gcd(mk, n) \neq 1$ let $p \mid d$, p a prime. Then $p \mid n$ and $p \mid mk$ But then $p \mid m$ or $p \mid k$, a contradiction either way because we have $\gcd(m, n) = 1 = \gcd(m, n)$.

19. Write $d = \gcd(m, n)$ and $d' = \gcd(km, kn)$. We must show $kd = d'$. First, $d \mid m$ and $d \mid n$, so $kd \mid km$ and $kd \mid kn$. Hence, $kd \mid d'$. On the other hand, write $km = qd'$ and $kn = pd'$. We have $d = xm + yn$, $x, y \in \mathbb{Z}$, so

$$kd = xkm + ykn = xqd' + ypd'.$$

Thus $d' \mid kd$. As $k \geq 1$ it follows that $d' = kd$.

21. If p is not a prime, then assume $p = mn$ with $m \geq 2$ and $n \geq 2$. But then $p \mid m$ or $p \mid n$ by hypothesis, so $p \leq m < p$ or $p \leq n < p$, a clear contradiction.

23. No. If $a = 18$ and $n = 12$ then $d = 6$ so $\frac{a}{d} = 3$ is not relatively prime to $n = 12$.

25. Let them be $2k + 1$, $2k + 3$, $2k + 5$. We have $k = 3q + r$, $r = 0, 1, 2$. If $r = 0$ then $3 \mid (2k + 3)$; if $r = 1$, then $3 \mid (2k + 1)$; and if $r = 2$, then $3 \mid (2k + 5)$. Thus one of these primes is a multiple of 3, and so is 3.

27. Let $d = \gcd(m, p^k)$, then $d \mid m$ and $d \mid p^k$. Thus $d = p^j$, $j \leq k$. If $j > 0$, then $p \mid d$, so (since $d \mid m$) $p \mid m$. This contradicts $\gcd(m, p) = 1$. So $j = 0$ and $d = 1$.

29. We have $a \mid a_1 b_1$ and $(a, b_1) = 1$. Hence $a \mid a_1$ by Theorem 5. Similarly $a_1 \mid a$, so $a = a_1$ because both are positive. Similarly $b = b_1$.

30. (a) $27783 = 3^4 \cdot 7^3$

    (c) $2431 = 11 \cdot 13 \cdot 17$

    (e) $241 = 241$ (a prime)

31. (a) $735 = 2^0 \cdot 3^1 \cdot 5^1 \cdot 7^2 \cdot 11^0$ and $110 = 2^1 \cdot 3^0 \cdot 5^1 \cdot 7^0 \cdot 11^1$. Hence
$\gcd(735, 110) = 2^0 \cdot 3^0 \cdot 5^1 \cdot 7^2 \cdot 11^0 = 5$, and

$\operatorname{lcm}(735, 110) = 2^1 \cdot 3^1 \cdot 5^1 \cdot 7^2 \cdot 11^1 = 16170$.

(c) $139 = 2^0 \cdot 139^1$ and $278 = 2^1 \cdot 139^1$. Hence
$\gcd(139, 278) = 2^0 \cdot 139^1 = 139$, and $\operatorname{lcm}(139, 278) = 2^1 \cdot 139^1 = 278$.

33. (a) Use Theorem 8. In forming $d = p_1^{d_1} \ldots p_r^{d_r}$, there are $(n_1 + 1)$ choices for d_1 among $0, 1, 2, \ldots, n_i$; then there are $(n_2 + 1)$ choices for d_2 among $0, 1, 2, \ldots, n_2$; and so on. Thus there are $(n_1 + 1)(n_2 + 1) \cdots (n_r + 1)$ choices in all, and each leads to a different divisor by the uniqueness in the prime factorization theorem.

35. Let $m = p_1^{m_1} \ldots p_r^{m_r}$ and $n = q_1^{n_1} \ldots q_s^{n_s}$ be the prime factorizations of m and n. Since $\gcd(m, n) = 1$, $p_i \neq q_j$ for all i and j, so the prime factorization of mn is $mn = p_1^{m_1} \ldots p_r^{m_r} q_1^{n_1} \ldots q_s^{n_s}$. Since $d \mid mn$, we have $d = p_1^{d_1} \ldots p_r^{d_r} q_1^{e_1} \ldots q_s^{e_s}$ where $0 \leq d_i \leq m_i$ for each i and $0 \leq e_j \leq n_j$ for each j. Take $m_1 = p_1^{d_1} \ldots p_r^{d_r}$ and $n_1 = q_1^{e_1} \ldots q_s^{e_s}$.

37. Write $a = p_1^{a_1} p_2^{a_2} \ldots p_r^{a_r}$ and $b = p_1^{b_1} p_2^{b_2} \ldots p_r^{b_r}$ where the p_i are distinct primes, $a_i \geq 0$ and $b_i \geq 0$. Let $u_i = \begin{cases} 0 & \text{if } a_i < b_i \\ a_i & \text{if } a_i \geq b_i \end{cases}$ and $v_i = \begin{cases} b_i & \text{if } a_i < b_i \\ 0 & \text{if } a_i \geq b_i \end{cases}$, and then take $u = p_1^{u_1} p_2^{u_2} \ldots p_r^{u_r}$ and $v = p_1^{v_1} p_2^{v_2} \ldots p_r^{v_r}$. Then $u \mid a$, $v \mid b$ and $\gcd(u, v) = 1$. Moreover $uv = \operatorname{lcm}(a, b)$ by Theorem 9 because $u_i + v_i = \max(a_i, b_i)$ for each i.

39. (a) By the division algorithm, $p = 4k + r$ for $r = 0, 1, 2$ or 3. But $r = 0$ or 2 is impossible since p is odd (being a prime greater than 2).

41. (a) $28665 = 3^2 \cdot 5^1 \cdot 7^2 \cdot 11^0 \cdot 13^1$ and $22869 = 3^3 \cdot 5^0 \cdot 7^1 \cdot 11^2 \cdot 13^0$ so,

$\gcd(28665, 22869) = 3^2 \cdot 5^0 \cdot 7^1 \cdot 11^0 \cdot 13^0 = 63$

$\operatorname{lcm}(28665, 22869) = 3^3 \cdot 5^1 \cdot 7^2 \cdot 11^2 \cdot 13^1 = 10,405,395$

43. Let $X = \{x_1 a_1 + \cdots + x_k a_k \mid x_i \in \mathbb{Z}, \ x_1 a_1 + \cdots + x_k a_k \geq 1\}$. Then $X \neq \varnothing$ because $a_1^2 \cdots + a_k^2 \in X$, so let m be the smallest member of X. Then $m = x_1 a_1 + \cdots + x_k a_k$ for integers a_k, so we show $d = m$. Since $d \mid a_i$ for each i, it is clear that $d \mid m$. We can show $m \mid d$, if we can show that m is a common divisor of the a_i (by definition of $d = \gcd(a_1, \cdots, a_k)$). Write $a_1 = qm + r$, $0 \leq r < m$. Then

$$r = a_1 - qm = (1 - qx_1)a_1 + (-qx_2)a_2 + \cdots + (-qx_k)a_k,$$

and this contradicts the minimality if $r \geq 1$. So $r = 0$ and $m \mid a_1$. A similar argument shows $m \mid a_i$ for each i.

45. (a) Let $m = qn + r$, $0 \leq r < n$. If $m < n$, then $q = 0$ and $r = m$. If $m \geq n$, then $q \geq 1$. Thus $q \geq 0$. We want $x \in \mathbb{Z}$ such that $2^m - 1 = x(2^n - 1) + (2^r - 1)$. Solving for x (possibly in $\mathbb{Q}$):

$$x = \frac{2^m - 2^r}{2^n - 1} = 2^r \left(\frac{2^{m-r} - 1}{2^n - 1} \right) = 2^r \left(\frac{(2^n)^q - 1}{2^n - 1} \right).$$

If $q = 0$, take $x = 2^r = 2^m$; if $q > 0$, take $x = (2^n)^{q-1} + \cdots + 2^n + 1$.

1.3 INTEGERS MODULO n

1. (a) True. $40 - 13 = 3 \cdot 9$

   (c) True. $-29 - 6 = (-5)7$

   (e) True. $8 - 8 = 0 \cdot n$ for any n.

   (g) False. $8^4 \equiv (64)^2 \equiv (-1)^2 \equiv 1 \ (\mathrm{mod}\, 13)$.

2. (a) $2k - 4 = 7q$, so q is even. Thus $k = 2 + 7x$ for some integer x; that is $k \equiv 2 \ (\mathrm{mod}\ 7)$.

   (c) $2k \equiv 0 \ (\mathrm{mod}\ 9)$, so $2k = 9q$. Thus $2 \mid q$, so $k = 9x$ for some integer x; that is $k \equiv 0 \ (\mathrm{mod}\ 9)$.

3. (a) $10 \equiv 0 \ (\mathrm{mod}\ k)$, so $k \mid 10$: $k = 2, 5, 10$.

   (c) $k^2 - 3 = qk$, so $k \mid 3$. Thus $k = 1, 3$ so, (as $k \geq 2$ by assumption) $k = 3$.

5. (a) $a \equiv b \ (\mathrm{mod}\ 0)$ means $a - b = q \cdot 0$ for some q, that is $a = b$.

6. (a) $a \equiv a$ for all a because $n \mid (a - a)$. Hence if $n \mid (a - b)$, then $n \mid (b - a)$. Hence if $a - b = xn$ and $b - c = yn$, $x, y \in \mathbb{Z}$, then $a - c = (x + y)n$.

7. If $n = pm$ and $a \equiv b(\mathrm{mod}\ n)$, then $a - b = qn = qpm$. Thus $a \equiv b(\mathrm{mod}\ m)$.

8. (a) In $\mathbb{Z}_7 : \overline{10} = \bar{3}$,   so $\overline{10}^2 = \bar{9} = \bar{2}$, $\overline{10}^3 = \bar{6} = \overline{-1}$, $\overline{10}^6 = \bar{1}$.   Since $515 = 6 \cdot 85 + 5$ we get $\overline{10}^{515} = (\overline{10}^6)^{85} \cdot \overline{10}^5 = \bar{1}^{85} \cdot \overline{10}^2 \cdot \overline{10}^3 = \bar{2} \cdot \overline{(-1)} = \bar{5}$. Hence $10^{515} \equiv 5(\mathrm{mod}\, 7)$.

9. (a) In $\mathbb{Z}_{10} : \bar{3}^2 = \bar{9} = -1$,   so $\bar{3}^4 = \bar{1}$. Since $1027 = 4 \cdot 256 + 3$,   we get $\bar{3}^{1027} = (\bar{3}^4)^{256} \cdot \bar{3}^3 = \bar{1}^{256} \cdot \overline{27} = \bar{7}$. The unit decimal is 7.

11. $\bar{p} = \bar{0}, \bar{1}, \bar{2}, \bar{3}, \bar{4}, \bar{5}$ in $\mathbb{Z}_6$. If $\bar{p} = \bar{0}, \bar{2}, \bar{4}$ then $2 \mid p$; if $\bar{p} = \bar{3}$, then $3 \mid p$. So $\bar{p} = \bar{1}$ or $\bar{p} = \bar{5}$.

12. (a) $\bar{a} = \bar{0}, \bar{1}, \bar{2}, \bar{3}$ in $\mathbb{Z}_4$, so $\bar{a}^2 = \bar{0}, \bar{1}, \bar{0}, \bar{1}$ respectively.

13. $\bar{a} = \bar{0}, \bar{1}, \ldots, \overline{10}$ in $\mathbb{Z}_{11}$. Taking each case separately:

$$\bar{0}^5 = \bar{0} \qquad\qquad\qquad \bar{6}^5 = \overline{(-5)}^5 = \overline{-5}^5 = \overline{-1}$$
$$\bar{1}^5 = \bar{1} \qquad\qquad\qquad \bar{7}^5 = \overline{(-4)}^5 = \overline{-4}^5 = \overline{-1}$$
$$\bar{2}^5 = \overline{32} = \overline{-1} \qquad\qquad \bar{8}^5 = \overline{(-3)}^5 = \overline{-3}^5 = \overline{-1}$$
$$\bar{3}^5 = \bar{9} \cdot \overline{27} = \bar{9} \cdot \bar{5} = \bar{1} \qquad \bar{9}^5 = \overline{(-2)}^5 = \overline{-2}^5 = \bar{1}$$
$$\bar{4}^5 = \overline{16} \cdot \overline{64} = \bar{5} \cdot \bar{9} = \bar{1} \qquad \overline{10}^5 = \overline{(-1)}^5 = \overline{-1}$$
$$\bar{5}^5 = \overline{25} \cdot \overline{25} \cdot \bar{5} = \bar{3} \cdot \bar{3} \cdot \bar{5} = \bar{1}$$

15. One of $a, a+1$ must be even so $2 \mid a(a+1)(a+2)$; similarly, one of a, $a+1, a+2$ is a multiple of 3 [in fact $a \equiv 0$ means $3 \mid a$, $a \equiv 1$ means $3 \mid a+2$, and $a \equiv 2$ means $3 \mid a+1$]. Hence $3 \mid a(a+1)(a+2)$. But 2 and 3 are relatively prime so $2 \cdot 3 = 6$ also divides $a(a+1)(a+2)$. Hence

$$\bar{a}(\bar{a} + \bar{1})(\bar{a} + \bar{2}) = \overline{a(a+1)(a+2)} = \bar{0} \text{ in } \mathbb{Z}_6.$$

17. Since $\bar{a} = \bar{0}, \bar{1}, \ldots, \bar{5}$ in $\mathbb{Z}_6$, we examine every case.

$$\bar{0}^3 = \bar{0} \qquad \bar{3}^3 = \overline{27} = \bar{3}$$
$$\bar{1}^3 = \bar{1} \qquad \bar{4}^3 = \overline{(-2)}^3 = -(\bar{2})^3 = \overline{-2} = \bar{4}$$
$$\bar{2}^3 = \bar{8} = \bar{2} \qquad \bar{5}^3 = \overline{(-1)}^3 = \overline{-1} = \bar{5}$$

Hence $\bar{a}^3 = \bar{a}$ in all cases.

18. (a) Since $\bar{a} = \bar{0}, \bar{1}, \ldots, \bar{4}$ in $\mathbb{Z}_5$, it suffices to show each of these is a cube in $\mathbb{Z}_5$. Look at the cubes in $\mathbb{Z}_5 : \bar{0}^3 = \bar{0}$, $\bar{1}^3 = \bar{1}$, $\bar{2}^3 = \bar{3}$, $\bar{3}^3 = \bar{2}$, and $\bar{4}^3 = \overline{(-1)}^3 = -\bar{1} = \bar{4}$. Thus every residue $\bar{0}, \bar{1}, \bar{2}, \bar{3}, \bar{4}$ is a cube in $\mathbb{Z}_5$.

19. (a) Since $\bar{k} = \bar{0}, \bar{1}, \bar{2}, \bar{3}, \bar{4}, \bar{5}, \bar{6}$ in $\mathbb{Z}_7$, we get $\bar{k}^2 + \bar{1} = \bar{1}, \bar{2}, \bar{5}, \bar{3}, \bar{3}, \bar{5}, \bar{2}$ respectively. Clearly $\bar{k}^2 + \bar{1} = \bar{0}$ does not occur in $\mathbb{Z}_7$.

21. We have $n = d_0 + 10d_1 + 10^2 d_2 + \cdots + 10^k d_k$.

(a) $\overline{10} = \bar{1}$ in $\mathbb{Z}_3$, so $\bar{n} = \overline{d_0} + \bar{1} \cdot \overline{d_1} + \bar{1}^2 \overline{d_2} + \cdots + \bar{1}^k \overline{d_k} = \overline{d_0 + d_1 + \cdots + d_k}$. Thus $\bar{n} = \overline{d_0} + \overline{d_1} + \cdots + \overline{d_k} \pmod 3$.

22. (a) By the euclidean algorithm, $13 = 1 \cdot 9 + 4$ so
$$9 = 2 \cdot 4 + 1$$

$$35 = 2 \cdot 13 + 9 \qquad 1 = 9 - 2(13 - 9)$$
$$= 3(35 - 2 \cdot 13) - 2 \cdot 13$$
$$= 3 \cdot 35 - 8 \cdot 13$$

Hence $(-8) \cdot 13 \equiv 1 \pmod{35}$, so $\overline{-8} = \overline{27}$ is the inverse of $\overline{13}$ in $\mathbb{Z}_{35}$. Then $\overline{13} \cdot \bar{x} = \bar{9}$ gives $\bar{x} = \overline{27} \cdot \overline{13} \cdot \bar{x} = \overline{27} \cdot \bar{9} = \overline{-8} \cdot \bar{9} = \overline{-72} = \overline{-2} = \overline{33}$.

(c) Euclidean algorithm: $11 = 9 + 2$ so
$$20 = 11 + 9$$
$$9 = 4 \cdot 2 + 1$$

$$1 = 9 - 4(11 - 9)$$
$$= 5 \cdot 9 - 4 \cdot 11$$
$$= 5(20 - 11) - 4 \cdot 11$$
$$= 5 \cdot 20 - 9 \cdot 11$$

Hence the inverse of $\overline{11}$ is $\overline{-9} = \overline{11}$, so $\overline{11} \cdot \bar{x} = \overline{16}$ gives $\bar{x} = \overline{11} \cdot \overline{16} = \overline{16}$.

23. (a) Let $\bar{d}$ be the inverse of $\bar{a}$ in $\mathbb{Z}_n$, so $\bar{d} \cdot \bar{a} = \bar{1}$ in $\mathbb{Z}_n$, then multiply $\bar{a} \cdot \bar{b} = \bar{a} \cdot \bar{c}$ by $\bar{d}$ to get $\bar{d} \cdot \bar{a} \cdot \bar{b} = \bar{d} \cdot \bar{a} \cdot \bar{c}$, that is $\bar{1} \cdot \bar{a} = \bar{1} \cdot \bar{c}$, that is $\bar{a} = \bar{c}$.

24. (a) If $\bar{c}$ and $\bar{d}$ are the inverses of $\bar{a}$ and $\bar{b}$ respectively in $\mathbb{Z}_n$, then $\bar{c} \cdot \bar{a} = \bar{1}$ and $\bar{d} \cdot \bar{b} = \bar{1}$. Multiplying, we find $\bar{c} \cdot \bar{a} \cdot \bar{d} \cdot \bar{b} = \bar{1}$, that is $(\bar{c} \cdot \bar{d})(\bar{a} \cdot \bar{b}) = \bar{1}$. Hence $\bar{c} \cdot \bar{d}$ is the inverse of $\bar{a} \cdot \bar{b} = \overline{ab}$ in $\mathbb{Z}_n$.

25. (a) Multiply equation 2 by $\bar{2}$ to get $\overline{10}x + \bar{2}y = \bar{2}$. Subtract this from equation 1: $\bar{7}x = \bar{1}$. But $\bar{8} \cdot \bar{7} = \bar{1}$ in $\mathbb{Z}_{11}$, so $x = \bar{8} \cdot \bar{1} = \bar{8}$. Then equation 2 gives $y = \bar{1} - \bar{5} \cdot \bar{8} = \bar{5}$.

(c) Multiply equation 2 by $\bar{2}$ to get $\bar{3}x + \bar{2}y = \bar{2}$. Comparing this with the first equation gives $\bar{1} = \bar{3}x + \bar{2}y = \bar{2}$, an impossibility. So there is no solution to these equations in $\mathbb{Z}_7$. (Compare with (a)).

(e) Multiply equation 2 by $\bar{2}$ to get $\bar{3}x + \bar{2}y = \bar{1}$, which is just equation 1. Hence, we need only solve equation 2. If $x = \bar{r}$ is arbitrary in $\mathbb{Z}_7$ (so $\bar{r} = \bar{0}, \bar{1}, \ldots, \bar{6}$), then $y = \bar{4} - \bar{5}x = \overline{4 - 5r}$. Thus the solutions are:

x	$\bar{0}$	$\bar{1}$	$\bar{2}$	$\bar{3}$	$\bar{4}$	$\bar{5}$	$\bar{6}$
y	$\bar{4}$	$\bar{6}$	$\bar{1}$	$\bar{3}$	$\bar{5}$	$\bar{0}$	$\bar{2}$

27. If an expression $x^2 + ax$ is given where a is a number, we can *complete the square* by adding $\left(\frac{1}{2}a\right)^2$. Then $x^2 + ax + \left(\frac{1}{2}a\right)^2 = (x + \frac{1}{2}a)^2$. The same thing works in $\mathbb{Z}_n$ except $\frac{1}{2}$ is replaced by the inverse of $\bar{2}$ if it exists.

(a) $x^2 + \bar{5}x + \bar{4} = \bar{0}$ means $x^2 + \bar{5}x = \bar{3}$ in $\mathbb{Z}_7$. The inverse of $\bar{2}$ is $\bar{4}$ in $\mathbb{Z}_7$, so the square is completed by adding $\left(\bar{4} \cdot \bar{5}\right)^2 = \bar{1}$ to both sides. The result is
$$(x + \bar{6})^2 = x^2 + \bar{5}x + \bar{1} = \bar{3} + \bar{1} = \bar{4}.$$
The only members of $\mathbb{Z}_7$ which square to $\bar{4}$ are $\bar{2}$ and $\overline{-2} = \bar{5}$. (See Exercise 26.) Hence $x + \bar{6} = \bar{2}$ or $\bar{5}$; that is $x = \bar{3}$ or $\bar{6}$.

(c) $x^2 + x + \bar{2} = \bar{0}$ gives $x^2 + x = \bar{3}$ in $\mathbb{Z}_5$. The inverse of $\bar{2}$ is $\bar{3}$ in $\mathbb{Z}_5$, so add $\bar{3}^2 = \bar{4}$ to both sides
$$(x + \bar{3})^2 = x^2 + x + \bar{4} = \bar{3} + \bar{4} = \bar{2}.$$
But $\bar{2}$ is not a square in $\mathbb{Z}_5$ $[\bar{0}^2 = \bar{0}, \bar{1}^2 = \bar{4}^2 = \bar{1}, \bar{2}^2 = \bar{3}^2 = \bar{4}]$, so there is no solution.

(e) Since n is odd, $\gcd(2, n) = 1$, so $\bar{2}$ has an inverse in $\mathbb{Z}_n$; call it $\bar{r}$. Now $x^2 + \bar{a}x + \bar{b} = \bar{0}$ in $\mathbb{Z}_n$ means $x^2 + \bar{a}x = \overline{-b}$. Complete the square by adding $(\bar{r} \cdot \bar{a})^2 = \overline{ra}^2$ to both sides. The result is
$$(x + \overline{ra})^2 = x^2 + \bar{a} + ra^2 = -b + ra^2 = (\bar{r}^2\bar{a}^2 - \bar{b}).$$
Thus, there is a solution if and only if $(\bar{r}^2\bar{a}^2 - \bar{b})$ is a square in $\mathbb{Z}_n$.

29. (a) Let $\bar{a} \cdot \bar{b} = \bar{0}$ in $\mathbb{Z}_n$. If $\gcd(a, n) = 1$, then a has an inverse in $\mathbb{Z}_n$, say $\bar{c} \cdot \bar{a} = \bar{1}$. Then $\bar{b} = \bar{1}\bar{b} = \bar{c} \cdot \bar{a} \cdot \bar{b} = \bar{c} \cdot \bar{0} = \bar{0}$.

31. $(1) \Rightarrow (2)$. Assume (1) holds but n is not a power of a prime. Then $n = p^k a$ where p is a prime, $k \geq 1$, and $a > 1$ has $p \nmid a$. Then $\gcd(n, a) = a > 1$, so $\bar{a}$ has no inverse in $\mathbb{Z}_n$. But $\bar{a}^n \neq \bar{0}$ too. In fact $\bar{a}^n = \bar{0}$ means $n \mid a^n$ whence $p \mid a^n$. By Euclid's lemma, this implies $p \mid a$, contrary to choice.

33. In $\mathbb{Z}_{223}$, $\bar{2}^8 = \overline{256} = \overline{33}$. Thus $\bar{2}^{16} = \overline{33}^2 = \overline{197}$, $\bar{2}^{32} = \overline{197}^2 = \bar{7}$, and finally $\bar{2}^{37} = \bar{2}^{32} \cdot \bar{2}^5 = \bar{7} \cdot \overline{32} = \overline{224} = \bar{1}$. Similarly, in $\mathbb{Z}_{641}$,
$$\bar{2}^8 = \overline{256}, \ \bar{2}^{16} = \overline{256}^2 = \overline{154}, \ \bar{2}^{32} = \overline{154}^2 = \overline{640} = \overline{-1}.$$

34. (a) If $ax \equiv b$ has a solution x in $\mathbb{Z}_n$, then $b - ax = qn$, q an integer, so $b = ax + qn$. It follows that $d = \gcd(a, n)$ divides b. Conversely, if $d \mid b$ write $b = qd$, q an integer. Now $d = ra + sn$ for integers r and s (Theorem 3 §1.2), so $b = qd = (qr)a + (qs)n$. Thus, $(qr)a \equiv b \pmod{n}$ and we have our solution.

35. Working modulo p, $x^2 = \bar{1}$ means $x^2 - \bar{1} = \bar{0}$. Thus $(x - \bar{1})(x + \bar{1}) = \bar{0}$ in $\mathbb{Z}_p$, so $x = \bar{1}$ or $x = -\bar{1}$ by Theorem 7.

37. (a) If $n = p^2 m$ and $a = pm$, then $a \not\equiv 0 \pmod{n}$ and $a^2 \equiv 0 \pmod{n}$. Hence $a^n \not\equiv a$.

1.4 PERMUTATIONS

1. (a) $\tau\sigma = \begin{pmatrix} 1 & 2 & 3 & 4 & 5 \\ 2 & 3 & 5 & 1 & 4 \end{pmatrix}$ (c) $\tau^{-1} = \tau = \begin{pmatrix} 1 & 2 & 3 & 4 & 5 \\ 3 & 2 & 1 & 5 & 4 \end{pmatrix}$

(e) $\mu\tau\sigma^{-1} = \begin{pmatrix} 1 & 2 & 3 & 4 & 5 \\ 4 & 5 & 2 & 3 & 1 \end{pmatrix}$

3. (a) $\chi = \sigma^{-1}\tau = \begin{pmatrix} 1 & 2 & 3 & 4 \\ 4 & 2 & 3 & 1 \end{pmatrix}$ (c) $\chi = \sigma\tau = \begin{pmatrix} 1 & 2 & 3 & 4 \\ 1 & 3 & 2 & 4 \end{pmatrix}$

(e) $\chi = \tau^{-1}\varepsilon\sigma^{-1} = \tau^{-1}\sigma^{-1} = \begin{pmatrix} 1 & 2 & 3 & 4 \\ 1 & 3 & 2 & 4 \end{pmatrix}$

5. *Solution 1.* We must have $\sigma1 = 1, 2, 3$ or 4; in each case we find $\sigma1 = \sigma3$, a contradiction.

If $\sigma1 = 1$:

$\tau1 = \tau\sigma1 = 2$
$\sigma2 = \sigma\tau1 = 2$
$\tau2 = \tau\sigma2 = 3$
$\sigma3 = \sigma\tau2 = 1$

If $\sigma1 = 3$:

$\tau3 = \tau\sigma1 = 2$
$\sigma2 = \tau\sigma3 = 4$
$\tau4 = \tau\sigma2 = 3$
$\sigma3 = \sigma\tau4 = 3$

If $\sigma1 = 2$:

$\tau2 = \tau\sigma1 = 2$
$\sigma2 = \sigma\tau2 = 1$
$\tau1 = \tau\sigma2 = 3$
$\sigma3 = \sigma\tau1 = 2$

If $\sigma1 = 4$:

$\tau4 = \tau\sigma1 = 2$
$\sigma2 = \sigma\tau4 = 3$
$\tau3 = \tau\sigma2 = 3$
$\sigma3 = \sigma\tau3 = 4$

Solution 2. Let $\sigma = \begin{pmatrix} 1 & 2 & 3 & 4 \\ a & b & c & d \end{pmatrix}$. Then we show $\sigma\tau = (a\ b\ c\ d)$ is a cycle, contrary to $\sigma\tau = (1\ 2)(3\ 4)$:

$\sigma1 = a \Rightarrow \tau a = \tau\sigma1 = 2 \Rightarrow \sigma\tau a = \sigma2 = b$
$\sigma2 = b \Rightarrow \tau b = \tau\sigma2 = 3 \Rightarrow \sigma\tau b = \sigma3 = c$
$\sigma3 = c \Rightarrow \upsilon\tau c = \tau\sigma3 = 4 \Rightarrow \sigma\tau c = \sigma4 = d$
$\sigma4 = d \Rightarrow \tau d = \tau\sigma4 = 1 \Rightarrow \sigma\tau d = \sigma1 = a$

6. If $\sigma k = k$, then $\sigma^{-1}k = \sigma^{-1}(\sigma k) = k$. If also $\tau k = k$, then $(\tau\sigma)k = \tau(\sigma k) = \tau k = k$.

7. (a) Here $\sigma = \begin{pmatrix} 1 & 2 & 3 & 4 & 5 \\ 1 & a & b & c & d \end{pmatrix}$ where a, b, c, d are 2, 3, 4, 5 in some order. Thus there are 4 choices for a, 3 for b, 2 for c, and 1 for d; and so we have $4 \cdot 3 \cdot 2 \cdot 1 = 4! = 24$ choices in all for σ.

(b) Now $\sigma = \begin{pmatrix} 1 & 2 & 3 & 4 & 5 \\ 1 & 2 & a & b & c \end{pmatrix}$ where a, b, c are 3, 4, 5 in some order. As in (a), there are $3 \cdot 2 \cdot 1 = 3! = 6$ choices in all for σ.

8. (a) If $\sigma\tau = \varepsilon$, then $\sigma = \sigma\varepsilon = \sigma(\tau\tau^{-1}) = (\sigma\tau)\tau^{-1} = \tau^{-1}$.

9. If $\sigma = \tau$, then $\sigma\tau^{-1} = \tau\tau^{-1} = \varepsilon$; if $\sigma\tau^{-1} = \varepsilon$, then

$$\tau = \varepsilon\tau = (\sigma\tau^{-1})\tau = \sigma(\tau^{-1}\tau) = \sigma\varepsilon = \sigma.$$

11. (a) $\begin{pmatrix} 1 & 2 & 3 & 4 & 5 & 6 & 7 & 8 & 9 \\ 8 & 2 & 6 & 1 & 9 & 4 & 5 & 7 & 3 \end{pmatrix}$

12. (a) $\varepsilon, \sigma = (1\ \ 2\ \ 3), \sigma^2 = (1\ \ 3\ \ 2), \tau = (1\ \ 2), \sigma\tau = (1\ \ 3), \sigma^2\tau = (2\ \ 3)$. These are all six elements of S_3. We have $\sigma^3 = \sigma\sigma^2 = \varepsilon$, $\tau^2 = \varepsilon$ and hence $\tau\sigma = (2\ \ 3) = \sigma^2\tau$.

13. (a) $\sigma = (1\ 4\ 8\ 3\ 9\ 5\ 2\ 7\ 6); \sigma^{-1} = (1\ 6\ 7\ 2\ 5\ 9\ 3\ 8\ 4)$
    (c) $\sigma = (1\ 2\ 8)(3\ 6\ 7)(4\ 9\ 5); \sigma^{-1} = (1\ 8\ 2)(3\ 7\ 6)(4\ 5\ 9)$
    (e) $\sigma = (1\ 3\ 8\ 7\ 2\ 5); \sigma^{-1} = (1\ 5\ 2\ 7\ 8\ 3)$

15. (a) $\varepsilon, (1\ 2\ 3\ 4\ 5), (1\ 2\ 3\ 4), (1\ 2\ 3), (1\ 2\ 3)(4\ 5), (1\ 2), (1\ 2)(3\ 4)$

17. (a) $\sigma^{-1} = (4\ 3\ 2\ 1)(7\ 6\ 5)$.

19. They are factored into disjoint cycles in the solution to Exercise 13, so the parities are:

    (a) even   (c) even + even + even = even   (e) odd

21. (a) We have $\gamma_i^2 = \varepsilon$ for all i because the γ_i are transpositions. Hence
$$(\gamma_1\gamma_2 \cdots \gamma_m)(\gamma_m\gamma_{m-1} \cdots \gamma_2\gamma_2) = (\gamma_1\gamma_2 \cdots \gamma_{m-1})(\gamma_{m-1} \cdots \gamma_2\gamma_1) = \ldots = \varepsilon.$$
Now use Exercise 8(a).

   (c) If σ and τ are products of k and m transpositions respectively, then τ^{-1} is also a product of m transpositions (by (a)) so $\tau\sigma\tau^{-1}$ is a product of $k + 2m$ transpositions. This has the same parity as k.

23. Let $\sigma k = 1$ for some $k \neq 1$. Then, as $n \geq 3$, choose an $m \notin \{k, 1\}$. Now let $\gamma = (k, m)$. This gives $\gamma\sigma k = \gamma 1 = 1$, but $\sigma\gamma k = \sigma m \neq 1$, since if $\sigma m = 1 = \sigma k$, then $m = k$ as σ is one-to-one, contrary to assumption.

25. It suffices to show that any pair of transpositions is a product of 3-cycles. If k, l, m and n are distinct, this follows from
$$(k\ l)(m\ n) = (k\ m\ l)(k\ m\ n), (k\ l)(k\ m) = (k\ m\ l), \text{ and } (k\ l)^2 = \varepsilon.$$

27. (a) Both sides have the same effect on each k_i, and both sides fix each $k \notin \{k_1, k_2, \ldots k_r\}$.

   (c) Using Exercise 26, we have for all $a = 1, 2, \ldots, n - 1$:
$$(1\ a+1) = (1\ a)(a\ a+1)(1\ a) \tag{*}$$
Now if $\sigma \in S_n$, write it as a product of factors $(1\ n)$. Use (*) to write each $(1\ n)$ as a product of $(1\ 2), \ldots, (1\ n-1)$, and $(n-1\ n)$. Then write each $(1, n-1)$ in terms of $(1\ 2), \ldots, (1\ n-2)$ and $(n-2, n-1)$. Continue. The result is (c).

28. (a) $\sigma = (1\ 2\ 3\ 4\ \ldots\ 2k-1\ 2k)$ so $\sigma^2 = (1\ 3\ 5\ \ldots\ 2k-1)(2\ 4\ 6\ \ldots\ 2k)$.

   (c) The action of σ is depicted in the diagram, and carries $k \to k+1 \to k+2 \ldots$. If $k+m > n$, the correct location on the circle is given by the remainder r when $k+m$ is divided by n, That is $k+m \equiv 4 \pmod{n}$. Now the action of σ^m is $\sigma^m k = k+m$, so $\sigma^m k \equiv k+m \bmod n$.

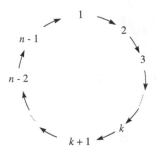

29. Each of σ and τ may be either even or odd, so four cases arise. They are the rows of the following table. The parity of $\sigma\tau$ in each case is clear, and so the result follows

σ	τ	$\sigma\tau$	sgn $\sigma\tau$	sgn σ	sgn τ
E	E	E	1	1	1
E	O	O	-1	1	-1
O	E	O	-1	-1	1
O	O	E	1	-1	-1

by verifying, sgn $\sigma\cdot$sgnτ = sgn(τ) in every case.

Chapter 2

Groups

2.1 BINARY OPERATIONS

1. (a) This is not commutative: $1 * 2 = -1$ while $2 * 1 = 1$. It is not associative: $(2 * 1) * 3 = 1 * 3 = -2$, while $2 * (1 * 3) = 2 * (-2) = 4$. There is no unity: If $e * a = a$ for all a, then $e - a = a$ so $e = 2a$ for all a. This is impossible.

(c) This is commutative: $a * b = a + b - ab = b + a - ba = b * a$. It is associative:

$$a * (b * c) = a + (b * c) - a(b * c) = a + b + c - (ab + ac + bcc) + abc$$

and, similarly, this equals $(a * b) * c$. The unity is 0:

$$a * 0 = a + 0 - a \cdot 0 = a.$$

Every $a \neq 1$ has an inverse

$$\frac{a}{a-1} = a + \frac{a}{a-1} - \frac{a^2}{a-1} = \frac{a(a-1) + a - a^2}{a-1} = 0.$$

(e) This is not commutative $(p, q) * (p', q') = (p, q')$ while $(p', q') * (p, q) = (p', q)$. It is associative:

$$(p, q) * [(p', q') * (p'', q'')] [= (p, q'') = [(p, q) * (p', q')] * (p'', q'').$$

There is no unity: If $(a, b) * (p, q) = (p, q)$ for all p, q, then $a = p$ for all p.

(g) This is commutative: $\gcd(n, m) = \gcd(m, n)$. It is associative: Write $d = \gcd(k, m)$, $d' = \gcd(m, n)$. Then $d_1 = (k * m) * n = \gcd(d, n)$, so $d_1 \mid d$ and $d_1 \mid n$. But then, $d_1 \mid k$, $d_1 \mid m$ and $d_1 \mid n$. It follows that $d_1 \mid k$ and $d_1 \mid d'$, so $d_1 \mid \gcd(k, d') = k * (m * n)$. A similar argument shows $k * (m * n) \mid (k * m) * n$, so these are equal. Finally, there is no unity: If $e * n = n$ for all n, then $n = \gcd(e, n)$ so $n \mid e$ for all n. This is impossible.

Student Solution Manual to Accompany Introduction to Abstract Algebra, Fourth Edition. W. Keith Nicholson.
© 2012 John Wiley & Sons, Inc. Published 2012 by John Wiley & Sons, Inc.

(i) If we write $(x, y, z) = \begin{bmatrix} x & y \\ 0 & z \end{bmatrix}$ as a matrix, this operation is matrix multiplication, and so is associative. It is not commutative $\begin{bmatrix} 1 & 1 \\ 0 & 0 \end{bmatrix} \begin{bmatrix} 0 & 0 \\ 0 & 1 \end{bmatrix} = \begin{bmatrix} 0 & 1 \\ 0 & 0 \end{bmatrix}$ while $\begin{bmatrix} 0 & 0 \\ 0 & 1 \end{bmatrix} \begin{bmatrix} 1 & 1 \\ 0 & 0 \end{bmatrix} = \begin{bmatrix} 0 & 0 \\ 0 & 0 \end{bmatrix}$. The unity is the identity matrix $\begin{bmatrix} 1 & 0 \\ 0 & 1 \end{bmatrix}$ $= (1\ 0\ 1)$, and $\begin{bmatrix} x & y \\ 0 & z \end{bmatrix}$ has an inverse if and only if $x \neq 0$ and $z \neq 0$. The inverse then is $\begin{bmatrix} 1/x & -y/xz \\ 0 & 1/z \end{bmatrix}$.

2. (a) $1(yz) = yz = (1y)z$; $x(1z) = xz = (x1)z$; and $x(y1) = xy = (xy)1$.

(c) M is clearly closed and 1 is the unity. M is associative by (a).

3. (a) We have $ab = b$, $b^2 = a$. Hence $a^2 = ab^2 = (ab)b = b^2 = a$, and hence $ba = bb^2 = b^2b = ab = b$. Hence a is the unity and the operation is associative by Exercise 2.

5. This is associative:
$$a_1 \cdots a_n \cdot [(b_1 \cdots b_m) \cdot (c_1 \cdots c_k)] = a_1 \cdots a_n \cdot b_1 \cdots b_m \cdot c_1 \cdots c_k.$$
Clearly this equals $[(a_1 \cdots a_n) \cdot (b_1 \cdots b_m)] \cdot c_1 \cdots c_k$. The unity is the empty word λ (with no letters). It is not commutative if $|A| > 1$: $a \cdot b \neq b \cdot a$ if $a \neq b$. Note that if $A = \{a\}$, then $W = \{1, a, aa, aaa, \ldots\}$ is commutative. If $wv = \lambda$, w, r words, it is clear that $w = \lambda = r$. So λ is the only unit.

7. It is associative:
$$(m, n)[(m', n')(m'', n'')][= (m, n)(m'm'', n'n'') = (mm'm'', nn'n''),$$
and this equals $[(m, n)(m'n')](m'', n'')$. The unity is $(1, 1)$. $M \times N$ is commutative if and only if M and N are both commutative. Finally, (m, n) is a unit if and only if m and n are units in M and N respectively, then $(m, n)^{-1} = (m^{-1}, n^{-1})$.

8. (a) Given $a^m = a^{m+n}$, we have $a^m = a^m a^n = a^{m+n} a^n = a^{m+2n}$. Continue to get $a^m = a^{m+kn}$ for all $k \geq 0$. Then multiply by a^r to get $a^{m+r} = a^{m+kn+r}$ for all $r \geq 0$. Hence a^{m+r} is an idempotent if $r \geq 0$ and $k \geq 0$ satisfy $2(m + r) = m + kn + r$, that is $m + r = kn$. So choose $k \geq 0$ such that $kn \geq m$; and then take $r = kn - m$. One choice: $k = m$, $r = m(n - 1)$. Then $m + r = m + m(n - 1)$, so a^{mn} is an idempotent.

9. (a) $a^{24}a = a^{25} = (a^5)^5 = (b^5)^5 = b^{25} = b^{24}b = a^{24}b$. Cancel a^{24} by cancelling a 24 times.

11. Let $e \neq f$ be left unities ($ex = x = fx$ for all x). If g is a right unity ($xg = x$), then $g = eg = e$ and $g = fg = f$, so $e = f$, contrary to hypothesis.

12. (a) If $au = bu$, then $(au)u^{-1} = (bu)u^{-1}$, that is $a(uu^{-1}) = b(uu^{-1})$; $a1 = b1$; $a = b$.

13. Let $(uv)w = 1 = w(uv)$. Then $u(vw) = 1$, and we claim that $(vw)u = 1$ too (so u is a unit). In fact $[(vw)u]v = (vw)(uv) = v[w(uv)] = v = 1v$, so $(vw)u = 1$ by hypothesis. Thus u^{-1} exists. But then $v = u^{-1}(uv)$ is a unit by Theorem 5 (since u and uv are both units.)

15. (a) If σ is a bijection, let $1 = \sigma(v)$ for some $v \in M$ (σ is onto). This means $1 = uv$. But $\sigma(vu) = u(vu) = (uv)u = u = \sigma 1$, so $vu = 1$ because σ is one-to-one.

Conversely, let u be a unit. If $\sigma a = \sigma b$, then $ua = ub$, so $a = u^{-1}ua = u^{-1}ub = b$. This shows σ is one-to-one. If $b \in M$, then $b = u(u^{-1}b) = \sigma(u^{-1}b)$, so σ is onto. Thus σ is a bijection.

17. (a) If $u^{-1} = v^{-1}$, then $u = (u^{-1})^{-1} = (v^{-1})^{-1} = v$ by Theorems 4 and 5.

(c) Use (b) twice: $uv = vu$ gives $u^{-1}v = vu^{-1}$, so (since v is a unit) $v^{-1}u^{-1} = u^{-1}v^{-1}$, as required. Alternatively, if $uv = vu$ then $(uv)^{-1} = (vu)^{-1}$ by Theorem 4, whence $v^{-1}u^{-1} = u^{-1}v^{-1}$ by Theorem 5.

18. (1) $\Rightarrow$ (2). If $ab = 1$ then a^{-1} exists by (1) so $b = 1b = a^{-1}ab = a^{-1}$. Hence b is a unit by Theorem 5.

19. Let $M = \{a_1, a_2, \ldots, a_n\}$, and consider $X = \{a_1 u, a_2 u, \ldots, a_n u\}$. If $a_i u = a_j u$, then $a_i = a_i uv = a_j$, so $i = j$. Thus $|X| = n = |M|$, so since $X \subseteq M$ we have $X = M$. In particular $1 \in X$, say $1 = wu$, $w \in M$. Then
$$w = w1 = w(uv) = (wu)v = v,$$
so $1 = wu = vu$. This means v is an inverse of u.

20. (a) $a \sim a$ for all a because $a = a \cdot 1$; if $a \sim b$, then $a = bu$, so $b = au^{-1}$, that is $b \sim a$. If $a \sim b$ and $b \sim c$, let $a = bu$, $b = cv$, u, v units. Then $a = (cv)u = c(vu)$, and so $a \sim c$ because vu is a unit. Note that M need not be commutative here.

(c) M is associative because $\bar{a}(\bar{b} \cdot \bar{c}) = \bar{a} \cdot \overline{bc} = \overline{a(bc)} = \overline{(ab)c} = (\overline{ab}) \cdot \bar{c} = (\bar{a} \cdot \bar{b})\bar{c}$. Since 1 is the unity of M, we obtain $\bar{1} \cdot \bar{a} = \overline{1a} = \bar{a}$; and similarly, $\bar{a} \cdot \bar{1} = \bar{a}$. Hence $\bar{1}$ is the unity of $\bar{M}$. Next $\bar{a} \cdot \bar{b} = \overline{ab} = \overline{ba} = \bar{b} \cdot \bar{a}$ so M is commutative. Finally, if $\bar{a}$ is a unit in M, let $\bar{a} \cdot \bar{b} = \bar{1}$. Then $ab \sim 1$ so $1 = abu$. Thus a is a unit in M, so $a \sim 1$. Hence $\bar{a} = \bar{1}$, as required.

21. (a) $E(M)$ is closed under composition since, if $\alpha, \beta \in E(M)$, then
$$\alpha\beta(xy) = \alpha[\beta(xy)] = \alpha[\beta(x) \cdot y] = \alpha\beta(x) \cdot y$$
for all $x, y \in M$. We have $1_M(xy) = xy = 1_M x \cdot y$, so $1_M \in E(M)$ and 1_M is the unity of $E(M)$. Finally, composition is always associative, so $E(M)$ is a monoid.

2.2 GROUPS

1. (a) Not a group. Only 0 has an inverse so G4 fails.

(c) Group. It is clearly closed and
$$a \cdot (b \cdot c) = a + (b + c + 1) + 1 = a + b + c + 2 = (a + b + 1) + c + 1$$
$$= (a \cdot b) \cdot c$$
proves associativity. The unity is -1, and the inverse of a is $-a - 2$. Note that G is also abelian.

(e) Not a group. It is not closed: $(1\ 2)(1\ 3) = (1\ 3\ 2)$ is not in G. Note that ε is a unity and each element is self inverse, so only G1 fails.

(g) Group. The unity is 16; associativity from $\mathbb{Z}_{20}$. For inverses and closure — see the Cayley table:

	16	12	8	4
16	16	12	8	4
12	12	4	16	8
8	8	16	4	12
4	4	8	12	16

(i) Not a group. It is closed (by Theorem 3 §0.3), and associative, and ε is the unity. However G4 fails. If $\sigma : \mathbb{N} \to \mathbb{N}$ has $\sigma n = 2n$ for all $n \in \mathbb{N}$, then σ has no inverse because it is not onto.

3. (a) First $ad = c$, $a^2 = d$ by the Corollary to Theorem 6. Next $ba \neq b, a, d$; and $ba = c \Rightarrow b = ac = a(ba) = (ab)a = 1a = a$, a contradiction. So $ba = 1$. Then $bd = a$, $bc = d$, $b^2 = c$. Next, $ca = b$, $cd = 1$, $c^2 = a$, $cb = d$. Finally, $da = c$, $db = a$, $dc = 1$, $d^2 = b$.

	1	a	b	c	d
1	1	a	b	c	d
a	a	d	1	b	c
b	b	1	c	d	a
c	c	b	d	a	1
d	d	c	a	1	b

5. A monoid is a group if each element is invertible. So check that every row and column contains exactly one 1.

7. The unity is I_3 and $\begin{bmatrix} 1 & a & b \\ 0 & 1 & c \\ 0 & 0 & 1 \end{bmatrix} \begin{bmatrix} 1 & a' & b' \\ 0 & 1 & c' \\ 0 & 0 & 1 \end{bmatrix} = \begin{bmatrix} 1 & a+a' & b'+ac'+b \\ 0 & 1 & c+c' \\ 0 & 0 & 1 \end{bmatrix}$ shows that G is closed. Since matrix multiplication in general is associative, it remains to show that each matrix in G has an inverse in G. But

$$\begin{bmatrix} 1 & a & b \\ 0 & 1 & c \\ 0 & 0 & 1 \end{bmatrix}^{-1} = \begin{bmatrix} 1 & -a & ac-b \\ 0 & 1 & -c \\ 0 & 0 & 1 \end{bmatrix}$$

as is easily verified.

8. (a) Write $\sigma = (1\ 2)(3\ 4)$, $\tau = (1\ 3)(2\ 4)$ and $\varphi = (1\ 4)(2\ 3)$. Then $\sigma^2 = \tau^2 = \varphi^2 = \varepsilon$ and $\sigma\tau = \tau\sigma = \varphi$, $\sigma\varphi = \varphi\sigma = \tau$ and $\varphi\tau = \tau\varphi = \sigma$. Hence G is closed and every element is self inverse. Since permutation multiplication in general is associative, G is a group. Here $x^2 = \varepsilon$ for all four elements x of G.

9. It is easy to show that

$$\sigma^2 = (1\ 3\ 5)(2\ 4\ 6), \sigma^3 = (1\ 4)(2\ 5)(3\ 6), \sigma^4 = (1\ 5\ 3)(2\ 6\ 4), \sigma^5 = (1\ 6\ 5\ 4\ 3\ 2)$$

and $\sigma^6 = \varepsilon$. Hence $G = \{\varepsilon, \sigma, \sigma^2, \sigma^3, \sigma^4, \sigma^5\}$ is closed by the exponent laws and $\sigma^{-1} = \sigma^5$, $(\sigma^2)^{-1} = \sigma^4$, $(\sigma^3)^{-1} = \sigma^3$, $(\sigma^4)^{-1} = \sigma^2$ and $(\sigma^5)^{-1} = \sigma$. Since permutation multiplication is associative, G is a group. Also, G is abelian because $\sigma^k \sigma^l = \sigma^{k+l} = \sigma^l \sigma^k$ for all k, l. Finally, there are two elements τ satisfying $\tau^2 = \varepsilon : \tau = \varepsilon$ and $\tau = \sigma^3$; the three with $\tau^3 = \varepsilon$ are $\tau = \varepsilon$, $\tau = \sigma^2$ and $\tau = \sigma^4$.

10. (a) $ab = ba^2$ gives $aba^2 = ba^4 = b$. Hence $a^2ba^2 = ab$, that is $a^2ba^2 = ba^2$. Cancellation gives $a^2 = 1$. Then $ab = ba^2 = b$, whence $a = 1$ by cancellation.

(c) $ab = ba^2$ gives $aba^4 = ba^6 = b$. Hence $a^2ba^4 = ab = ba^2$, so $a^2ba^2 = b$ by cancellation. Finally $a^3ba^2 = ab = ba^2$ so $a^3 = 1$. Hence $b = aba^4 = aba$.

11. (a) We claim that $b(ab)^n a = (ba)^{n+1}$ for all $n \geq 0$. It is clear if $n = 0$. If it holds for some $n \geq 0$, then

$$b(ab)^{n+1}a = b(ab)(ab)^n a = ba(ba)^{n+1} = (ba)^{n+2}.$$

Hence this holds for all $n \geq 0$ by induction. Now suppose $(ab)^n = 1$. Then $(ba)^{n+1} = b(ab)^n a = b1a = ba$. Cancelling ba gives $(ba)^n = 1$.

13. α is onto because $g = (g^{-1})^{-1} = \alpha(g^{-1})$ for all $g \in G$. If $\alpha(g) = \alpha(g_1)$, then $g^{-1} = g_1^{-1}$, so $g = (g^{-1})^{-1} = (g_1^{-1})^{-1} = g_1$. This shows that α is one-to-one.

15. Define $\sigma : X \to Xa$ by $\sigma(x) = xa$. This is clearly onto and $\sigma(x) = \sigma(x_1)$ implies $xa = x_1a$, so $x = x_1$ by cancellation. Hence σ is one-to-one.

17. If $e^2 = e$, then $ee = e1$, so $e = 1$ by cancellation. Thus 1 is the only idempotent.

19. If G is abelian, then $gh = hg$, so $(gh)^{-1} = (hg)^{-1} = g^{-1}h^{-1}$ by Theorem 3. Conversely, given $x, y \in G$, we are assuming $(xy)^{-1} = x^{-1}y^{-1}$. By Theorem 3, this is $y^{-1}x^{-1} = x^{-1}y^{-1}$; that is any two inverses commute. But this means that G is abelian because *every* element g of G is an inverse [in fact $g = (g^{-1})^{-1}$].

21. If G is abelian, then $(gh)^2 = g(hg)h = g(gh)h = g^2h^2$ for all g, h. Conversely, if $(gh)^2 = g^2h^2$, then $g(hg)h = g(gh)h$. Thus $hg = gh$ by cancellation (twice).

23. (a) If $g = g^{-1}$, then $g^2 = gg^{-1} = 1$; if $g^2 = 1$, then $g^{-1} = g^{-1}1 = g^{-1}g^2 = g$.

25. Let $a^5 = 1$ and $a^{-1}ba = b^m$. Then

$$a^{-2}ba^2 = a^{-1}b^m a = (a^{-1}ba)^m = (b^m)^m = b^{m^2}.$$

Next $a^{-3}ba^3 = a^{-1}b^{m^2}a = (a^{-1}ba)^{m^2} = (b^m)^{m^2} = b^{m^3}$. This continues to give $a^{-4}ba^4 = b^{m^4}$ and finally $b = a^{-5}ba^5 = b^{m^5}$. Hence $1 = b^{m^5 - 1}$ by cancellation.

27. In multiplicative notation, $a^1 = a$, $a^2 = a \cdot a$, $a^3 = a \cdot a \cdot a, \ldots$; in additive notation $a + a = 2a$, $a + a + a = 3a, \ldots$. In $\mathbb{Z}_n$, $\bar{k} = \bar{1} + \bar{1} + \cdots + \bar{1} = k\bar{1}$, so $\mathbb{Z}_n$ is generated by 1.

29. (a) We first establish left cancellation: If $gx = gy$ in G, then $x = y$. In fact, let $hg = e$. Then $gx = gy$ implies $x = ex = hgx = hgy = ey = y$. Thus $hg = e = ee = hge$, so $g = ge$ by left cancellation. This shows that e is the unity. Finally, $h(gh) = (hg)h = eh = h = he$, so $gh = e$, again by left cancellation. Thus h is the inverse of g.

(c) Choose $g \in G$ and let $ge = g$, $e \in G$ (by hypothesis). If $zg = e$, $z \in G$, then $e = zg = zge = ee = e^2$. Now, given $h \in G$, let $h = ex$. Then, $eh = e^2x = ex = h$. Similarly, $h = ye$, $y \in G$, implies $he = h$. Thus e is the unity for G. But now, given h, we can find c, d such that $ch = e = hd$. Then $c = ce = c(hd) = (ch)d = ed = d$, so $ch = e = hc$. Thus h has an inverse.

2.3 SUBGROUPS

1. (a) No, $1 + 1 \notin H$.

(c) No, $3^2 = 9 \notin H$.

(e) No, $(1\ 2)(3\ 4) \cdot (1\ 3)(2\ 4) = (1\ 4)(2\ 3) \notin H$.

(g) Yes, $0 = 6 \in H$. H is closed because it consists of the even residues in $\mathbb{Z}_6$; $-4 = 2$, $-2 = 4$, so it is closed under inverses.

(i) Yes, the unity $(0,0) \in H$. If (m,k) and (m',k') are in H, then so is $(m,k) + (m',k') = (m+m', k+k')$ and $-(m,k) = (-m,-k)$.

3. Yes. If H is a subgroup of G and K is a subgroup of H, then $1 \in K$ (it is the unity of H). If $a,b \in K$, then $ab \in K$ because this is their product in H. Finally, a^{-1} is the inverse of a in H, hence in K.

5. (a) We have $1 \in H$ because $1 = 1^2$. If $a,b \in H$, then $a^{-1} = a$ (because $a^2 = 1$), so $a^{-1} \in H$. Finally, the fact that $ab = ba$ gives $(ab)^2 = a^2 b^2 = 1 \cdot 1 = 1$, so $ab \in H$.

6. (a) We have $1 \in H$ because $1 = 1^2$. If $x, y \in H$, write $x = g^2$, $y = h^2$. Then $x^{-1} = (g^{-1})^2 \in H$ and (since G is abelian) $xy = g^2 h^2 = (gh)^2 \in H$.

(c) The set of squares in A_4 consists of ε and all the 3-cycles. This is not a subgroup since $(1\ 2\ 3)(1\ 2\ 4) = (1\ 3)(2\ 4)$.

7. (a) We have $1 = g^0 \in \langle g \rangle$. If x and y are in $\langle g \rangle$, write $x = g^k$, $y = g^m$, $k, m \in \mathbb{Z}$. Then $x^{-1} = g^{-k} \in \langle g \rangle$ and $xy = g^{k+m} \in \langle g \rangle$. Use the subgroup test.

8. (a) If $x \in X$ (since X is nonempty), $1 = xx^{-1} \in \langle X \rangle$. Clearly $\langle X \rangle$ is closed, and if $g = x_1^{k_1} \ldots x_m^{k_m} \in X$, then $g^{-1} = x_m^{-k_m} \ldots x_1^{-k_1} \in X$. Hence $\langle X \rangle$ is a subgroup; clearly $X \subseteq \langle X \rangle$.

9. We have $1 \in C(g)$ because $1g = g1$. If $z, w \in C(g)$, then
$$(zw)g = z(wg) = z(gw) = (zg)w = (gz)w = g(zw).$$
This shows $zw \in C(g)$. Finally $zg = gz$ implies $g = z^{-1}gz$, so $gz^{-1} = z^{-1}g$. Thus $z^{-1} \in C(g)$. Use the subgroup test.

11. We have $\begin{bmatrix} 1 & 0 \\ 0 & 1 \end{bmatrix} \in G$. If $X, Y \in G$, write $X = \begin{bmatrix} a & b \\ 0 & a \end{bmatrix}$, $Y = \begin{bmatrix} c & d \\ 0 & c \end{bmatrix}$. Then $XY = \begin{bmatrix} ac & ad+bc \\ 0 & ac \end{bmatrix} \in G$ and $X^{-1} = \begin{bmatrix} a^{-1} & -a^{-1}ba^{-1} \\ 0 & a^{-1} \end{bmatrix} \in G$. Use Theorem 1.

13. (a) Clearly the unity $(1,1)$ of $G \times G$ is in H. If $x, y \in H$, write $x = (f,f)$ and $y = (g,g)$. Then $xy = (fg, fg) \in H$ and $x^{-1} = (f^{-1}, f^{-1}) \in H$. So H is a subgroup of G by Theorem 1.

15. (a) $C_5 = \{1, g, g^2, g^3, g^4\}$, $g^5 = 1$. If $H \neq \{1\}$ is a subgroup, one of g, g^2, g^3, g^4 is in H. If $g \in H$, then $H = C_5$. But $g = (g^2)^3 = (g^3)^2 = (g^4)^4$, so $H = C_5$ in any case. Hence $\{1\}$ and C_5 are the only subgroups.

C_5

|

|

$\{1\}$

(c) $S_3 = \{1, \sigma, \sigma^2, \tau, \tau\sigma, \tau\sigma^2\}$, $\sigma^3 = 1 = \tau^2$, $\sigma\tau = \tau\sigma^2$. We claim $\{1\}, \{1, \sigma, \sigma^2\}, \{1, \tau\}$, $\{1, \tau\sigma\}$ and $\{1, \tau\sigma^2\}$ are all the proper subgroups. They are subgroups by Theorem 2. Suppose a subgroup H is not one of these:

Case 1. $\sigma \in H$ or $\sigma^2 \in H$. Then $\{1, \sigma, \sigma^2\} \subseteq H$, so H contains one of $\tau, \tau, \sigma, \tau\sigma^2$. But $\tau = (\tau\sigma)\sigma^2 = \tau(\tau\sigma^2)\sigma$, so $\sigma \in H$ and $\tau \in H$. This means $H = S_3$.

Case 2. $\sigma \notin H$ and $\sigma^2 \notin H$. Then H contains two of $\tau, \tau\sigma, \tau\sigma^2$. But $\tau(\tau\sigma) = \sigma$, $\tau(\tau\sigma^2) = \sigma^2$ and $(\tau\sigma)(\tau\sigma^2) = \tau(\tau\sigma^2)\sigma^2 = \sigma$, so this case cannot occur.

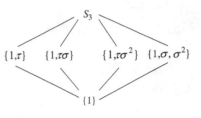

16. (a) $1 \in H \cap K$ because $1 \in H$ and $1 \in K$. If $a \in H \cap K$, then $a \in H$ and $a \in K$. Thus $a^{-1} \in H$ and $a^{-1} \in K$, so $a^{-1} \in H \cap K$. If $b \in H \cap K$ also, then $b \in H$, and $b \in K$, so $ab \in H$ and $ab \in K$. Thus $ab \in H \cap K$.

17. If $H \subseteq K$ or $K \subseteq H$, then $H \cup K$ is K or H respectively, so $H \cup K$ is a subgroup. Conversely, suppose $H \cup K$ is a subgroup and $H \nsubseteq K$. We show $K \subseteq H$. If $k \in K$, we must show that $k \in H$. Choose $h \in H \smallsetminus K$. Then $kh \notin K$ (if $kh = k_1 \in K$, then $h = k^{-1}k_1 \in K$). Since kh is in $H \cup K$ (because $H \cup K$ is a subgroup), this gives $kh \in H$. But $kh = h_1$ implies $k = h^{-1}h_1 \in H$, as required.

19. (a) $g^{-1}Hg = g^{-1}gH = 1H = H$. So the only conjugate of H is H itself.

20. (a) If $g \in G$, then $gh = hg$ for all $h \in H$ (since $H \subseteq Z(G)$), so $gH = Hg$. Hence $g^{-1}Hg = g^{-1}gH = 1H = H$.

21. Let $\begin{bmatrix} x & y \\ 0 & z \end{bmatrix} \in Z(G)$. Then $\begin{bmatrix} a & b \\ 0 & c \end{bmatrix}\begin{bmatrix} x & y \\ 0 & z \end{bmatrix} = \begin{bmatrix} x & y \\ 0 & z \end{bmatrix}\begin{bmatrix} a & b \\ 0 & c \end{bmatrix}$ for all a, b, c. This means $ay + bz = xb + yc$ for all a, b, c. If we take $a = c = 0$ and $b = 1$, we get $x = z$. Take $a = 1$ and $b = c = 0$ to get $y = 0$. Hence

$$Z(G) = \left\{ \begin{bmatrix} x & 0 \\ 0 & x \end{bmatrix} \,\middle|\, 0 \neq x \in \mathbb{R} \right\}.$$

22. Let $Z = \begin{bmatrix} x & y \\ z & w \end{bmatrix}$ be in $Z[GL_2(\mathbb{R})]$. Then $ZA = AZ$ for all $A \in GL_2(\mathbb{R})$. Taking $A = \begin{bmatrix} 0 & 1 \\ 1 & 0 \end{bmatrix}$ leads to $y = z$ and $x = w$; taking $A = \begin{bmatrix} 1 & 1 \\ 0 & 1 \end{bmatrix}$ leads to $z = 0$ (and $x = w$). Hence $X = \begin{bmatrix} x & 0 \\ 0 & x \end{bmatrix}$. Each matrix $\begin{bmatrix} x & 0 \\ 0 & x \end{bmatrix} = xI$ is central because $(xI)A = xA = A(xI)$ for all A.

23. Yes. If $\sigma = (1\ 2\ 3)$, then $H = \{\varepsilon, \sigma, \sigma^2\}$ is an abelian subgroup of S_3, but $Z(S_3) = \{\varepsilon\}$.

25. Assume that $KH \subseteq HK$. Then $1 = 11 \in HK$. If $h \in H$ and $k \in K$, then $(hk)^{-1} = k^{-1}h^{-1} \in KH \subseteq HK$, and

$$(hk)(h_1k_1) \in h(KH)k_1 \subseteq h(HK)k_1 \subseteq HK.$$

Conversely, if HK is a subgroup then $kh = (h^{-1}k^{-1})^{-1} \in HK$.

2.4 CYCLIC GROUPS AND THE ORDER OF AN ELEMENT

1. If $o(g) = n$, we use Theorem 8: g^k generates $G = \langle g \rangle$ if and only if $\gcd(k, n) = 1$.

   (a) $o(g) = 5$. Then $G = \langle g^k \rangle$ if $k = 1, 2, 3, 4$.

   (c) $o(g) = 16$. Then $G = \langle g^k \rangle$ if $k = 1, 3, 5, 7, 9, 11, 13, 15$.

2. Since $\mathbb{Z}_n$ is cyclic and $\mathbb{Z}_n = \langle \bar{1} \rangle$, the solution to Exercise 1 applies.

   (a) $\mathbb{Z}_5$ has generators $\bar{1}, \bar{2}, \bar{3}, \bar{4}$.

   (c) $\mathbb{Z}_{16}$ has generators $\bar{1}, \bar{3}, \bar{5}, \bar{7}, \bar{9}, \overline{11}, \overline{13}, \overline{15}$.

3. (a) $G = \langle g \rangle$, $o(g) = \infty$. We claim g and g^{-1} are the only generators. Note that $g^k = (g^{-1})^{-k}$ for all $k \in \mathbb{Z}$, so $G = \langle g^{-1} \rangle$. Suppose $G = \langle g^m \rangle$. Then $g \in \langle g^m \rangle$, say $g = (g^m)^k$. Thus $g^1 = g^{mk}$ so $1 = mk$ by Theorem 3. Since m and k are integers, this shows $m = \pm 1$.

4. (a) $G = \mathbb{Z}_7^* = \{\bar{1}, \bar{2}, \bar{3}, \bar{4}, \bar{5}, \bar{6}\}$. We have $\bar{2}^3 = \bar{1}$, so $o(\bar{2}) \leq 3$. As to $\bar{3}$: $\bar{3}^0 = \bar{1}$, $\bar{3}^1 = \bar{3}$, $\bar{3}^2 = \bar{2}$, $\bar{3}^3 = \bar{2}$, $\bar{3}^3 = \bar{6}$, $\bar{3}^4 = \bar{4}$, $\bar{3}^5 = \bar{5}$. Thus $\mathbb{Z}_7^* = \langle \bar{3} \rangle$.

   (c) $G = \mathbb{Z}_{16}^* = \{\bar{1}, \bar{3}, \bar{5}, \bar{7}, \bar{9}, \overline{11}, \overline{13}, \overline{15}\}$. Here $\bar{3}^4 = \overline{81} = \bar{1}$, so $\overline{13}^4 = \overline{(-3)}^4 = \bar{1}$. Similarly $\bar{5}^4 = \bar{1} = \overline{11}^4$, $\bar{7}^2 = \bar{1} = \bar{9}^2$, and $\overline{15}^2 = \bar{1} = \bar{1}^2$. Thus $\mathbb{Z}_{16}^*$ is not cyclic.

5. (a) No, If $\mathbb{Q}^*$ is cyclic, suppose $\mathbb{Q}^* = \langle \frac{n}{m} \rangle$ where $\gcd(m, n) = 1$. Then $-1 = \langle \frac{n}{m} \rangle^k$ for $0 \neq k \in \mathbb{Z}$. Now if $k < 0$, then $(-1) = \left(\frac{m}{n} \right)^{-k}$, so we may assume $k > 0$. Then $-m^k = n^k$, and this is impossible for relatively prime m and n, unless $n = \pm 1$, $m = \pm 1$. Then $\frac{n}{m} = \pm 1$ generates $\mathbb{Q}^*$, a contradiction.

7. Given $o(g) = 20$:

   (a) $o(g^2) = \frac{20}{2} = 10$ by Theorem 5.

   (c) $o(g^5) = \frac{20}{5} = 4$ by Theorem 5.

8. (a) Each element σ of S_5 factors into disjoint cycles in one of the following ways:

$$\begin{aligned}
\sigma &= (a \quad b \quad c \quad d \quad e) & o(\sigma) &= 5 \\
\sigma &= (a \quad b \quad c \quad d) & o(\sigma) &= 4 \\
\sigma &= (a \quad b \quad c) & o(\sigma) &= 3 \\
\sigma &= (a \quad b \quad c)(d \quad e) & o(\sigma) &= 6 \\
\sigma &= (a \quad b) & o(\sigma) &= 2 \\
\sigma &= (a \quad b)(c \quad d) & o(\sigma) &= 2
\end{aligned}$$

Hence, by Theorem 4, any permutation of the form $(a \quad b \quad c)(d \quad e)$ has maximum order 6.

9.

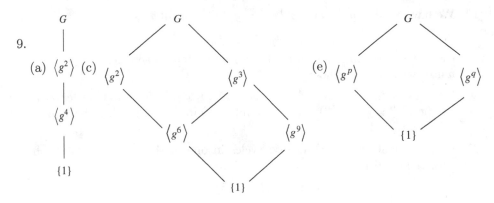

(a) $\langle g^2 \rangle$ (c) $\langle g^2 \rangle$ ... $\langle g^3 \rangle$ ... (e) $\langle g^p \rangle$ ... $\langle g^q \rangle$

$\langle g^4 \rangle$ ... $\langle g^6 \rangle$ ... $\langle g^9 \rangle$ ... $\{1\}$

$\{1\}$ ... $\{1\}$

10. (a) If $o(g) = n$ and $o(h) = m$, then $(gh)^{nm} = (g^n)^m(h^n)^m = 1$ because $gh = hg$.

11. (a) If $G = \langle a \rangle$ where $o(a) = n$, let $g = a^k$. Then
$$g^n = (a^k)^n = a^{kn} = (a^n)^k = 1^k = 1.$$

13. (a) Observe first that $g^{-1} = g$ if and only if $g = 1$ or $o(g) = 2$. Thus all the elements in the product $a = g_1 g_2 \cdots g_n$ which are not of order 2 (if any) cancel in pairs because G is abelian. Since $a^2 = g_1^2 g_2^2 \cdots g_n^2$, and since 1 and the elements of order 2 (if any) all square to 1, the result follows.

15. We have $\langle a, ab \rangle \subseteq \langle a, b \rangle$ by Theorem 10 because $a \in \langle a, b \rangle$ and $b \in \langle a, b \rangle$. The reverse inclusion follows because $a \in \langle a, ab \rangle$ and $b = a^{-1}(ab) \in \langle a, ab \rangle$. Similarly, $\langle a, b \rangle = \langle a^{-1}, b^{-1} \rangle$ because $a^{-1}, b^{-1} \in \langle a, b \rangle$, and $a = (a^{-1})^{-1}$ and $b = (b^{-1})^{-1}$ are both in $\langle a^{-1}, b^{-1} \rangle$.

16. (a) We have $a = a^4(a^3)^{-1} \in H$, so $G = \langle a \rangle \subseteq H$. Thus $H = G$.

(c) We have $d = xm + yk$ with $x, y \in \mathbb{Z}$, so $a^d = (a^m)^x(a^k)^y \in H$. Thus $\langle a^d \rangle \subseteq H$. But $d \mid m$, say $m = qd$, so $a^m = (a^d)^q \in \langle a^d \rangle$. Similarly $a^k \in \langle a^d \rangle$, so $H = \langle a^d \rangle$ by Theorem 10.

(e) $\{(1,1), (a,b), (a^2, b^2), (a^3, b^3)\} = \langle (a,b) \rangle \subseteq H$ and

$$\{(1,1), (a^3, b), (a^2, b^2), (a, b^3)\} = \langle (a^3, b) \rangle \subseteq H.$$

Then

$$(a^2, 1) = (a,b)(a, b^3) \in H, (1, b^2) = (a,b)(a^3, b) \in H.$$

Hence $K \subseteq H$ where

$$K = \{(1,1), (a,b), (a^2, b^2), (a^3, b^3), (a^3, b), (a, b^3), (a^2, 1), (1, b^2)\}.$$

Since $K = \{(a^k, b^m) \mid k + m \text{ even}\}$, it is a subgroup containing (a,b) and (a^3, b). Hence $H \subseteq K$, so $K = H$.

17. (a) Since $X \subseteq Y$ and $Y \subseteq \langle Y \rangle$, we have $X \subseteq \langle Y \rangle$. But $\langle Y \rangle$ is a subgroup, so $\langle X \rangle \subseteq \langle Y \rangle$ by Theorem 10.

19. We have $xy^{-1} = y^{-1}x$ and $x^{-1}y^{-1} = y^{-1}x^{-1}$ for all $x, y \in X$. If
$$g = x_1^{k_1} x_2^{k_2} \cdots x_m^{k_m} \in \langle X \rangle,$$
then each $x_i^{k_i}$ commutes with all the others. Hence each element of $\langle X \rangle$ commutes with all the others.

20. (a) If $C_6 = \langle a \rangle$ and $C_{15} = \langle b \rangle$, then $(a^3, b), (a, b^3), (a, b)$ all have order 30. Since $(x, y)^{30} = (x^{30}, y^{30}) = (1, 1)$ for all (x, y) in $C_6 \times C_{15}$, these have maximal order.

21. Each element of S_5 factors into cycles in one of the following ways (shown with their orders).

$$
\begin{array}{ll}
(1 \quad 2 \quad 3 \quad 4 \quad 5) & 5 \\
(1 \quad 2 \quad 3 \quad 4) & 4 \\
(1 \quad 2 \quad 3) & 3 \\
(1 \quad 2 \quad 3)(4 \quad 5) & 6 \\
(1 \quad 2) & 2 \\
(1 \quad 2)(3 \quad 4) & 2
\end{array}
$$

Since $\mathrm{lcm}(5, 4, 3, 6, 2, 2) = 60$, we have $\sigma^{60} = \varepsilon$ for all $\sigma \in S_5$. On the other hand, if $\sigma^n = \varepsilon$ for all $\sigma \in S_5$, then $o(\sigma)$ divides n for all σ, and so n is a common multiple of $5, 4, 3, 6, 2, 2$. Thus $60 \leq n$.

23. (a) We have $(ghg^{-1})^k = gh^k g^{-1}$ for all $k \geq 1$. Hence $h^k = 1$ if and only if $(ghg^{-1})^k = 1$. It follows that $o(h) = o(ghg^{-1})$ as in Example 10.

24. (a) If h is the only element of order 2 in G, then $h = g^{-1}hg$ for all $g \in G$ since $(g^{-1}hg)^2 = g^{-1}h(gg^{-1})hg = g^{-1}h^2g = g^{-1}g = 1$. Thus $gh = hg$ for all $g \in G$, that is $h \in Z(G)$. Note that $C_4 = \langle a \rangle$, $o(a) = 4$, has such an element: a^2.

25. Let $G = \langle g \rangle$ and $H = \langle h \rangle$ where $o(g) = m$ and $o(h) = n$. Since we have $|G \times H| = |G| \, |H| = mn$, it suffices to show that $o((g, h)) = nm$. We have $(g, h)^{nm} = (g^{nm}, h^{nm}) = (1, 1)$. If $(g, h)^k = (1, 1)$, then $g^k = 1$ and $h^k = 1$, so $m \mid k$ and $n \mid k$. But $\gcd(n, m) = 1$, then implies $nm \mid k$ (Theorem 5 §1.2)), so $o((g, h)) = mn$, as required.

26. (a) Write $o(gh) = d$. Since $gh = hg$, we have
$$(gh)^{mn} = g^{mn}h^{mn} = (g^m)^n(h^m)^n = 1.$$
This means $d \mid mn$. To prove $mn \mid d$, it suffices to show $m \mid d$ and $n \mid d$ (by Theorem 5 §1.2 because $\gcd(m, n) = 1$). This in turn follows if we can show $g^d = 1$ and $h^d = 1$. We have $1 = (gh)^d = g^d h^d$, so $g^d = h^{-d} \in \langle g \rangle \cap \langle h \rangle$. But $\langle g \rangle \cap \langle h \rangle = \{1\}$ because $\gcd(m, n) = 1$. Thus $g^d = 1$ and $h^{-d} = 1$, as required. If $\gcd(m, n) \neq 1$, nothing can be said (for example $h = g^{-1}$).

27. (a) If $A \subseteq B$, then $g^a \in B = \langle g^b \rangle$, say $g^a = g^{bq}$, $q \in \mathbb{Z}$. Since $o(g) = \infty$, $a = qb$. Conversely, if $a = qb$, then $g^a \in B$, so $A \subseteq B$.

29. Write $o(g^k) = m$. Then $(g^k)^{n/d} = (g^n)^{k/d} = 1^{k/d} = 1$ implies that $m \mid (n/d)$. On the other hand, write $d = xk + yn$ with $x, y \in \mathbb{Z}$ (by Theorem 3 §1.2).

Then $(g^k)^m = 1$ implies $g^{dm} = (g^{km})^x \cdot (g^n)^{ym} = 1$, so $n \mid dm$. If $qn = dm$, $q \in \mathbb{Z}$, then $q \cdot \frac{n}{d} = m$, so $(n/d) \mid m$. This shows $(n/d) = m$, as required.

31. (a) We have $a \mid m$ and $b \mid m$, so $g^m \in A$ and $g^m \in B$. Thus $g^m \in A \cap B$, whence $\langle g^m \rangle \subseteq A \cap B$. Conversely, write $A \cap B = \langle g^c \rangle$. Then $g^c \in A$, say $g^c = (g^a)^x$. Since $o(g) = \infty$, this implies $c = ax$. Similarly, $g^c \in B$ implies $c = by$. Thus c is a common multiple of a and b, so $m \mid c$ by the definition of the least common multiple. This implies $A \cap B = \langle g^c \rangle \subseteq \langle g^m \rangle$.

32. $(1) \Rightarrow (2)$. Let H and K be subgroups of $G = \langle g \rangle$ where $o(g) = p^n$. By Theorem 9, let $H = \langle g^a \rangle$ and $K = \langle g^b \rangle$ where a and b are divisors of p^n. Since p is a prime, this means $a = p^l$ and $b = p^m$. If $l \leq m$, this says $a \mid b$, whence $K \subseteq H$. The other alternative is $m \leq l$, so $H \subseteq K$.

33. If G is cyclic, it is finite (because infinite cyclic groups have infinitely many subgroups). So assume G is not cyclic. Use induction on the number n of distinct subgroups of G. If $n = 1$, $G = \{1\}$ is finite. If it holds for $n = 1, 2, \ldots, k$, let $H_1 = \{1\}$, $H_2, \ldots, H_k, H_{k+1} = G$ be all the subgroups of G. If $1 \leq i \leq k$ then $H_i \subseteq G$ so H_i is finite by induction. So it suffices to show $G = H_1 \cup H_2 \cup \cdots \cup H_k$. But if $g \in G$, then $\langle g \rangle \neq G$ because we are assuming that G is not cyclic. Hence $\langle g \rangle = H_i$ for some i, so $g \in H_i$.

35. (a) Let $m = p_1^{m_1} p_2^{m_2} \cdots p_r^{m_r}$ and $n = p_1^{n_1} p_2^{n_2} \cdots p_r^{n_r}$ where the p_i are distinct primes and $m_i \geq 0$, $n_i \geq 0$ for each i. For each i, define x_i and y_i by

$$x_i = \begin{cases} m_i & \text{if } m_i \geq n_i \\ 0 & \text{if } m_i < n_i \end{cases}, \qquad y_i = \begin{cases} 0 & \text{if } m_i \geq n_i \\ n_i & \text{if } m_i < n_i \end{cases}.$$

If $x = p_1^{x_1} p_2^{x_2} \cdots p_r^{x_r}$ and $y = p_1^{y_1} p_2^{y_2} \cdots p_r^{y_r}$, then $x \mid m$, $y \mid n$ and x and y are relatively prime. Thus $o(a^{m/x}) = x$ and $o(b^{n/y}) = y$ by Theorem 10, so $o(a^{m/x} \cdot b^{n/y}) = xy$ by Exercise 26(a). But $x_i + y_i = \max(m_i, n_y)$ for each i, so $xy = \operatorname{lcm}(m, n)$ by Theorem 9 §1.2.

37. Let cards numbered $1, 2, 3, \ldots$ be initially in position $1, 2, 3, \ldots$ in the deck. Then after a perfect shuffle, position 1 contains card 1, position 2 contains card $n + 1$, position 3 contains card 2, position 4 contains cards $n + 2, \ldots$. In general,

$$\text{position } k \text{ in the deck contains card } \sigma k = \begin{cases} \dfrac{k+1}{2} & \text{if } k \text{ is odd} \\ n + \dfrac{k}{2} & \text{if } k \text{ is even.} \end{cases}$$

Thus $\sigma = \begin{pmatrix} 1 & 2 & 3 & 4 & 5 & 6 & \cdots & 2n-1 & 2n \\ 1 & n+1 & 2 & n+2 & 3 & n+3 & \cdots & n & 2n \end{pmatrix}$. Note that σ fixes 1 and $2n$. The number of shuffles required to regain the initial order is $o(\sigma)$. Use Example 9.

(a)
$\begin{aligned} n &= 4 & \sigma &= (2\ 5\ 3)(4\ 6\ 7) & o(\sigma) &= 3 \\ n &= 5 & \sigma &= (2\ 6\ 8\ 9\ 5\ 3)(4\ 7) & o(\sigma) &= 6 \\ n &= 6 & \sigma &= (2\ 7\ 4\ 8\ 10\ 11\ 6\ 9\ 5\ 3) & o(\sigma) &= 10 \\ n &= 7 & \sigma &= (2\ 8\ 11\ 6\ 10\ 12\ 13\ 7\ 4\ 9\ 5\ 3) & o(\sigma) &= 12 \end{aligned}$

2.5 HOMOMORPHISMS AND ISOMORPHISMS

1. (a) It is a homomorphism because

$$\alpha(r)\alpha(s) = \begin{bmatrix} 1 & r \\ 0 & 1 \end{bmatrix}\begin{bmatrix} 1 & s \\ 0 & 1 \end{bmatrix} = \begin{bmatrix} 1 & r+s \\ 0 & 1 \end{bmatrix} = \alpha(r+s).$$

It is clearly not onto, but it is one-to-one because $\alpha(r) = \alpha(s)$ implies $\begin{bmatrix} 1 & r \\ 0 & 1 \end{bmatrix} = \begin{bmatrix} 1 & s \\ 0 & 1 \end{bmatrix}$, so $r = s$.

3. If α is an automorphism, then $a^{-1}b^{-1} = \alpha(a)\cdot\alpha(b) = \alpha(ab) = (ab)^{-1} = b^{-1}a^{-1}$ for all a, b. Thus G is abelian. Conversely, if G is abelian,

$$\alpha(ab) = (ab)^{-1} = b^{-1}a^{-1} = a^{-1}b^{-1} = \alpha(a)\cdot\alpha(b),$$

so α is a homomorphism; α is a bijection because $\alpha^{-1} = \alpha$.

5. $\sigma_a = 1_G$ if and only if $aga^{-1} = g$ for all $g \in R$, if and only if $ag = ga$ for all $g \in R$.

7. Let $\alpha : \mathbb{Z} \to \mathbb{Z}_n$ be given by $\sigma(k) = \bar{k}$. Then $o(1) = \infty$ in $\mathbb{Z}$, but $o(\alpha(\bar{1})) = o(1) = n$ in $\mathbb{Z}_n$.

8. (a) If $\alpha : \mathbb{Z} \to \mathbb{Z}$ is a homomorphism, let $\alpha(1) = m$. Then

$$\alpha(k) = \alpha(k\cdot 1) = k[\alpha(1)] = km.$$

Thus α is multiplication by m, and each such map is a homomorphism $\mathbb{Z} \to \mathbb{Z}$.

9. $1 \in K$ because $\alpha(1) = 1$. If $g, h \in K$ then $\alpha(ab) = \alpha(g)\alpha(h) = 1\cdot 1 = 1$ and $\alpha(g^{-1}) = \alpha(g)^{-1} = 1^{-1} = 1$. Thus $gh \in K$ and $g^{-1} \in K$, so the subgroup test applies.

11. It is not difficult to show this from the formuls $\sigma_a(g) = a^{-1}ga$ for all $g \in G$. Howevef, we show that $\sigma_a : G \to G$ is a bijection by showing that it has an inverse. Indeed, $\sigma_a\sigma_{a^{-1}} = 1_G$ because $\sigma_a\sigma_{a^{-1}}(g) = a[a^{-1}ga]a = g = 1_G(g)$ for all $g \in G$. Similarly, $\sigma_{a^{-1}}\sigma_a = 1_G$ so $\sigma_{a^{-1}}$ is the inverse of σ_a.

12. (a) Yes. Bijection since

$$\alpha^{-1}(x) = \tfrac{1}{2}x\cdot\alpha(x+y) = 2(x+y) = 2x + 2y = \alpha(x) + \alpha(y).$$

(c) No. $\sigma(1) = 1^2 = 1$ and $\alpha(4) = 4^2 = 1$, so α is not one-to-one.

(e) Yes. $\alpha(g+h) = 2(g+h) = 2g + 2h = \alpha(g)\alpha(h)$; α is a bijection:

$$\alpha(\bar{0}) = \bar{0}, \alpha(\bar{1}) = \bar{2}, \alpha(\bar{2}) = \bar{4}, \alpha(\bar{3}) = \bar{6}, \alpha(\bar{4}) = \bar{1}, \alpha(\bar{5}) = \bar{3}, \alpha(\bar{6}) = \bar{5}.$$

(g) Yes. $\alpha(gh) = (gh)^2 = \alpha(g)\cdot\alpha(h)$; bijection because $\alpha^{-1}(g) = \sqrt{g}$.

(i) Yes. $\alpha(2k + 2m) = 3(k + m) = \alpha(2k) + \alpha(2m)$; α is one-to-one because $\alpha(2k) = \alpha(2m) \Rightarrow 3k = 3m \Rightarrow k = m$.

13. If $A = \begin{bmatrix} 0 & -1 \\ 1 & 0 \end{bmatrix}$, then $A^2 = \begin{bmatrix} -1 & 0 \\ 0 & -1 \end{bmatrix}$, $A^3 = \begin{bmatrix} 0 & 1 \\ -1 & 0 \end{bmatrix}$ and $A^4 = \begin{bmatrix} 1 & 0 \\ 0 & 1 \end{bmatrix}$. Thus $G = \{I, A, A^2, A^3\}$ and $A^4 = I$. Similarly $\{1, i, -1, -i\} = \{1, i, i^2, i^3\}$ and $i^4 = 1$. They are both cyclic of order 4.

15. We have $G = \langle a \rangle$ where $o(a) = n$, so we define $\sigma : \mathbb{Z}_n \to G$ by $\sigma(\bar{k}) = a^k$. To see that this mapping is well defined, recall that $\bar{k} = \bar{m}$ in $\mathbb{Z}_n \Leftrightarrow a^k = a^m$ by

Theorem 2, §2.4. Hence σ is well defined (by $\Rightarrow$), and as a bonus σ is one-to-one (by $\Leftarrow$). Since σ is clearly onto, it remains to verify that it is a homomorphism:

$$\sigma(\bar{k} + \bar{m}) = \sigma(\overline{k + m}) = a^{k+m} = a^k a^m = \sigma(\bar{k}) \cdot \sigma(\bar{m}).$$

Hence σ is an isomorphism.

17. Let $\sigma_g : G \to G$ be the inner automorphism given by $\sigma_g(a) = g^{-1}ag$. Then $\sigma_g(gh) = g^{-1}(gh)g = hg$. Hence $\sigma_g : \langle gh \rangle \to \langle hg \rangle$ is onto. It is clearly one-to-one and a homomorpism. Hence $\langle gh \rangle \cong \langle hg \rangle$, so $o(hg) = o(gh)$ by Theorem 4.

19. Let $z \in Z(G)$. If $g_1 \in G_1$, write $g_1 = \sigma(g)$, $g \in G$. Then

$$g_1 \cdot \sigma(z) = \sigma(g) \cdot \sigma(z) = \sigma(gz) = \sigma(zg) = \sigma(z) \cdot g_1,$$

so $\sigma(z) \in Z(G_1)$. Thus $\sigma[Z(G)] \subseteq Z(G_1)$. Now let $w \in Z(G_1)$ and write $w = \sigma(z)$, $z \in G$. Given $g \in G$, $\sigma(zg) = w \cdot \sigma(g) = \sigma(g) \cdot w = \sigma(gz)$, so $zg = gz$ because σ is one-to-one. Hence $z \in Z(G)$, so $w = \sigma(z) \in \sigma[Z(G)]$.

21. $\mathbb{Z}_{10}^* = \{\bar{1}, \bar{3}, \bar{7}, \bar{9}\} = \langle \bar{3} \rangle$ is cyclic because $\bar{3}^2 = \bar{9}$, $\bar{3}^3 = \bar{7}$, and $\bar{3}^4 = \bar{1}$. However $\mathbb{Z}_{12}^* = \{\bar{1}, \bar{5}, \bar{7}, \overline{11}\}$ is not cyclic ($\bar{5}^2 = \bar{7}^2 = \overline{11}^2 = \bar{1}$).

23. Suppose $\sigma : \mathbb{C}^\circ \to \mathbb{R}^*$ is an isomorphism. Write $w = e^{2\pi i/3} \in \mathbb{C}^\circ$, so $w^3 = 1$. Write $\sigma(w) = r$. Then $r^3 = [\sigma(w)]^3 = \sigma(w^3) = \sigma(1) = 1$. This means $r = 1$, so $\sigma(w) = 1$. Thus $w = 1$, a contradiction. So no such σ exists.

25. $\mathbb{Z}$ is infinite cyclic, so $\mathbb{Z} \cong \mathbb{Q}$ means $\mathbb{Q}$ is infinite cyclic too. Suppose that $\mathbb{Q} = \langle q \rangle = \{kq \mid k \in \mathbb{Z}\}$. In particular $q^2 = k_0 q$, so $q = k_0 \in \mathbb{Z}$. Thus $\mathbb{Q} = \{kk_0 \mid k \in \mathbb{Z}\} \subseteq \mathbb{Z}$, a contradiction.

27. Let $\sigma : G \to G_1$ be an automorphism. Then $o(\sigma(a)) = o(a) = 6$, so $\sigma(a) = b$ or $\sigma(a) = b^5 = b^{-1}$. If $\sigma(a) = b$, then $\sigma(a^k) = b^k$ for all $k \in \mathbb{Z}$, while $\sigma(a) = b^{-1}$ gives $\sigma(a^k) = b^{-k}$ for all $k \in \mathbb{Z}$. Thus these are the only possible isomorphisms. If we define λ and $\mu : G \to G_1$ by $\lambda(a^k) = b^k$ and $\mu(a^k) = b^{-k}$, we have $a^k = a^m \Leftrightarrow k \equiv m(\text{mod } 6) \Leftrightarrow b^k = b^m$, so λ is well-defined and one-to-one. It is clearly onto, and is easily checked to be an isomorphism. Similarly μ is an isomorphism.

29. $\tau_{a',b'}(\tau_{a,b}(x)) = a'(ax + b) + b' = a'ax + (a'b + b') = \tau_{a'a,a'b+b'}(x)$. Thus G_1 is closed, clearly $1_\mathbb{R} = \tau_{1,0}$, and $\tau_{a,b}^{-1} = \tau_{a^{-1},-a^{-1}b}$ as is easily verified. So G_1 is a subgroup of $S_\mathbb{R}$. Similarly, $\begin{bmatrix} a & b \\ 0 & 1 \end{bmatrix}\begin{bmatrix} a' & b' \\ 0 & 1 \end{bmatrix} = \begin{bmatrix} aa' & ab'+b \\ 0 & 1 \end{bmatrix}$, so G is closed; $\begin{bmatrix} 1 & 0 \\ 0 & 1 \end{bmatrix} \in G$; $\begin{bmatrix} a & b \\ 0 & 1 \end{bmatrix}^{-1} = \begin{bmatrix} a^{-1} & -a^{-1}b \\ 0 & 1 \end{bmatrix} \in G$, so G is a subgroup of $GL_2(\mathbb{R})$. Now define $\sigma : G \to G_1$ by $\sigma\begin{bmatrix} a & b \\ 0 & 1 \end{bmatrix} = \tau_{a,b}$. This is clearly onto; it is one-to-one because $\tau_{a,b} = \tau_{a_1,b_1} \Rightarrow ax + b = a_1 x + b_1$ for all x, $\Rightarrow a = a_1$, $b = b_1$. Finally

$$\sigma\left[\begin{bmatrix} a & b \\ 0 & 1 \end{bmatrix}\begin{bmatrix} a' & b' \\ 0 & 1 \end{bmatrix}\right] = \sigma\begin{bmatrix} aa' & ab'+b \\ 0 & 1 \end{bmatrix} = \tau_{aa',ab'+b} = \tau_{a,b} \cdot \tau_{a',b'} = \sigma\begin{bmatrix} a' & b' \\ 0 & 1 \end{bmatrix}\sigma\begin{bmatrix} a' & b \\ 0 & 1 \end{bmatrix}.$$

So σ is an isomorphism.

31. (a) If $o(a) = 2$, let $\sigma : G \to G$ be an automorphism. Then $o(\sigma(a)) = o(a) = 2$, so $\sigma(a) = a$. This means $\sigma = 1_G$, so $\text{aut } G = \{1_G\}$.

33. Define $\theta : G \to \text{inn } G$ by $\theta(a) = \sigma_a$ for each $a \in G$, where $\sigma_a(g) = a^{-1}ga$ for all g. Then θ is an onto homomorphism by Example 17. So, if $Z(G) = \{1\}$, we must show that θ is one-to-one. If $\theta(a) = \theta(b)$, then $\sigma_a = \sigma_b$, so $a^{-1}ga = b^{-1}gb$ for all

$g \in G$. Thus $g(ab^{-1}) = (ab^{-1})g$ for all g, whence $ab^{-1} \in Z(G) = \{1\}$. This gives $b = a$, and shows θ is one-to-one, as required.

35. (a) If $\sigma, \tau \in S(g)$, then $(\sigma\tau)g = \sigma(\tau g) = \sigma(g) = g$, so $\sigma\tau \in S(g)$. Since $\sigma(g) = g$ we get $g = \sigma^{-1}(g)$, so $\sigma^{-1} \in S(g)$; $\varepsilon \in S(g)$ is clear.

36. (a) $a \sim a$ because $a = 1a1^{-1}$; if $a \sim b$, then $b = gag^{-1}$, so $a = g^{-1}bg$, $b \sim a$. If $a \sim b$, $b \sim c$, then $b = gag^{-1}$, $c = hbh^{-1}$, so

$$c = h(gag^{-1})h^{-1} = (hg)a(hg)^{-1}.$$

Hence $c \sim a$.

37. If $g_1 \in G_1$, write $g_1 = \sigma(g)$, $g \in G$. Now $G = \langle X \rangle$, so

$$g = x_1^{k_1}x_2^{k_2}\cdots x_m^{k_m}, x_i \in X, k_i \in \mathbb{Z}.$$

Thus $g_1 = \sigma(x_1^{k_1}x_2^{k_2}\cdots x_m^{k_m}) = \sigma(x_1)^{k_1}\sigma(x_2)^{k_2}\cdots\sigma(x_m)^{k_m} \in \langle \sigma(X) \rangle$.   Thus $G_1 = \langle \sigma(X) \rangle$.

2.6   COSETS AND LAGRANGE'S THEOREM

1. (a) $1H = H1 = \{1, a^4, a^8, a^{12}, a^{16}\}$
      $aH = Ha = \{a, a^5, a^9, a^{13}, a^{17}\}$
      $a^2H = Ha^2 = \{a^2, a^6, a^{10}, a^{14}, a^{18}\}$
      $a^3H = Ha^3 = \{a^3, a^7, a^{11}, a^{15}, a^{19}\}$
      $1K = K1 = \{1, a^2, a^4, a^6, a^8, a^{10}, a^{12}, a^{14}, a^{16}, a^{18}\}$
      $aK = Ka = \{a, a^3, a^5, a^7, a^9, a^{11}, a^{13}, a^{15}, a^{17}, a^{19}\}$

   (c) $0 + H = H + 0 = H = \{2k \mid k \in \mathbb{Z}\}$ $0 + K = K + 0 = \{3k \mid k \in \mathbb{Z}\}$
      $1 + H = H + 1 = \{2k + 1 \mid k \in \mathbb{Z}\}$ $1 + K = K + 1 = \{3k + 1 \mid k \in \mathbb{Z}\}$
      $2 + K = K + 2 = \{3k + 2 \mid k \in \mathbb{Z}\}$

   (e) $1H = H1 = \{1, a^2\}$ $aH = Ha = \{a, a^3\}$
      $bH = Hb = \{b, ba^2\}$ $baH = Hba = \{ba, ba^3\}$
      $K1 = \{1, b\}$ $1K = \{1, b\}$
      $Ka = \{a, ba\}$ $aK = \{a, ba^3\}$
      $Ka^2 = \{a^2, ba^2\}$ $a^2K = \{a^2, ba^2\}$
      $Ka^3 = \{a^3, ba^3\}$ $a^3K = \{a^3, ba\}$

3. No. If $H = \{1, b\} \subseteq D_3$, then $Ha = \{a, ba) = Hba$, but $aH = \{a, ba^2\} \neq baH$.

5. (a) $a \equiv a$ because $a^{-1}a = 1 \in H$. If $a \equiv b$ then $b^{-1}a \in H$, whence $a^{-1}b = (b^{-1}a)^{-1} \in H$, so $b \equiv a$. Finally, if $a \equiv b$ and $b \equiv c$ then $b^{-1}a \in H$ and $c^{-1}b \in H$, so $c^{-1}a = (c^{-1}b)(b^{-1}a) \in H$. Thus $a \equiv c$.

7. If $Ha = bH$, then $a \in Ha$ gives $a \in bH$, so $aH = bH$. Thus $aH = Ha$. Similarly $bH = Hb$, so $aH = Ha = bH = Hb$.

9. (a) If $x \in \mathbb{R}^*$, then $\mathbb{R}^+x$ equals $\mathbb{R}^+$ or $\mathbb{R}^+(-1) = \{r \mid r < 0\}$, according as $x > 0$ or $x < 0$. Here $\mathbb{R}^+$ is the set of positive real numbers, and $\mathbb{R}^+(-1)$ is the set of negative real numbers.

(c) If $x \in \mathbb{R}$, write $x = n + t$, $n \in \mathbb{Z}$, $0 \leq t < 1$. Then $\mathbb{Z} + x = \mathbb{Z} + t$ consists of all points on the line at distance t to the right of an integer.

10. (a) Write $H = \langle a^6 \rangle$. Then $|H| = o(a^6) = \frac{30}{6} = 5$, so $|G : H| = \frac{|G|}{|H|} = \frac{30}{5} = 6$.

11. (a) $(H \cap K)a \subseteq Ha \cap Ka$ is clear. If $x \in Ha \cap Ka$, write $x = ha = ka$, $h \in H$, $k \in K$. Then $h = k$ by cancellation, so $h \in H \cap K$. Thus $x = ha \in (H \cap K)a$.

12. (a) If $o(g) = m$, we show $m = 12$. We have $m \mid 12$ by Lagrange's theorem, so m is one of $1, 2, 3, 4, 6$ or 12. If $m \neq 12$, then $m|4$ or $m|6$, so $g^4 = 1$ or $g^6 = 1$, contrary to hypothesis.

    (c) Now $o(g)$ divides 60, so is one of $1, 2, 3, 4, 5, 6, 10, 12, 15, 20, 30, 60$. Again each of these except 60 divides one of 12, 20 or 30, so $o(g) = 60$.

13. If $|H| = m$, then $m|12$ and $4|m$ (because $K \subseteq H$). Thus $|H| = 4$ or $|H| = 12$; that is $H = K$ or $H = A_4$.

15. Since $H \cap K \subseteq H$, we have that $|H \cap K|$ divides the prime $|H|$. So either $|H \cap K| = 1$ (and $H \cap K = \{1\}$) or $|H \cap K| = |H|$ (so $H \cap K = H$, that is $H \subseteq K$).

16. (a) We have $g^n = 1$ by Lagrange's theorem. Since $\gcd(m, n) = 1$, write $1 = xn + ym$; $x, y \in \mathbb{Z}$. If $g^m = 1$, then $g = g^1 = (g^m)^x (g^n)^y = 1^x 1^y = 1$.

17. Let H be a subgroup of G, $H \neq G$. Then $|H| = 1$ or p by Lagrange's Theorem. If $|H| = 1$, then $H = \langle 1 \rangle$. If $|H| = p$, then H is cyclic by Corollary 3 of Lagrange's Theorem.

19. If $x = a^k = b^k$, then $x \in \langle a \rangle \cap \langle b \rangle$. But $\langle a \rangle \cap \langle b \rangle = \{1\}$ by Corollary 4 of Lagrange's Theorem because $|\langle a \rangle| = m$ and $|\langle b \rangle| = n$ are relatively prime. Thus $a^k = 1$ and $b^k = 1$, so $m \mid k$ and $n \mid k$. Thus $mn \mid k$, again because m and n are relatively prime.

21. If $n = 1$ then $|\mathbb{Z}, n\mathbb{Z}| = |\mathbb{Z}, \mathbb{Z}| = 1$. If $n \geq 2$ and $k \in \mathbb{Z}$, then $k = qn + r$, $0 \leq r \leq n - 1$, so $k - r \in n\mathbb{Z}$. Thus $n\mathbb{Z} + k = n\mathbb{Z} + r$; that is the cosets are

$$\{n\mathbb{Z} + r \mid 0 \leq r \leq n - 1\}.$$

These are distinct. For if $n\mathbb{Z} + r = n\mathbb{Z} + s$ with $0 \leq r \leq s \leq n - 1$ then $s - r \in n\mathbb{Z}$, so $0 \leq s - r \leq s \leq n - 1$. But $s - r \geq 0$ is a multiple of n; so $s - r = 0$. Thus there are exactly n cosets.

23. Let $g \in G$, $g \neq 1$. Then $o(g)$ divides $|G| = p^k$, say $o(g) = p^m$, $m \leq k$. Since $m \neq 0$, $o(g^{p^{m-1}}) = p$ by Theorem 5 §2.4.

25. (a) For $k \geq 1$ we induct on k. It is given for $k = 1$. If $a^k b a^k = b$ for some $k \geq 0$, then $a^{k+1} b a^{k+1} = aba = a$. Thus $a^k b a^k = b$ for all $k \geq 1$ by induction. It is clear if $k = 0$. But $a^{-1} b a^{-1} = a^{-1}(aba)a^{-1} = b$, and a similar argument shows $a^{-k} b a^{-k} = b$ if $k \geq 1$.

27. No. $D_5 \times C_3$ has no element of order 10, while $D_3 \times C_5$ has 12 elements of order 10.

29. (a) $H \cap K$ is a subgroup of H so $H \cap K = \{1\}$ or $H \cap K = H$ by Lagrange's Theorem. Similarly, $H \cap K \subseteq K$ implies that $H \cap K = \{1\}$ or $H \cap K = K$. Thus $H \cap K \neq \{1\}$ implies $H = H \cap K = K$.

31. If $|H:K| = n$, let $Kh_1, \ldots, Kh_n$ be the distinct cosets of K in H. Thus $H = Kh_1 \cup \cdots \cup Kh_n$, a disjoint union. Then $Hg \subseteq Kh_1g \cup \cdots \cup Kh_ng$ is clear, and it is equality because $K \subseteq H$. Thus each H-coset in G is the union of n K-cosets. If $|G:H| = m$ this gives $|G:K| = mn = |G:H|\,|H:K|$. Conversely, if $|G:K|$ is finite, then $|H:K|$ is clearly finite and $|G:H|$ is finite by the hint since each H-coset is a union of K-cosets.

32. (a) $(H \cap K)g \subseteq Hg \cap Kg$ is clear. If $x \in Hg \cap Kg$, write $x = hg = kg$, $h \in H$, $k \in K$. Then $h = k \in H \cap K$ by cancellation, so $x = hg \in (H \cap K)g$. Hence $(H \cap K)g = Hg \cap Kg$, so each $(H \cap K)g$ coset is the intersection of one of the m H-cosets with one of the n K-cosets. There are thus at most mn $H \cap K$-cosets.

33. If $H_1, \cdots, H_n$ are all of finite index in G, we show $H_1 \cap \cdots \cap H_k$ is of finite index for each $k = 1, 2, \ldots, n$. This is clear if $k = 1$. If it holds for some k, then $H_1 \cap \cdots \cap H_{k+1} = (H_1 \cap \cdots \cap H_k) \cap H_{k+1}$ is a finite index by part (a) of the preceding exercise.

35. (a) $a \equiv a$ for all a because $a = 1a1$. If $a \equiv b$, then $a = hbk$, $h \in H$, $k \in K$, so $b = h^{-1}ak^{-1}$, that is $b \equiv a$. If $a \equiv b$ and $b \equiv c$, then $a = hbk$, $b = h_1ck_1$, so $a = (hh_1)c(kk_1)$. Thus $a \equiv c$.

2.7 GROUPS OF MOTIONS AND SYMMETRIES

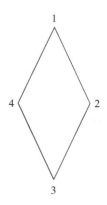

1. Label the figure as shown. Clearly (1 3) and (2 4) are motions, as is their product. Hence the group of motions is $\{\varepsilon, (1\ 3), (2\ 4), (1\ 3)(2\ 4)\}$, isomorphic to the Klein group K_4.

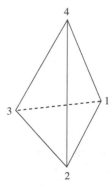

3. Label the figure as shown. Then (1 2 3) and (1 3 2) are motions (rotations of 120° and 240° about a line through vertex 4 and the center of the triangle base). Clearly every motion (indeed every symmetry) must fix vertex 4. Hence the group of motions is $G = \{\varepsilon, (1\ 2\ 3), (1\ 3\ 2)\}$. However (1 2), (1 3) and (2 3) are all symmetries (which are not motions), so the group of symmetries if S_3.

5. Label the figure as shown. Clearly $(1\,2)(3\,4)$ and $(1\,4)(2\,3)$ are such symmetries, and hence their product is $(1\,3)(2\,4)$. The rest of the symmetries of the square do not preserve blue edges, so the group is $\{\varepsilon, (1\ 2)(3\ 4), (1\ 3)(2\ 4), (1\ 4)(2\ 3)\} \cong K_4$.

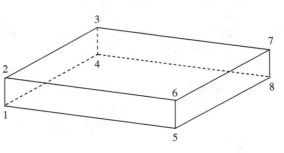

7. Label the vertices as shown. Let $\lambda = (1\,3)(2\,4)(5\,7)(6\,8)$ and $\mu = (1\,6)(2\,5)(3\,8)(4\,7)$. These are motions (rotations of π radians about axes through the sides). Also $\lambda\mu = \mu\lambda = (1\,8)(2\,7)(3\,6)(4\,5)$ is the sides). Also $\lambda\mu = \mu\lambda = (1\,8)(2\,7)(3\,6)(4\,5)$ is the rotation about a vertical axis. The group of motions is $\{\varepsilon, \lambda, \mu, \lambda\mu\} \cong K_4$.

However, there are symmetries which are not motions. We have
$$\sigma = (1\,5)(2\,6)(3\,7)(4\,8),$$
$$\tau = (1\,4)(2\,3)(5\,8)(6\,7), \text{ and}$$
$$\gamma = (1\,2)(3\,4)(5\,6)(7\,8)$$
which are reflections in various planes of symmetry. Now compute
$$\sigma\tau = \tau\sigma = \lambda\mu$$
$$\lambda\tau = (1\,2)(3\,4)(5\,6)(7\,8) = \tau\lambda = \mu\sigma = \sigma\mu = \gamma$$
$$\lambda\sigma = (1\,7)(2\,8)(3\,5)(4\,6) = \sigma\lambda = \mu\tau = \tau\mu = \delta$$
Call these last two $\gamma = \lambda\tau$ and $\delta = \lambda\sigma$. Then the group of symmetries is
$$G = \{\varepsilon, \lambda, \mu, \lambda\mu, \sigma, \tau, \gamma, \delta\}.$$
The fact that $x^2 = \varepsilon$ for all $x \in G$ gives the following multiplication table:

	ε	λ	μ	$\lambda\mu$	σ	τ	γ	δ
ε	ε	λ	μ	$\lambda\mu$	σ	τ	γ	δ
λ	λ	ε	$\lambda\mu$	μ	δ	γ	τ	σ
μ	μ	$\lambda\mu$	ε	λ	γ	δ	σ	τ
$\lambda\mu$	$\lambda\mu$	μ	λ	ε	τ	σ	δ	γ
σ	σ	δ	γ	τ	ε	$\lambda\mu$	μ	λ
τ	τ	γ	δ	σ	$\lambda\mu$	ε	λ	μ
γ	γ	τ	σ	δ	μ	λ	ε	$\lambda\mu$
δ	δ	σ	τ	γ	λ	μ	$\lambda\mu$	ε

The group is abelian and $x^2 = \varepsilon$ for all x. These are called elementary abelian groups.

2.8 NORMAL SUBGROUPS

1. (a) $H = \{1, a^6, b, ba^6\}$ is a subgroup because
$$a^6 b = ba^{12-6} = ba^6, b^2 = (ba^6)^2 = 1 = (a^6)^2.$$
It is not normal in D_4 because $Ha = \{a, a^7, ba, ba^7\}$ while $aH = \{a, a^7, ba^{11}, ba^5\}$.

   (c) This is closed because $a^2 b = ba^{10}$, $a^4 b = ba^8$, $a^6 b = ba^6$; $(ba^k)^2 = 1$ for each k. Hence H is a subgroup. It is normal, being of index 2.

3. A_4 has no subgroup of order 6 [Exercise 34 §2.6] so the only subgroups (normal or not) have order 1, 2, 3 or 4. If $|H| = 4$, it contains no 3-cycle and so equals K. Thus $K \lhd A_4$ being unique of its order. If $|H| = 3$, let $H = \langle (1, 2, 3) \rangle$ without loss of generality. But $(1\ 4)^{-1}(1\ 2\ 3)(1\ 4) = (2\ 3\ 4)$ shows H is not normal. Finally $|H| = 2 \Rightarrow H = \{\varepsilon, (1\ 2)\}$ without loss of generality. But $(1\ 3)^{-1}(1\ 2)(1\ 3) = (2\ 3)$ shows H is not normal.

5. First aKa^{-1} is a subgroup, by Theorem 5 §2.3, and $aKa^{-1} \subseteq aHa^{-1} \subseteq H$ because $H \lhd G$. If $h \in H$, we must show $h(aKa^{-1})h^{-1} \subseteq aKa^{-1}$. We have $h^{-1}Kh = K$ because $K \lhd H$, so
$$h(aKa^{-1})h^{-1} = ha(h^{-1}Kh)a^{-1}h^{-1} = (hah^{-1})K(hah^{-1})^{-1} = K$$
because $K \lhd G$.

7. Let $H = \{1, h\}$. Given $g \in G$, we have $g^{-1}hg \in H$ because $H \lhd G$. But $g^{-1}hg \neq 1$ because $h \neq 1$, so $g^{-1}hg = h$. Thus $hg = gh$ for all $g \in G$; that is $h \in Z(G)$. If $H = \{1, a, a^2\} \subseteq D_3 = \{1, a, a^2, b, ba, ba^2\}$ then $H \lhd D_3$ but $Z(D_3) = \{1\}$.

9. $(1, 1) \in D$, $(g, g)^{-1} = (g^{-1}, g^{-1}) \in D$; $(g, g)(g_1, g_1) = (gg_1, gg_1) \in D$. So D is always a subgroup of $G \times G$. If G is abelian, then $G \times G$ is abelian so every subgroup is normal. Conversely, if D is normal, let $g, a \in G$. Then $(a, g)^{-1}(g, g)(a, g) \in D$, that is $(a^{-1}ga, g) \in D$. This means $a^{-1}ga = g$, so $ga = ag$. Hence G is abelian.

11. Let H and K be subgroups of G with $|H| = p$ and $|K| = q$. Then $H \cap K = \{1\}$ by Lagrange's theorem. Moreover $H \lhd G$ because it is unique of its order, and similarly $K \lhd G$. Hence $G \cong H \times K$ by Corollary 2 of Theorem 6. Since p and q are primes, H and K are cyclic of relatively prime orders. Hence $H \times K$ is cyclic by Exercise 25 §2.4.

13. (a) Let $g \in G$, $h \in H$. Write $g = x_1^{k_1} \cdots x_r^{k_r}$, $x_i \in X$, $k_i \in \mathbb{Z}$. Then
$$ghg^{-1} = x_1^{k_1} \cdots x_r^{k_r} h x_r^{-k_r} \cdots x_1^{-k_1},$$
so it suffices to show $x^k h x^{-k} \in H$ for all $x \in X$, $k \in \mathbb{Z}$. But then $x^k h x^{-k} = (xhx^{-1})^k \in H$ because $xhx^{-1} \in H$ by hypothesis. The converse is clear.

15. (a) Since $a \notin K$, $Ka \neq K$, so $Ka = G \backslash K$ (because $|G : K| = 2$). Similarly $b^{-1} \notin K$ gives $Kb^{-1} = G - K$. Thus $Ka = Kb^{-1}$, so $ab \in K$.

17. (a) If $H = \langle a^d \rangle$, $d \mid n$, let $n = md$. Since ba^k is selfinverse for all k, we have
$$(ba^k)^{-1} a^{dt}(ba^k) = ba^k a^{dt} ba^k = b(a^{k+dt})ba^k = b \cdot b \cdot a^{-k-dt} \cdot a^k$$
$$= b^2 a^{-k} a^{-dt} a^k = a^{-dt} \in H.$$

19. Assume that $HK \subseteq KH$. If $g \in KH$, say $g = kh$, then $g^{-1} = h^{-1}k^{-1} \in HK \subseteq$ KH. If $g^{-1} = k_1 h_1$, then $g = h_1^{-1}k_1^{-1} \in HK$, proving that $KH \subseteq HK$. Hence $HK = KH$ and Lemma 2 applies.

21. $HK \subseteq \langle H \cup K \rangle$ always holds because $\langle H \cup K \rangle$ is a subgroup containing H and K. But HK is itself a subgroup (Lemma 2) and contains H and K, so $\langle H \cup K \rangle \subseteq KH$.

23. (a) Let $D_n = \{1, a, a^2, \dots, a^{n-1}, b, ba, ba^2, \dots, ba^{n-1}\}$ where $o(a) = n$, $o(b) = 2$ and $aba = b$. If $n = 2m$, let $H = \{1, a^m\}$. We have $a^m b = ba^{n-m} = ba^m$, so $H \subseteq Z(D_n)$. Thus $H \lhd D_n$. Let $K = \{1, a^2, a^4, \dots, a^{n-2}, b, ba^2, \dots, ba^{n-2}\}$. We have $o(a^2) = m$, $o(b) = 2$, and $a^2 ba^2 = a(aba)a = aba = b$. Thus $K \cong D_m$. Moreover, $K \lhd D_n$ because it is of index 2. But then we have $G \cong H \times K \cong C_2 \times D_m$ by Corollary 2 of Theorem 6 since
$$|C_2| \, |D_n| = 2 \cdot 2m = 2n = |D_n|.$$

24. (a) If H is characteristic in G then $aHa^{-1} = \sigma_a(H) \subseteq H$ where σ_a is the inner automorphism.

(c) Let $G = \langle a \rangle \times \langle a \rangle$ where $o(a) = 2$, and define $\sigma : G \to G$ by $\sigma(x,y) = (y,x)$. It is easy to verify that σ is an automorphism of G and, if $H = \langle (a,1) \rangle = \{(1,1), (a,1)\}$, then $H \lhd G$, but $\sigma(H) \nsubseteq H$, so H is not characteristic in G.

(e) If $a \in G$, let $\sigma_a : G \to G$ denote the inner automorphism. Then $\sigma_a(K) = K$ because $K \lhd G$, so $\sigma_a : K \to K$ is an automorphism of K. Hence $\sigma_a(H) = H$ because H is characteristic in K, that is $aHa^{-1} = H$.

(g) If $G = \langle a \rangle$ and $H = \langle a^m \rangle$, let $\sigma \in \text{aut } G$ and let $\sigma(a) = a^n$. If $h = a^{mk} \in H$, then $\sigma(h) = [\sigma(a)]^{mk} = a^{nmk} = (a^{nk})^m \in H$. Thus $\sigma(H) \subseteq H$. If G is finite it is easier: $\sigma(H)$ is a subgroup of G of the same order as H so (Theorem 7 §2.4) $\sigma(H) = H$. It fails if G is abelian by (c).

(i) Clearly $K \subseteq 1_G(H) = H$. If $\tau \in \text{aut } G$, then $K \subseteq \tau^{-1}\sigma(H)$ for all $\sigma \in \text{aut } G$ by the definition of K, so $\tau(K) \subseteq \sigma(H)$ for all σ. Thus $\tau(K) \subseteq K$. Similarly, $\tau^{-1}(K) \subseteq K$, so $K \subseteq \tau(K)$. Thus $K = \tau(K)$ and K is characteristic. Finally, let $C \subseteq H$, C a characteristic subgroup of G. If $\sigma \in \text{aut } G$, then $C = \sigma(C) \subseteq \sigma(H)$, so $C \subseteq K$ by the definition of K.

25. (a) We have $1 \in N(X)$ because $1X1^{-1} = X$. If $a, b \in N(X)$, then we have $aXa^{-1} = X = bXb^{-1}$. Thus $a^{-1} \in N(X)$ because
$$a^{-1}X(a^{-1})^{-1} = a^{-1}Xa = a^{-1}(aXa^{-1})a = X,$$
and $ab \in N(X)$ because $(ab)X(ab)^{-1} = a[bXb^{-1}]a^{-1} = a[X]a^{-1} = X$. Hence $N(X)$ is a subgroup of G.

(c) Suppose K is a subgroup of G and $H \lhd K$. If $k \in K$, then $k^{-1}Hk = H$ by Theorem 3, so $k \in N(H)$. Thus $K \subseteq N(H)$.

26. (a) Write $K = \text{core } H = \cap_a aHa^{-1}$. Clearly $1 \in K$. If $g, g_1 \in K$, then $gg_1 \in aHa^{-1}$ for all a, so $gg_1 \in K$. Also, $g^{-1} \in a^{-1}H(a^{-1})^{-1}$ for all a, so $g \in aHa^{-1}$ and $g \in K$. Hence K is a subgroup. If $g \in G$ and $k \in K$ then $gkg^{-1} \in g[(g^{-1}a)H(g^{-1}a)^{-1}]g^{-1} = aHa^{-1}$ for all a, as required.

(c) core $H \cap K \subseteq H \cap K \subseteq H$ and core $H \cap K \lhd G$ gives

$$\text{core } H \cap K \subseteq \text{core } H.$$

Similarly, core $(H \cap K) \subseteq$ core K, so core $(H \cap K) \subseteq$ core $H \cap$ core K. On the other hand, core $H \subseteq H$ and core $K \subseteq K$ gives core $H \cap$ core K $\subseteq$ core $H \cap K$. Since core $H \cap$ core $K \lhd G$, this gives core $H \cap$ core K $\subseteq$ core $(H \cap K)$.

27. (a) First, $\bar{X}$ is a subgroup of G. If $g, g_1 \in \bar{X}$ and $X \subseteq N \lhd G$, then $1 \in H$ and the fact that $g, g_1 \in N$ implies that $g^{-1} \in N$ and $gg_1 \in N$. Thus $1, g^{-1}, gg_1 \in \bar{X}$, so $\bar{X}$ is a subgroup. But if $a \in G$ and $g \in \bar{X}$, then we have $aga^{-1} \in aNa^{-1} = N$ for all $N \lhd G$, $X \subseteq N$, so $aga^{-1} \in \bar{X}$. Thus $a\bar{X}a^{-1} \subseteq \bar{X}$ for all $a \in G$, so $\bar{X} \lhd G$. Clearly $X \subseteq \bar{X}$.

(c) $H \cap K \subseteq \bar{H} \cap \bar{K} \lhd G$ so $\overline{H \cap K} \subseteq \bar{H} \cap \bar{K}$. If $G = S_3$, $H = \{\varepsilon, \sigma, \sigma^2\}$ and $K = \{\varepsilon, \tau\}$, then $\bar{H} = H$, $\bar{K} = G$, so $\overline{H \cap K} = \{1\} \subset H = \bar{H} \cap \bar{K}$.

28. (a) If $c, c_1 \in C(X)$, then $(cc_1)x = c(c_1x) = c(xc_1) = (cx)c_1 = xcc_1$, for all $x \in X$, so $cc_1 \in C(X)$. Since $cx = xc$ for all $x \in X$, it follows that $xc^{-1} = c^{-1}x$; that is $c^{-1} \in C(X)$. Finally, $1 \in C(X)$ is clear.

2.9 FACTOR GROUPS

1. (a) $D_6 = \{1, a, a^2, a^3, a^4, a^5, b, ba, ba^2, ba^3, ba^4, ba^5\}$; $K = Z(D_6) = \{1, a^3\}$ by Exercise 26 §2.6. The cosets are

$$K = \{1, a^3\} \qquad Ka = \{a, a^4\}$$
$$Ka^2 = \{a^3, a^5\} \qquad Kb = \{b, ba^3\}$$
$$Kba = \{ba, ba^4\} \qquad Kba^2 = \{ba^2, ba^5\}$$

	K	Ka	Ka^2	Kb	Kba	Kba^2
K	K	Ka	Ka^2	Kb	Kba	Kba^2
Ka	Ka	Ka^2	K	Kba^2	Kb	Kba
Ka^2	Ka^2	K	Ka	Kba	Kba^2	Kb
Kb	Kb	Kba	Kba^2	K	Ka	Ka^2
Kba	Kba	Kba^2	Kb	Ka^2	K	Ka
Kba^2	Kba^2	Kb	Kba	Ka	Ka^2	K

We have $|Ka| = 3$, $|Kb| = 2$ and $KaKbKa = Kaba = Kb$. Hence $D_6/K \cong D_3$.

(c) $G = A \times B$, $K = \{(a, 1) \mid a \in A\}$. Note $K \lhd G$ because $(x, y)^{-1}(a, 1)(x, y) = (x^{-1}ax, 1) \in K$. If $b \in B$, $a \in A$, note that $(ab)(1, b)^{-1} \in K$, so

$$K(a, b) = K(1, b).$$

Thus $G/K = \{K(1, b) \mid b \in B\}$. Observe that $K(1, b)K(1, b') = K(1, bb')$ gives the Cayley table. Define $\theta : B \to G/K$ by $\theta(b) = K(1, b)$. Then θ is onto, and it is $1 : 1$ because $\theta(b) = \theta(b_1)$ gives $K(1, b) = K(1, b_1)$, so $(1, bb_1^{-1}) = (1, b)(1, b_1)^{-1} \in K$. Hence $bb_1^{-1} = 1$, $b = b_1$. Finally, $\theta(b) \cdot \theta(b_1) = K(1, b)K(1, b_1) = K(1, bb_1) = \theta(bb_1)$. So θ is an isomorphism; $G/K \cong B$.

3. (a) We have $G = \langle a \rangle$, so $G/K = \langle Ka \rangle$. We claim $|Ka| = 12$. Certainly $(Ka)^{12} = Ka^{12} = K$. But $|G/K| = |G|/|K| = 24/2 = 12$, so, $|Ka| = 12$ because Ka generates G/K. Now $|(Ka)^2| = 12/2 = 6$ by Theorem 7 §2.4. Similarly, $|Ka^3| = |(Ka)^3| = 12/3 = 4$, and $|Ka^4| = |(Ka^4)| = 12/4 = 3$. Finally, $Ka^5 = (Ka)^5$ is a generator of G/K because $\gcd(5,12) = 1$ (Theorem 6 §2.4). Thus $|Ka^5| = 12$.

4. (a) $G = \langle a \rangle \times \langle b \rangle$ with $o(a) = 8$ and $o(b) = 12$, and $K = \langle (a^2, b^3) \rangle$. We want $|K(a^4, b)|$ in G/K. Then, as $(a^4, b)^{12} = (1,1) \in K$, $|K(a^4, b)| \mid 12$, so $|K(a^4, b)| \in \{2, 3, 4, 6\}$. But since none of
$$(a^4, b)^2 = (1, b^2), \ (a^4, b)^3 = (a^4, b^3), \ (a^4, b)^4 = (1, b^4)$$
or $(a^4, b)^6 = (1, b^6)$ are in K, $|K(a^4, b)| = 12$.

5. (a) We have $Ka^2 = K$ because $a^2 \in K$. So $o(Ka^2) = 1$. Next we have $(Ka^3)^2 = Ka^6 = K$, so $o(Ka^3)$ divides 2. Since
$$a^3 \notin K = \{1, a^2, a^4, a^6, a^8, a^{10}\}, \ \text{we have} \ |Ka^3| = 2.$$
Similarly $(Kba)^2 = K(ba)^2 = K$ and $ba \notin K$, so $o(Kba) = 2$. Finally $(Ka^5)^2 = Ka^{10} = K$, so $o(Ka^5) = 2$ because $a^5 \notin K$.

7. If $0 < n < m$ in $\mathbb{Z}$, then $\mathbb{Z} + \frac{1}{n} \neq \mathbb{Z} + \frac{1}{m}$ because $\frac{1}{n} - \frac{1}{m} \notin \mathbb{Z}$. Hence $\mathbb{Q}/\mathbb{Z}$ contains the infinite set $\{\mathbb{Z} + \frac{1}{n} \mid n \geq 1\}$. Now let $\mathbb{Z} + \frac{m}{n}$ be any element of $\mathbb{Q}/\mathbb{Z}$. Then $n(\mathbb{Z} + \frac{m}{n}) = \mathbb{Z} + m = \mathbb{Z}$, so $\mathbb{Z} + \frac{m}{n}$ has finite order.

9. If $o(g) = n$, then $g^n = 1$, so $(Kg)^n = Kg^n = K1 = K$. Thus $o(Kg)$ divides n.

11. Let $o(g) = n$, $g \in G$. Then $(Kg)^n = Kg^n = K$ in G/K. On the other hand, $\left| \frac{G}{K} \right| = |G : K| = m$, so $(Kg)^m = K$ by Lagrange's theorem. But $\gcd(m, n) = 1$ implies $1 = nx + my$, $x, y \in \mathbb{Z}$, so
$$Kg = (Kg)^{nx}(Kg)^{my} = KK = K.$$
Thus $g \in K$.

13. (a) If $z \in Z(G)$, then $zK \in Z(G/K)$, so $z \in K$. Then $z \in Z(K)$, so $z = 1$.

(c) If $g \in G$, then $(g^{p^n}K) = (gK)^{p^n} = K$ for some n, so $g^{p^n} \in K$. Thus $(g^{p^n})^{p^m} = 1$, so $g^{p^{m+n}} = 1$, so $o(g) = p^k$ for some $k \leq n + m$.

15. If $Kg \in G/K$, then $g \in \langle X \rangle$, say $g = x_1^{k_1} x_2^{k_2} \cdots x_r^{k_r}$, $x_i \in X$, $k_i \in \mathbb{Z}$. Then $Kg = (Kx_1)^{k_1}(Kx_2)^{k_2} \cdots (Kx_r)^{k_r} \in \langle \{Kx \mid x \in X\} \rangle$.

17. (a) If $o(g) = n$, $o(g_1) = m$, then $(gg_1)^{nm} = g^{nm}g_1^{nm} = 1$ because G is abelian, and $(g^{-1})^n = 1$.

(c) If G is torsion, it is clear that H and G/H are torsion. Conversely, if $g \in G$ then $(Hg)^n = H$ for some n, so $g^n \in H$. Thus $(g^n)^m = 1$ for some $m \neq 0$, that is $g \in T(G)$. Hence $T(G) = G$.

19. (a) If G is abelian, then $G/\{1\}$ is abelian, so $G' \subseteq \{1\}$. Thus $G' = \{1\}$.

(c) Write
$$D_6 = \{1, a, a^2, a^3, a^4, a^5, b, ba, ba^2, ba^3, ba^4, ba^5\}, o(a) = 6, o(b) = 2, aba = a.$$
Let $K = \langle a^2 \rangle = \{1, a^2, a^4\}$. Then $a^2b = ba^4$ and $a^4b = ba^2$, so $b^{-1}Kb \subseteq K$. Hence $K \triangleleft D_6$ and D_6/K is abelian (it has order 4). This means $D_6' \subseteq K$. But $D_6' \neq \{1\}$ because D_6 is not abelian. Thus $D_6' = K$.

21. Given commutators $[a, b]$ in G' and $[x, y]$ in H', we have
$$([a, b], [x, y]) = [(a, x), (b, y)] \in (G \times H)'.$$
If $(g, h) \in G' \times H'$, then $(g, h) = (g, 1)(1, h)$ and each of $(g, 1)$ and $(1, h)$ are products of commutators of the form $([a, b], [x, y])$. Hence $G' \times H' \subseteq (G \times H)'$. Conversely $[(a, x), (b, y)] = ([a, x], [b, y])$ shows that every commutator of $G \times H$ lies in $G' \times H'$, so $(G \times H)' \subseteq G' \times H'$.

23. (a) The unity $K = K1$ of G/K is in H/K because $1 \in H$. If Kh and Kh' are in H/K where $h, h' \in H$, then $Kh \cdot Kh' = Khh' \in H/K$ and also $(Kh)^{-1} = Kh^{-1} \in H/K$. Use the subgroup test.

25. (a) $[Ka, Kb] = (Ka)(Kb)(Ka^{-1})(Kb^{-1}) = K(aba^{-1}b^{-1}) = K[a, b]$.

27. Let $\sigma : G \to G$ be an automorphism. We must show $\sigma(H) = H$. Define $\bar{\sigma} : G/K \to G/K$ by $\bar{\sigma}(Kg) = K[\sigma(g)]$. Since K is characteristic in G,
$$Kg = Kg_1 \Leftrightarrow gg_1^{-1} \in K \Leftrightarrow \sigma(g)[\sigma(g_1)]^{-1} \in \sigma(K) = K \Leftrightarrow K[\sigma(g)] = K[\sigma(g_1)].$$
Hence $\bar{\sigma}$ is well-defined and one-to-one; it is onto because σ is onto. Finally,
$$\bar{\sigma}(Kg \cdot Kg_1) = K\sigma(gg_1) = K[\sigma(g)] \cdot K[\sigma(g_1)] = \bar{\sigma}(Kg) \cdot \bar{\sigma}(Kg_1).$$
Hence $\bar{\sigma}$ is an automorphism of G/K, so $\bar{\sigma}(H/K) = HK$ by hypothesis. Hence, if $h \in H$, $\bar{\sigma}(Kh) = Kh_1$, $h_1 \in H$, so $[\sigma(h)]h_1^{-1} \in K \subseteq H$. Thus $\sigma(h) \in H$, whence $\sigma(H) \subseteq H$. Similarly $\sigma^{-1}(H) \subseteq H$ and so $H \subseteq \sigma(H)$. Thus $H = \sigma(H)$.

28. (a) Suppose $|G : Z(G)| = p$ is a prime. Then $|G/Z(G)| = p$, so the group $G/Z(G)$ is cyclic by Lagrange's theorem (Corollary 3). Then Theorem 2 shows G is abelian. Thus $Z(G) = G$, so $|G : Z(G)| = 1$, a contradiction.

29. Let $n = kd$ and $D_n = \{1, a, \dots, a^{n-1}, b, ba, \dots, ba^{n-1}\}$, $o(a) = n$, $o(b) = 2$, $aba = b$. Let $K = \langle a^k \rangle$. Then $K \triangleleft G$ because $b^{-1}a^k b = a^{-k} \in K$. In D_n/K, let $\bar{a} = Ka$, $\bar{b} = Kb$. Then $o(\bar{a}) = k$, $o(\bar{b}) = 2$ and
$$D_n/K = \{\bar{1}, \bar{a}, \dots, \bar{a}^{k-1}, \bar{b}, \bar{b} \cdot \bar{a}, \dots, \bar{b} \cdot \bar{a}^{k-1}\}.$$
Since $\bar{a} \cdot \bar{b} \cdot \bar{a} = \overline{aba} = \bar{b}$, $D_n/K \cong D_k$.

31. Let $A \subseteq C_4$ and $B \subseteq C \subseteq C_8$ where $|A| = |B| = 2$ and $|C| = 4$.
    (a) Let $K = A \times C$ and $H = C_4 \times B$. Then $K \cong C_2 \times C_4 \cong H$, but $\frac{G}{K} \cong \frac{C_4}{A} \times \frac{C_8}{C} \cong C_2 \times C_2$, while $\frac{G}{H} \cong \frac{C_4}{C_4} \times \frac{C_8}{B} \cong \{1\} \times C_4$. So $\frac{G}{K} \ncong \frac{G}{H}$.

2.10   THE ISOMORPHISM THEOREM

1. Define $\alpha : G \to \mathbb{R}^* \times \mathbb{R}^*$ by $\alpha\left(\begin{bmatrix} a & b \\ 0 & c \end{bmatrix}\right) = (a, c)$. This is a group homomorphism by direct calculation. We have $\begin{bmatrix} a & b \\ 0 & c \end{bmatrix} \in \ker\alpha$ if and only if $a = 1 = c$, so $\ker\alpha = K$.

3. Since $|G : H| = 2$, $H \triangleleft G$ and $G/H = \{H, G \smallsetminus H\}$ is a group with H as unity. Thus $\sigma : G/H \to \{1, -1\}$ is an isomorphism if $\sigma(H) = 1$ and $\sigma(G - H) = -1$. Let $\varphi : G \to G/H$ be the coset map $\varphi(g) = Hg$, and let $\alpha = \sigma\varphi : G \to \{1, -1\}$. Then α is a homomorphism and: If $g \in H$, then $\alpha(g) = \sigma\varphi(g) = \sigma(H) = 1$; if $g \notin H$, then $\alpha(g) = \sigma\varphi(g) = \sigma(G - H) = -1$.

4. (a) We have $1 \in \alpha^{-1}(X)$ because $\alpha(1) = 1 \in X$. If $g, h \in \alpha^{-1}(X)$, then $\alpha(g) \in X$ and $\alpha(h) \in X$, so $\alpha(g^{-1}) = [\alpha(g)]^{-1} \in X$ and $\alpha(gh) = \alpha(g) \cdot \alpha(h) \in X$. Thus $g^{-1} \in \alpha^{-1}(X)$ and $gh \in \alpha^{-1}(X)$, so $\alpha^{-1}(X)$ is a subgroup of G. If $X \lhd \alpha(G)$, let $g \in \alpha^{-1}(X)$, $a \in G$. Then
$$\alpha(a^{-1}ga) = [\alpha(a)]^{-1} \cdot \alpha(g) \cdot \alpha(a) \in [\alpha(a)]^{-1}X\alpha(a) \subseteq X.$$
Thus $a^{-1}ga \in \alpha^{-1}(X)$ for all $a \in G$; that is $\alpha^{-1}(X) \lhd G$.

   (c) Since $X \cap Y \subseteq X$ and $X \cap Y \subseteq Y$, $\alpha^{-1}(X \cap Y) \subseteq \alpha^{-1}(X) \cap \alpha^{-1}(Y)$ by (b). If $g \in \alpha^{-1}(X) \cap \alpha^{-1}(Y)$, then $\alpha(g) \in X$ and $\alpha(g) \in Y$, so $\alpha(g) \in X \cap Y$. Hence $g \in \alpha^{-1}(X \cap Y)$.

5. (a) If $g^d = 1$, then since $d \mid m$, $1 = g^m = \rho_m g$, so $g \in \ker \rho_m$. Conversely, if $\rho_m g = 1$, then $g^m = 1$. We have also that $g^n = 1$ (since $|G| = n$). Since $d = xm + yn$, $x, y \in \mathbb{Z}$, this gives $g^d = (g^m)^x(g^n)^y = 1^x 1^y = 1$.

   (c) Let $G = \langle a \rangle$ where $o(a) = n$. If $\sigma : G \to G$ is an automorphism then $o(\sigma(a)) = o(a) = n$. Hence Theorem 8, §2.4 gives $\sigma(a) = a^m$ where $gcd(m, n) = 1$. Now let $g \in G$, say $g = a^k$. Then
$$\sigma(g) = \sigma(a^k) = \sigma(a)^k = (a^m)^k = (a^k)^m = g^m = \rho_m(g).$$
It follows that $\sigma = \rho_m$.

7. Let $\alpha : G \to G$ and let $\ker \alpha = \langle X \rangle$ and $\alpha(G) = \langle Y \rangle$ where X and Y are finite sets. Since $Y \subseteq \alpha(G)$, let $Z \subseteq G$ be a finite set such that, if $y \in Y$, $y = \alpha(z)$ for some $z \in Z$. If $g \in G$, then $\alpha(g) \in \langle Y \rangle$, so
$$\begin{aligned}
\alpha(g) &= y_1^{k_1} y_2^{k_2} \cdots y_n^{k_n}, \ y_i \in Y \\
&= [\alpha(z_1)]^{k_1}[\alpha(z_2)]^{k_2} \cdots [\alpha(z_n)]^{k_n}, \ z_i \in Z \\
&= \alpha(z_1^{k_1} z_2^{k_2} \cdots z_n^{k_n}).
\end{aligned}$$
If we write $h = z_1^{k_1} z_2^{k_2} \cdots z_n^{k_n}$, then $gh^{-1} \in \ker \alpha$, so $gh^{-1} = x_1^{n_1} x_2^{n_2} \cdots x_m^{n_m}$, $x_i \in X$. Hence
$$g = \{x_1^{n_1} x_2^{n_2} \cdots x_m^{n_m})h = x_1^{n_1} x_2^{n_2} \cdots x_m^{n_m} z_1^{k_1} z_2^{k_2} \cdots z_n^{k_n} \in \langle X \cup Z \rangle.$$
Hence $G = \langle X \cup Z \rangle$, so since $X \cup Z$ a finite set, G is finitely generated.

8. (a) Let $C_6 = \langle g \rangle$, $o(g) = 6$, and write $K_4 = \{1, a, b, ab\}$, where $a^2 = b^2 = 1$ and $ab = ba$. If $\alpha : C_6 \to K_4$ is a homomorphism, then α is determined by the choice of $\alpha(g)$ in K_4. If $\alpha(g) = 1$ then α is trivial ($\alpha(x) = 1$ for all $x \in G$). If $\alpha(g) = a$ (say), then $\alpha(g^k) = a^k$. If we *define* α by $\alpha(g^k) = a^k$, it is well defined because
$$g^k = g^m \Rightarrow k \equiv m \pmod 6 \Rightarrow k \equiv m \pmod 2 \Rightarrow a^k = a^m.$$
Thus α is well defined, it is clearly a homomorphism. In the same way, there is a homomorphism carrying g to $1, a, b$ and ab; so there are four in all.

   (c) Let $D_3 = \{1, a, a^2, b, ba, ba^2\}$ where $o(a) = 3$, $o(b) = 2$, $aba = b$; and let $C_4 = \langle c \rangle$, $o(c) = 4$. If $\alpha : D_3 \to C_4$, then $o(\alpha(a))$ divides $o(a) = 3$, so $\alpha(a) = 1$ because C_4 has no element of order 3. Similarly $o(\alpha(b))$ divides $o(b) = 2$, so $\alpha(b) = 1, c^2$. If $\alpha(b) = 1$, then α is trivial. If $\alpha(b) = c^2$, then $\alpha(b^k a^m) = c^{2k} 1^m = c^{2k}$. If we now *define* α by this formula, it is possible (but tedious) to check it is well defined and a homomorphism. There is another way. Write $H = \langle a \rangle$ in D_3. Then $H \lhd D_3$ being of index 2, so

$D_3/H = \{H, bH\}$. Then there is an isomorphism $\{H, bH\} \xrightarrow{\sigma} \{1, c^2\} \subseteq C_4$ where $\sigma(H) = 1$, $\sigma(bH) = c^2$. If $\varphi : D_3 \to D_3/H$ is the coset map, we get

$$D_3 \xrightarrow{\varphi} D_3/H \xrightarrow{\sigma} C_4.$$

Then: $\sigma\varphi(b^k a^m) = \sigma(b^k a^m H) = \sigma(b^k H) = \sigma[(bH)^k] = [\sigma(bH)]^k = c^{2k}$. Hence the map we want is $\alpha = \sigma\varphi$. This is the only non-trivial homomorphism.

9. No. If $\alpha : S_4 \to A_4$ has $\ker \alpha = K = \{\varepsilon, (1\ 2)(3\ 4), (1\ 3)(2\ 4), (1\ 4)(2\ 3)\}$, then $A_4/K \cong \alpha(A_4)$. Now $S_4/K \cong D_3$ [if $a = K(1\ 2\ 3)$ and $b = K(1\ 2)$, then $o(g) = 3$, $o(b) = 2$ and $aba = b$]. Hence $\alpha(S_4)$ would be a subgroup of A_4 which is isomorphic to D_3. But there is no such subgroup: If $o(\sigma) = 3$ and $o(\tau) = 2$ and $\sigma\tau\sigma = \tau$, then σ is a 3-cycle, say $\sigma = (1\ 2\ 3)$, and τ is one of $(1\ 2)(3\ 4), (1\ 3)(2\ 4)$ or $(1\ 4)(2\ 3)$. It is easy to check that $\sigma\tau\sigma \neq \tau$ in each case.

10. (a) No. If $\alpha : S_3 \to K_4$ is onto, then $K_4 \cong S_3/\ker \alpha$, so $|K_4| = 4$ would divide $|S_3| = 6$, a contradiction.

    (c) Yes. If $S_3 = \{\varepsilon, \sigma, \sigma^2, \tau, \tau\sigma, \tau\sigma^2\}$, let $K = \langle \sigma \rangle$. Then $|S_3/K| = 2$, so $S_3/K = \{K, \tau(K)\}$. If $C_2 = \langle c \rangle$, $o(c) = 2$, then $S_3 \to S_3/K \to C_3$ is onto, where $\sigma(K) = 1$, $\sigma[\tau(K)] = c$. Thus $\alpha = \sigma\varphi : S_3 \to C_2$ is given by

$$\alpha(\lambda) = \begin{cases} 1 & \lambda \in K \\ c & \lambda \notin K \end{cases}.$$

11. (a) We have $\theta(gh) = (gh, gh) = (g, g)(h, h) = \theta(g) \cdot \theta(h)$ for all $g, h \in G$, so θ is a homomorphism. If $\theta(g) = (1, 1)$, then $g = 1$, so θ is one-to-one.

    (b) (1) $\Rightarrow$ (3). If G is abelian, define $\varphi : G \times G \to G$ by $\varphi(g, h) = gh^{-1}$. Then

$$\varphi(g, h)(g_1, h_1)] = (gg_1, hh_1) = gg_1(hh_1)^{-1} = gg_1 h_1^{-1} h^{-1}$$
$$= gh^{-1} g_1 h_1^{-1} = \varphi(g, h)\varphi(g_1, h_1).$$

    Thus φ is a homomorphism, and $\varphi\theta(g) = \varphi(g, g) = gg^{-1} = 1$ for all g. Hence $\theta(G) \subseteq \ker \varphi$. If $(g, h) \in \ker \varphi$, then $gh^{-1} = \phi(g, h) = 1$, so $h = g$. Thus $(g, h) \in \theta(G)$, so $a(G) = \ker \varphi$.

13. Let G be simple. If $\alpha : G \to G_1$ is a nontrivial homomorphism, then $\ker \alpha \neq G$. Since $\ker \alpha \lhd G$, $\ker \alpha = \{1\}$ by simplicity, that is α is one-to-one. Hence $G \cong \alpha(G) \subseteq G_1$.

    Conversely, if G_1 has a subgroup G_0 and $\sigma : G \to G_0$ is an isomorphism, then $\sigma : G \to G_1$ is a (one-to-one) homomorphism, which is nontrivial because $G_0 \neq \{1\}$, being simple.

15. We have $G/Z(G) \cong \text{inn}\, G$, and $\text{inn}\, G$ is cyclic by hypothesis (a subgroup of $\text{aut}\, G$). Hence G is abelian by Theorem 2 §2.9. Thus $\tau : G \to G$ is an automorphism where $\tau(g) = g^{-1}$ for all $g \in G$. Write $\text{aut}\, G = \langle \sigma \rangle$. Then $\sigma^k = \tau$ for some k, so $\sigma^{2k} = \tau^2 = \varepsilon$. Thus $o(\sigma)$ is finite, that is $|\text{aut}\, G|$ is finite. Since $o(\tau) = 2$ divides $|\text{aut}\, G|$, we are done.

17. (a) If $g \in G'$, $g = [a_1, b_1][a_2, b_2] \cdots [a_n, b_n]$ where $[a, b] = a^{-1}b^{-1}ab$ is a commutator. Then

$$\alpha(g) = \alpha[a_1, b_1] \cdots \alpha[a_n, b_n]$$
$$= [\alpha(a_1,), \alpha(b_1)][\alpha(a_2), \alpha(b_2)] \cdots [\alpha(a_n), \alpha(b_n)] \in G_1'.$$

19. (a) Define $\alpha : G \to \mathbb{R}^*$ by $\alpha(A) = \det A$. Then $\det A \in \mathbb{R}^*$ because $\det A \neq 0$ if A is invertible, and α is a homomorphism because $\det AB = \det A \cdot \det B$. Since $\ker \alpha = K$, we get $G/K \cong \alpha(G) = \mathbb{R}^*$.

21. Define $\alpha : \mathbb{C}^* \to \mathbb{R}^+$ by $\alpha(z) = |z|$ for $z \in \mathbb{C}$. [Note that $|z| > 0$ because $z \neq 0$.] Then α is a homomorphism because $|zw| = |z|\,|w|$, and $\ker \alpha = \{z \mid |z| = 1\} = \mathbb{C}^\circ$. Thus $\mathbb{C}^*/\mathbb{C}^\circ \cong \alpha(\mathbb{C}^*) \cong \mathbb{R}^+$ by the isomorphism theorem.

23. If $a \neq 0$, then $\tau_{a,b}$ is a bijection, so $G \subseteq S_\mathbb{R}$. We have $1_\mathbb{R} = \tau_{1,0} \in G$ and $(\tau_{a,b})^{-1} = \tau_{a^{-1},-a^{-1}b} \in G$, and $\tau_{a,b} \cdot \tau_{a_1,b_1} = \tau_{a_1 a, a_1 b + b_1} \in G$, so G is a subgroup. Now define $\alpha : G \to \mathbb{R}^*$ by $(\tau_{a,b}) = a$. Then α is well defined: If $\tau_{a,b} = \tau_{a',b'}$ then $ax + b = a'x + b'$ for all x, so $a = a'$. Now

$$\alpha(\tau_{a,b} \cdot \tau_{a_1,b_1}) = \alpha(\tau_{aa_1,ab_1+b}) = aa_1 = \alpha(\tau_{a,b})\alpha(\tau_{a_1,b_1})$$

so α is a homomorphism. Since $\ker \alpha = K$, we have $K \triangleleft G$ and $G/K \cong \alpha(G) = \mathbb{R}^*$.

25. Define $\alpha : G \times G \times G \to G \times G$ by $\alpha(a,b,c) = (ac^{-1}, bc^{-1})$. Then α is a homomorphism because G is abelian:

$$\alpha[(a,b,c)(a_1,b_1,c_1)] = [aa_1(cc_1)^{-1}, bb_1(cc_1)^{-1}] = (ac^{-1} \cdot a_1 c_1^{-1}, bc^{-1} \cdot b_1 c_1^{-1})$$
$$= (ac^{-1}, bc^{-1}) \cdot (a_1 c_1^{-1}, b_1 c_1^{-1}) = \alpha(a,b,c) \cdot \alpha(a_1,b_1,c_1).$$

Now

$$\ker \alpha = \{(a,b,c) \mid (ac^{-1}, bc^{-1}) = (1,1)\} = \{(a,b,c) \mid ac^{-1} = 1, bc^{-1} = 1\} = K.$$

Hence, since α is onto, we are done by the isomorphism theorem.

27. Define $\beta : \alpha(G) \to G/K$ by $\beta[\alpha(g)] = Kg$. This is well defined because $\alpha(g) = \alpha(g_1)$ implies $\alpha(gg_1^{-1}) = 1$, so $gg_1^{-1} \in \ker \alpha \subseteq K$, that is $Kg = Kg_1$. It is a homomorphism because

$$\beta[\alpha(g) \cdot \alpha(g_1)] = \beta[\alpha(gg_1)] = Kgg_1 = KgKg_1 = \beta[\alpha(g)] \cdot \beta[\alpha(g_1)].$$

Finally

$$\ker \beta = \{\alpha(g) \mid Kg = K\} = \{\alpha(g) \mid g \in K\} = \alpha(K).$$

Thus $\alpha(K) \triangleleft \alpha(G)$ and $\alpha(G)/\alpha(K) \cong \alpha(G) = G/K$. Note: This can also be solved by verifying $\alpha(K) \triangleleft \alpha(G)$ directly, defining $\gamma : G \to \alpha(G)/\alpha(K)$ by $\gamma(g) = [\alpha(K)]\alpha(g)$, and showing $\ker \gamma = K$.

29. (a) If $A = \begin{bmatrix} 1 & a & b \\ 0 & 1 & c \\ 0 & 0 & 1 \end{bmatrix}$ and $B = \begin{bmatrix} 1 & a' & b' \\ 0 & 1 & c' \\ 0 & 0 & 1 \end{bmatrix}$ then $AB = \begin{bmatrix} 1 & a+a' & b'+ac'+b \\ 0 & 1 & c'+c \\ 0 & 0 & 1 \end{bmatrix} \in G$

and $A^{-1} = \begin{bmatrix} 1 & -a & ac-b \\ 0 & 1 & -c \\ 0 & 0 & 0 \end{bmatrix} \in G$. Thus G is a subgroup of $M_3(\mathbb{R})^*$. The

center consists of all $\begin{bmatrix} 1 & a' & b' \\ 0 & 1 & c' \\ 0 & 0 & 1 \end{bmatrix}$ with $b' + ac' + b = b + a'c + b'$ for all a, b, c.

Taking $a = 1$, $c = 0$ gives $c' = 0$; taking $a = 0$, $c = 1$ gives $a' = 0$. Thus $Z(G) = \left\{ \begin{bmatrix} 1 & 0 & b \\ 0 & 1 & 0 \\ 0 & 0 & 1 \end{bmatrix} \middle| b \in R \right\}$. The map $b \mapsto \begin{bmatrix} 1 & 0 & b \\ 0 & 1 & 0 \\ 0 & 0 & 1 \end{bmatrix}$ is an isomorphism $\mathbb{R} \to Z(G)$.

31. Let $\bar{k} = k + m\mathbb{Z} \in \mathbb{Z}_m$ and $\tilde{k} = k + n\mathbb{Z} \in \mathbb{Z}_n$. Define $\alpha : \mathbb{Z} \to \mathbb{Z}_m \times \mathbb{Z}_n$ by $\alpha(k) = (\bar{k}, \tilde{k})$. Then $\alpha(k + l) = (\bar{k} + \bar{l}, \tilde{k} + \tilde{l}) = \alpha(k)\alpha(l)$ so α is a homomorphism. Now

$$k \in \ker \alpha \Leftrightarrow \bar{k} = \bar{0} \text{ and } \tilde{k} = \tilde{0} \Leftrightarrow m \mid k \text{ and } n \mid k \Leftrightarrow s \mid k \Leftrightarrow k \in s\mathbb{Z}.$$

Thus $\ker \alpha = s\mathbb{Z}$, so $\alpha(\mathbb{Z}) \cong \mathbb{Z}/s\mathbb{Z} = \mathbb{Z}_s$.

33. (a) $\mathbb{Z}_4$ has subgroups $\{0\}$, $\{0, 2\}$ and $\mathbb{Z}_4$. The factors are isomorphic to $\mathbb{Z}_4$, $\mathbb{Z}_2$, $\{0\}$ respectively, so these are the only possible images.

(c) The normal subgroups of A_4 are

$$\{1\}, K = \{\varepsilon, (1\ 2)(3\ 4), (1\ 2)(2\ 4), (1\ 4)(2\ 3)\},$$

and A_4 (Exercise 19 §2.8). Thus the factor groups are isomorphic to A_4, C_3 (since $|A_4/K| = 3$) and $\{1\}$.

35. If $\varphi : G_1 \to G_1/X$ is the coset map, we have $\beta = \varphi\alpha : G \to G_1/X$. This is onto because α and φ are both onto, and since

$$\ker \varphi = X, \ker \beta = \{g \mid \alpha(g) \in \ker \varphi\} = \alpha^{-1}(X).$$

The isomorphism theorem completes the proof.

36. (a) It is closed because $\alpha + \beta$ is a homomorphism:

$$(\alpha + \beta)(x + y) = \alpha(x + y) + \beta(x + y) = \alpha(x) + \alpha(y) + \beta(x) + \beta(y)$$
$$= (\alpha + \beta)x + (a + \beta)y.$$

We have $\alpha + \beta = \beta + \alpha$ because

$$(\alpha + \beta)x = \alpha(x) + \beta(x) = \beta(x) + \alpha(x) = (\beta + \alpha)x.$$

Similarly, $\alpha + (\beta + \gamma) = (\alpha + \beta) + \gamma$. The unity is $\theta : X \to Y$ where $\theta(x) = 0$ for all x. Finally, the negative of α is $-\alpha : X \to Y$ defined by $(-\alpha)(x) = -[\alpha(x)]$ for all x.

(c) For convenience, write $\tilde{x} = x + m\mathbb{Z} \in \mathbb{Z}_m$ and $\bar{x} = x + n\mathbb{Z} \in \mathbb{Z}_n$. Let $d = \gcd(m, n)$, and write $e = \frac{n}{d}$. Given $k \in \mathbb{Z}$, define $\alpha_k : \mathbb{Z}_m \to \mathbb{Z}_n$ by

$$\alpha_k(\tilde{x}) = ke\bar{x} \qquad \text{for all} \qquad \tilde{x} \in \mathbb{Z}_m.$$

This is well defined:

$$\tilde{x} = \tilde{y} \Rightarrow x - y = qm \Rightarrow ke(x - y) = k\frac{n}{d}qm = \left(k\frac{m}{d}e\right)n.$$

This means $ke\bar{x} = \overline{kex} = \overline{key} = ke\bar{y}$. So α_k is well defined. We have

$$\alpha_k(\tilde{x} + \tilde{y}) = ke(\overline{x + y}) = ke\bar{x} + ke\bar{y} = \alpha_k(\tilde{x}) + \alpha_k(\tilde{y}).$$

Hence $\alpha_k \in \hom(\mathbb{Z}_m, \mathbb{Z}_n)$ and we have a map

$$\alpha : \mathbb{Z} \to \hom(\mathbb{Z}_m, \mathbb{Z}_n) \qquad \text{given by} \qquad \alpha(k) = \alpha_k$$

α *is a homomorphism:*

$$\alpha_{k+1}(\tilde{x}) = (k + l)e\bar{x} = ke\bar{x} + le\bar{x} = \alpha_k(\tilde{x}) + \alpha_l(\tilde{x}) = (\alpha_k + \alpha_l)(\tilde{x}).$$

Hence we have $\alpha_{k+l} = \alpha_k + \alpha_l$; that is $\alpha(k + l) = \alpha(k) + \alpha(l)$.

α *is onto:* Let $\lambda : \mathbb{Z}_m \to \mathbb{Z}_n$, and write $\lambda\tilde{1} = \bar{y}$. Then

$$m\bar{y} = m(\lambda\tilde{1}) = \lambda(m\tilde{1}) = \lambda\tilde{0} = \bar{0}.$$

Thus $\overline{my} = 0$, so $n \mid my$. This gives $\frac{n}{d} \mid \frac{m}{d} y$. But $\frac{n}{d}$ and $\frac{m}{d}$ are relatively prime, so $\frac{n}{d} \mid y$, that is $e \mid y$, say $y = ek$. But then

$$\lambda(\tilde{x}) = \lambda(\tilde{1}x) = [\lambda(\tilde{1})]x = \bar{y}x = \overline{ekx} = ke\bar{x} = \alpha_k(\tilde{x})$$

Thus $\lambda = \alpha_k$ and α is onto.

$\ker \alpha = d\mathbb{Z}$: We have $\ker \alpha = \{k \mid \alpha_k = 0\}$. But

$$\alpha_k = \theta \Leftrightarrow \alpha_k(\tilde{1}) = 0 \Leftrightarrow ke\bar{1} = 0$$

$$\Leftrightarrow n \mid ke \Leftrightarrow qn = k\frac{n}{d} \text{ for some } q.$$

$$\Leftrightarrow qd = k \text{ for some } q.$$

$$\Leftrightarrow d \mid k.$$

Hence $\hom(\mathbb{Z}_m, \mathbb{Z}_n) = im\,\alpha \cong \mathbb{Z}/\ker \alpha = \mathbb{Z}/d\mathbb{Z} = \mathbb{Z}_d$.

37. (a) The unity of G^ω is $[1) = (1,1,1,\ldots)$. The inverse of $[g_i)$ is $[g_i)^{-1} = [g_i^{-1})$. Finally

$$\{[g_i)[h_i)\}[k_i) = [g_i h_i)[k_i) = [g_i h_i k_i) = [g_i)[h_i k_i) = [g_i)\{[h_i)[k_i)\}.$$

Thus G^ω is associative.

(c) $F \overset{\sigma}{\to} G^\omega$ is an isomorphism when $\sigma(f) = [f(i)) = [f(0), f(1), f(2), \ldots)$.

2.11 AN APPLICATION TO BINARY LINEAR CODES

1. (a) 5   (c) 6

2. (a) 3   (c) 7

3. By (1) of Theorem 1,

$$d(u + v, u + w) = wt[(u + v) - (u + w)] = wt(v - w) = d(v, w).$$

5. (a) It suffices to look at individual digits:

$$\overline{1 + 1} = 0 + 0 = 1 + 1; \bar{1} + \bar{0} = 0 + 1 = 1 + 0; \bar{0} + \bar{0} = 1 + 1 = 0 + 0.$$

6. The table lists the codewords across the top and the distances of w from them.

Code words		0000000	0101010	1010101	1110000
(a)	$w = 0110101$	4	5	2	3
(c)	$w = 1011001$	4	5	2	3

Code words		1011010	0100101	0001111	1111111
(a)	$w = 0110101$	6	1	4	3
(c)	$w = 1011001$	2	5	5	4

(a) The unique nearest neighbor of 0110101 is 0100101 (so it corrects the single error).

(c) 1011001 has both 1010101 and 1011010 at distance 2, so the 2 errors are detected but not corrected.

7. (a) The minimum weight of C is 4. Detects 3 errors, corrects 1 error.

9. (a) We can have $k = 4$, $n = 7$ for the code, and (Example 9), the minimum weight for the code is 3. Thus it corrects $t = 1$ errors. Hence $\binom{7}{0} + \binom{7}{1} = 2^{7-4}$ shows the code is perfect.

(c) If the minimum distance is $5 = 2 \cdot 2 + 1$, it corrects $t = 2$ errors. We have $\binom{7}{0} + \binom{7}{1} + \binom{7}{2} = 1 + 7 + 28 = 36$ while $2^{7-2} = 2^5 = 32$. So no such code exists.

10. (a) If $k = 2$, $t = 1$, the Hamming bound is $\binom{n}{0} + \binom{n}{1} \le 2^{n-2}$, that is $1 + n \le 2^{n-2}$. The first $n \ge 1$ for which this holds $n = 5$. The $(5, 2)$-code
$$C = \begin{cases} 00000 \\ 01110 \\ 10011 \\ 11101 \end{cases}$$ has minimum distance $3 = 2 \cdot 1 + 1$, so it corrects 1 error by
Theorem 2.

11. (a) Here $k = 3$ and $t = 2$, so n must be such that $\begin{bmatrix} n \\ 0 \end{bmatrix} + \begin{bmatrix} n \\ 1 \end{bmatrix} + \begin{bmatrix} n \\ 2 \end{bmatrix} \le 2^{n-3}$. If $n = 3, 4, \ldots, 8$ this would read respectively
$$7 \le 1, \ 11 \le 2, \ 16 \le 4, \ 22 \le 8, \ 29 \le 16, \ 37 \le 32.$$
Hence $n \ge 9$. [Note: $\begin{bmatrix} 9 \\ 0 \end{bmatrix} + \begin{bmatrix} 9 \\ 1 \end{bmatrix} + \begin{bmatrix} 9 \\ 2 \end{bmatrix} = 1 + 9 + 36 = 46 \le 64 = 2^6$.]

13. Suppose $w - c$ is the coset leader in $w + C$, and write it as $w - c = e$. Then coset decoding decodes w as $w - e = c$, and this is correct. Conversely, assume coset decoding decodes w correctly as c. If e is the coset leader in $w + C$, this means $w - e = c$. Hence $w - c = e$ is the coset leader in $w + C$.

15. (a) If a $(4, 2)$-code C corrects 1 error, the weight of C must be at least 3. Thus the nonzero words of C are all contained in $\{1111, 1110, 1101, 1011, 0111\}$. But the sum of two distinct words here is no longer in the set.

16. (a) If a $(6, 3)$-code C corrects 2 errors, the weight of C must be at least 5. Proceed as in (a) of the preceding exercise.

(c) If a $(7, 3)$-code C has minimum distance 5, the nonzero words in C have weight 7, 6 or 5. The table gives the possible weights of $w + v$ for various choices of w and v (nonzero) in C. For example, if w and v both have weight 5, we may take $w = 1111100$. The weight of $w + v$ depends upon how many 0-digits match. The three cases are illustrated as follows:
$$1111100 \quad 1111100 \quad 1111100$$
$$1111100 \quad 1111001 \quad 1110011.$$
The weight of $w + v$ is $0, 2, 4$ respectively The table shows that $w + v$ is not in C (unless $w = v$), so no such group C can exist.

w	v	$w + v$		
7	7	0		
7	6	1		
7	5	2		
6	6	0	2	
6	5	1	3	
5	5	0	2	4

17. (a) Write $V = \{i \mid v_i = 1\}$ and $W = \{i \mid w_i = 1\}$, so $wt(v) = |V|$ and $wt(w) = |W|$. Then

$$V \cap W = \{i \mid v_i = i = w_i\}$$

so $|V \cap W| = wt(vw)$. Similarly

$$(V \smallsetminus W) \cap (W \smallsetminus V) = \{i \mid v_i + w_i = 1\}$$

so $wt(v + w) = |V \smallsetminus W| + |W \smallsetminus V|$. Then (a) follows because

$$
\begin{aligned}
wt(v + w) + 2wt(vw) &= |V \smallsetminus W| + |W \smallsetminus V| + 2\,|V \cap W| \\
&= \{|V \smallsetminus W| + |V \cap W|\} + \{|W \smallsetminus V| + |V \cap W|\} \\
&= |V| + |W| \\
&= wt(v) + wt(w).
\end{aligned}
$$

(c) Using (a), equality holds in (b) if and only if $wt(vw) = 0$, that is if and only if $V \cap W = \emptyset$. If this holds, then $v_i = 1 \Rightarrow w_i = 0$ which is the condition in (c). Conversely, if $v_i = 1 \Rightarrow w_i = 0$, then $i \in V \cap W$ is impossible (it means $v_i = 1 = w_i$), so $V \cap W = \emptyset$. Thus $wt(vw) = 0$ and (a) implies equality in (b).

19. We have $|D| = |C| + |w + C| = 2^k + 2^k = 2^{k+1}$. Hence it suffices to show that D is a subgroup of B^n. Clearly $0 \in D$ and, if $x \in D$, then $-x = x \in D$. Finally, if $x, y \in D$ consider the following cases:

If $x, y \in C$ then $x + y \in C$ because C is a subgroup.

If $x \in C$ and $y \in w + C$ then $x + y \in x + w + C = w + (x + C)$ $= w + C$ because $x \in C$.

If $x \in w + C$ and $y \in C$ then $x + y \in w + C$ as in the preceding case.

If $x \in w + C$ and $y \in w + C$ then $x + y = w + w + C = C$ because $w + w = 0$.

Hence $x + y \in C \cup (w + C) = D$ in every case.

20. (a) $G = [1\,1\,1\,1\,1]$ $H = \begin{bmatrix} 1 & 1 & 1 & 1 \\ 1 & 0 & 0 & 0 \\ 0 & 1 & 0 & 0 \\ 0 & 0 & 1 & 0 \\ 0 & 0 & 0 & 1 \end{bmatrix}$

(c) $G = \begin{bmatrix} 1 & 0 & 0 & 1 & 0 & 1 & 1 \\ 0 & 1 & 0 & 0 & 1 & 1 & 1 \\ 0 & 0 & 1 & 1 & 1 & 1 & 0 \end{bmatrix}$ $H = \begin{bmatrix} 1 & 0 & 1 & 1 \\ 0 & 1 & 1 & 1 \\ 1 & 1 & 1 & 0 \\ 1 & 0 & 0 & 0 \\ 0 & 1 & 0 & 0 \\ 0 & 0 & 1 & 0 \\ 0 & 0 & 0 & 1 \end{bmatrix}$

21. (a) $\{0000, 1011, 0100, 1111\}$

(c) $\{000000, 100101, 010110, 001001, 110011, 101100, 011111, 111010\}$

23. Write $D = \{w \in B^n \mid wH = 0\}$. We have $C \subseteq D$ as before (by the Lemma). If $w \in D$, write $w = [u, v]$, $u \in B^k$, $v \in B^{n-k}$. Then

$$0 = wH = [u, v] \begin{bmatrix} A \\ I_{n-k} \end{bmatrix} = uA = vI_{n-k} = uA + v.$$

Thus $v = -uA = uA$ (because $-x = x$ in B^{n-k}). Now

$$uG = u[I_k, A] = [uI_k, uA] = [u, v] = w,$$

so $w = uG \in C$. Hence $D \subseteq C$.

24. (a) We have $C = \{uG \mid u \in B^k\}$ and $C' = \{uG' \mid u \in B^k\}$, so it is clear that $G = G' \Rightarrow C = C'$. If $C = C'$, write $G = \{I_k, A\}$ and $G' = \{I_{k'}, A'\}$. Given $u \in B^k$, $[u, uA] = uG \in C$, so there exists $u' \in B^k$ such that $[u, uA] = U'G' = [u', u'A']$. This gives $u = u'$ and $uA = u'A'$, so $uA = uA'$ for all $u \in B^k$. If $u = b_i$ is row i of I_k, then $b_iA = $ row i of A and $b_iA' = $ row i of A'. Thus $A = A'$, whence $G = G'$.

25. (a) Let $D = \{w \in B^n \mid wt(w) \text{ is even}\}$. Define $\varphi : B^n \to \mathbb{Z}_2$ by $\varphi(w) = \sum w_i$ where w_i is digit i of w. Then φ is onto and (if v_i is digit i of v)

$$\varphi(w) + \varphi(v) = \sum w_i + \sum v_i = \sum (w_i + v_i) = \varphi(w + v)$$

shows φ is a group homomorphism. Since

$$\ker \varphi = \{w \mid \sum w_i = 0 \text{ in } \mathbb{Z}_2\} = D,$$

the fact that $|\mathbb{Z}_2| = 2$ shows that D is a subgroup of B^n of index 2. If $C \subseteq D$, then each word in C has even weight. Otherwise, choose $c \in C \smallsetminus D$ and define $\sigma : C \cap D \to C \smallsetminus (C \cap D)$ by $\sigma(x) = x + c$. [Clearly $x + c \in C$; if $x + c \in D$, then $c \in D$ (as $x \in D$), a contradiction. So $x + c \in C \smallsetminus (C \cap D)$. Since σ is clearly 1:1, it remains to show that σ is onto. If $y \in C \smallsetminus (C \cap D)$, we want $y = x + c$, $x \in C \cap D$. Try $x = y - c$. Clearly $x \in C$. Since $y \notin D$ and $c \notin D$, we know $\varphi(y) = 1 = \varphi(c)$, so $\varphi(y - c) = 1 - 1 = 0$. Thus $x = y - c \in D$ too.

(c) Let D be any subgroup of B^n of index 2. Then $C \subseteq D$ or $[C : C \cap D] = 2$. The proof is in (a) where $\varphi : B^n \to \mathbb{Z}_2$ has $\varphi(w) = \begin{cases} 1 \text{ if } w \notin D \\ 0 \text{ if } w \in D \end{cases}$.

27. If $B = \begin{bmatrix} b_1 \\ \vdots \\ b_k \end{bmatrix}$ where $\{b_1, \ldots, b_k\}$ is a $\mathbb{Z}_2$-basis of C, and if $B \to R$ by the gaussian algorithm, let c_i be row i of R, $1 \le i \le k$. There exists an invertible $k \times k$ matrix U (the product of the elementary matrices used to carry $B \to R$) such that $UB = R$. If $U = [u_{ij}]$, then row i of R is

$$c_i = [u_{i1} u_{i2} \cdots u_{in}] \begin{bmatrix} b_1 \\ \vdots \\ b_n \end{bmatrix} = \sum_j u_{ij} b_j \in C.$$

Since $B = U^{-1}R$, we have $b_i \in \text{span}\{c_1, \ldots, c_n\}$ for each i, so $C = \text{span} \{c_1, \ldots, c_k\}$. It follows that $\{c_1, \ldots, c_k\}$ is a basis of C. Since $\text{rank } R = \text{rank } B = k$, and since R is a reduced row-echelon matrix, the proof is complete.

Chapter 3

Rings

3.1 EXAMPLES AND BASIC PROPERTIES

1. (a) Not an additive group; only 0 has an inverse.

   (c) It is an additive abelian group (as in Example 4), unity is $1_{\mathbb{R}}$, multiplication is associative (it is function composition). But $h \circ (f + g) = h \circ f + h \circ g$ fails: $h \circ (f + g)(x) = h[f(x) + g(x)]$ while

   $$[h \circ f + h \circ g](x) = h[f(x)] + h[g(x)].$$

   These may not be equal. Note that $(f + g) \circ h = f \circ h + g \circ h$ does hold.

3. (a) Clearly $\begin{bmatrix} 0 & 0 \\ 0 & 0 \end{bmatrix} \in S$ and $\begin{bmatrix} 1 & 0 \\ 0 & 1 \end{bmatrix} \in S$. If $\begin{bmatrix} a & b \\ c & d \end{bmatrix}$ and $\begin{bmatrix} a' & b' \\ c' & d' \end{bmatrix}$ are in S then $\begin{bmatrix} -a & -b \\ -c & -d \end{bmatrix} \in S$, and $\begin{bmatrix} a & b \\ c & d \end{bmatrix} + \begin{bmatrix} a' & b' \\ c' & d' \end{bmatrix} = \begin{bmatrix} a+a' & b+b' \\ c+c' & d+d' \end{bmatrix} \in S$ because

   $$(a + a') + (c + c') = (a + c) + (a' + d') = (b + d) + (b' + d')$$
   $$= (b + b') + (d + d').$$

   $\begin{bmatrix} a & b \\ c & d \end{bmatrix} \begin{bmatrix} a' & b' \\ c' & d' \end{bmatrix} = \begin{bmatrix} aa' + bc' & ab' + bd' \\ ca' + dc' & cb' + dd' \end{bmatrix} \in S$ because

   $$(aa' + bc') + (ca' + dc') = (a + c)a' + (b + d)c' = (a + c)(a' + c')$$
   $$(ab' + bd') + (cb' + dd') = (a + c)b' + (b + d)d' = (a + c)(b' + d')$$
   $$= (a + c)(a' + c').$$

   (c) Clearly $I, 0 \in S$ and $A, B \in S \Rightarrow A + B$ and $-A \in S$. To check multiplication

   $$\begin{bmatrix} a & 0 & b \\ 0 & c & d \\ 0 & 0 & a \end{bmatrix} \begin{bmatrix} a' & 0 & b' \\ 0 & c' & d' \\ 0 & 0 & a' \end{bmatrix} = \begin{bmatrix} aa' & 0 & ab' + ba' \\ 0 & cc' & cd' + da' \\ 0 & 0 & aa' \end{bmatrix} \in S.$$

Student Solution Manual to Accompany Introduction to Abstract Algebra, Fourth Edition. W. Keith Nicholson.
© 2012 John Wiley & Sons, Inc. Published 2012 by John Wiley & Sons, Inc.

5. We have 1, $0 \in C(X)$ as $1x = x = x1$ and $0x = 0 = x0$ for all $x \in X$. If $c, d \in C(x)$ then

$$(-c)x = -(cx) = -(xc) = x(-c),$$
$$(c + d)x = cx + dx = xc + xd = x(c + d),$$
$$\text{and } (cd)x = c(dx) = c(xd) = (cx)d = (xc)d = x(cd)$$

show that $-c$, $c + d$, and cd are in $C(x)$.

6. (a) If $ab = 0$ and $a \neq 0$ then a^{-1} exists so $b = a^{-1}ab = 0$.

7. If $A = \begin{bmatrix} a & b \\ c & d \end{bmatrix} \in Z[M_2(R)]$, then $\begin{bmatrix} a & b \\ c & d \end{bmatrix}\begin{bmatrix} 0 & 1 \\ 0 & 0 \end{bmatrix} = \begin{bmatrix} 0 & 1 \\ 0 & 0 \end{bmatrix}\begin{bmatrix} a & b \\ c & d \end{bmatrix}$, that is $\begin{bmatrix} 0 & a \\ 0 & c \end{bmatrix} = \begin{bmatrix} c & d \\ 0 & 0 \end{bmatrix}$. Hence $c = 0$ and $a = d$, so $A = \begin{bmatrix} a & b \\ 0 & a \end{bmatrix}$. Similarly the fact that A commutes with $\begin{bmatrix} 0 & 0 \\ 1 & 0 \end{bmatrix}$ forces $b = 0$. Finally $\begin{bmatrix} a & 0 \\ 0 & a \end{bmatrix}\begin{bmatrix} r & 0 \\ 0 & r \end{bmatrix} = \begin{bmatrix} r & 0 \\ 0 & r \end{bmatrix}\begin{bmatrix} a & 0 \\ 0 & a \end{bmatrix}$ shows $ar = ra$ for all $r \in R$; that is $a \in Z(R)$. Conversely if $a \in Z(R)$: $\begin{bmatrix} a & 0 \\ 0 & a \end{bmatrix}\begin{bmatrix} x & y \\ z & w \end{bmatrix} = \begin{bmatrix} ax & ay \\ az & aw \end{bmatrix} = \begin{bmatrix} xa & ya \\ za & wa \end{bmatrix} = \begin{bmatrix} x & y \\ z & a \end{bmatrix}\begin{bmatrix} a & 0 \\ 0 & a \end{bmatrix}$. Thus

$$Z[M_2(R)] = \left\{ \begin{bmatrix} a & 0 \\ 0 & a \end{bmatrix} | a \in Z(R) \right\}.$$

8. (a) If $(a + b)(a - b) = a^2 - b^2$. Then $a^2 + ba - ab - b^2 = a^2 - b^2$; whence $ba - ab = 0$, $ab = ba$. Conversely if $ab = ba$ then

$$(a + b)(a - b) = a^2 + ba - ab - b^2 = a^2 - b^2.$$

9. $$(a + b)(1 + 1) = (a + b)1 + (a + b)1 = a + b + a + b;$$

$$(a + b)(1 + 1) = a(1 + 1) + b(1 + 1) = a + a + b + b.$$

Hence $a + b + a + b = a + a + b + b$, so $b + a = a + b$ (additive cancellation on the left and on the right).

10. (a) $ab + ba = 1$ gives $aba + ba^2 = a$, so $ba^2 = a - aba$. Similarly $a^2b = a - aba$ so $a^2b = ba^2$. Then $ab = a^3b = a(a^2b) = aba^2 = (1 - ba)a^2 = a^2 - ba$. Hence $a^2 = ab + ba = 1$.

11. If $a^2 = 0$ implies $a = 0$, let $a^n = 0$, $n \geq 1$. If $a \neq 0$, let $a^m = 0$, $a^{m-1} \neq 0$. Then $m \geq 2$ and so $(a^{m-1})^2 = a^{2m-2} = 0$ since $2m - 2 \geq m$ ($m \geq 2$). This contradicts the hypothesis. So $a = 0$. The converse is clear.

13. By the hint: $u(u^{-1} + v^{-1})v = uu^{-1}v + uv^{-1}v = v + u = u + v$. Hence

$$u^{-1} + v^{-1} = u^{-1}(u + v)v^{-1}$$

is a unit, being a product of units [Theorem 2.5]. Thus

$$(u^{-1} + v^{-1})^{-1} = v(u + v)^{-1}u.$$

15. (3) $\Rightarrow$ (1). Let e be the unique right unity. Given $b \in R$,

$$r(e + eb - b) = re + reb - rb = r + rb - rb = r$$

for all $r \in R$. Hence $e + eb - b$ is a right unity too, so $e = e + eb - b$ by uniqueness. Thus $eb = b$ for all b, that is e is also a left unity.

16. (a) This is by the subring test since $0 = 01_R$; $1_R = 11_R$; $k1_R + m1_R = (k + m)1_R$; $-(k1_R) = (-k)1_R$ and $(k1_R)(m1_R) = (km)1_R$. To see that

$\mathbb{Z}1_R$ is central, let $s \in R$. Then $(k1_R)s = (k1_R)(1s) = (k1)(1_Rs) = ks$ by Theorem 2, and $s(k1_R) = ks$ in the same way.

(c) If char $R = 0$ define $\sigma : \mathbb{Z} \rightarrow \mathbb{Z}1_R$ by $\sigma(k) = k1_R$. Then $\sigma(km) = \sigma(k) \cdot \sigma(m)$ and $\sigma(k + m) = \sigma(k) + \sigma(m)$ as in (b), and $\sigma(k) = \sigma(m) \Rightarrow (k - m)1_R = 0 \Rightarrow k = m$ because $o(1_R) = \infty$. Since σ is clearly onto, it is an isomorphism.

17. If R has characteristic 1 then $1 \cdot 1_R = 0$, $1_R = 0$. Hence $R = 0$, the zero ring.

18. (a) $\mathbb{Z}_n \times \mathbb{Z}_m$. The unity is $(1, 1)$ so $k(1, 1) = 0 \Longleftrightarrow k1 = 0$ in $\mathbb{Z}_n$ and $k1 = 0$ in $\mathbb{Z}_m \Longleftrightarrow n|k$ and $m|k$. So the characteristic of $\mathbb{Z}_n \times \mathbb{Z}_m = lcm(n, m)$.

    (c) $\mathbb{Z} \times \mathbb{Z}_n$. $k(1, 1) = 0 \Longleftrightarrow k1 = 0$ in $\mathbb{Z}$ and $k1 = 0$ in $\mathbb{Z}_n \Longleftrightarrow k = 0$. So the characteristic of $\mathbb{Z} \times \mathbb{Z}_n$ is 0.

19. If $u \in R^*$, $k \in \mathbb{Z}$, then $k1 = 0$ in $R \Rightarrow ku = 0$. But
$$ku = 0 \Rightarrow k1 = k(uu^{-1}) = (ku)u^{-1} = 0.$$
So $o(u) = o(1) = \text{char } R$ (since char $R < \infty$).

21. (a) $(1 - 2e)^2 = (1 - 2e)(1 - 2e) = 1 - 2e - 2e + 4e^2 = 1$.

22. (a) If $a = (1 - e)re$ then $ea = 0$ and $ae = a$. Hence $a^2 = (ae)a = a(ea) = 0$. Similarly if $b = er(1 - e)$, then $eb = b$, $be = 0$, so $b^2 = 0$.

    (c) This follows from (a) and Example 17.

23. $(2) \Rightarrow (1)$. If $r \in R$, $f = e + (1 - e)re$ is an idempotent, so $ef = fe$ by (2). Hence $re = ere$. Similarly $re = ere$, so $re = re$ and e is central.

24. $(1) \Rightarrow (2)$. If $a^n = 0$ then $1 + a$ is a unit $[(1 + a)^{-1} = 1 - a + a^2 - \cdots]$. Hence $r(1 + a) = (1 + a)r$ for all $r \in R$, that is $ra = ar$ for all r.

25. If $r^3 = r$ then $e = r^2$ is idempotent. Iterating $r^3 = r$: $r^9 = r^3 = r$, $r^{27} = r^3 = r$, and in general $r^{3^k} = r$ for $k \geq 1$. Hence if $r^n = 0$ then $r = r^{3^k} = 0$ for any k such that if $3^k \geq n$. Thus R has no nonzero nilpotents. Hence idempotents in R are central by Example 18, so r^2 is central for all r. Finally, if $r, s \in R$, $rs = (rs)^3 = (rs)^2rs = r(rs)^2s = r^2srs^2 = s^2srr^2 = s^3r^3 = sr$.

26. (a) If $ab = 1$ and $R = \{r_1, r_2, \ldots, r_n\}$ then $br = bs \Rightarrow r = s$ in R. But then $br_1, br_2, \ldots, br_n$ are all distinct, so $\{br_1, br_2, \ldots, br_n\} = R$. In particular $bc = 1$ for some c. Then $a = a(bc) = (ab)c = c$ so $1 = bc = ba$.

27. (a) If $a^m = a^{m+n}$ then
$$a^{m+2n} = a^{m+n} = a^n, \; a^{m+3n} = a^{m+n} = a^m, \ldots, a^{m+kn} = a^m, \; k \geq 1.$$
Then $a^{m+t} = a^{m+t+kn}$ for all $t \geq 0$, so $a^{r+kn} = a^r$ for all $r \geq m$ and for all $k \geq 1$. We want r and k such that $r + kn = 2r$; that is $kn = r$. Since $n \geq 1$ choose k such that $kn \geq m$, and take $r = km$. Then a^r is an idempotent.

28. (a) $\mathbb{Z}$. Units: ± 1; idempotents: 0,1; nilpotents: 0.

    (c) $R_2(\mathbb{Z}_2)$. Units: $\left\{ \begin{bmatrix} 1 & 0 \\ 1 & 0 \end{bmatrix}, \begin{bmatrix} 0 & 1 \\ 1 & 0 \end{bmatrix}, \begin{bmatrix} 0 & 1 \\ 1 & 1 \end{bmatrix}, \begin{bmatrix} 1 & 1 \\ 1 & 0 \end{bmatrix}, \begin{bmatrix} 1 & 1 \\ 0 & 1 \end{bmatrix}, \begin{bmatrix} 1 & 0 \\ 1 & 1 \end{bmatrix} \right\}$;

    idempotents: $\begin{bmatrix} 0 & 0 \\ 0 & 0 \end{bmatrix}, \begin{bmatrix} 1 & 0 \\ 0 & 1 \end{bmatrix}, \begin{bmatrix} 1 & 1 \\ 0 & 0 \end{bmatrix}, \begin{bmatrix} 1 & 0 \\ 1 & 0 \end{bmatrix}, \begin{bmatrix} 0 & 0 \\ 1 & 1 \end{bmatrix}, \begin{bmatrix} 0 & 1 \\ 0 & 1 \end{bmatrix}, \begin{bmatrix} 1 & 0 \\ 0 & 0 \end{bmatrix}, \begin{bmatrix} 0 & 0 \\ 0 & 1 \end{bmatrix}$;

    nilpotents: $\left\{ \begin{bmatrix} 0 & 0 \\ 0 & 0 \end{bmatrix}, \begin{bmatrix} 0 & 1 \\ 0 & 0 \end{bmatrix}, \begin{bmatrix} 0 & 0 \\ 1 & 0 \end{bmatrix}, \begin{bmatrix} 1 & 1 \\ 1 & 1 \end{bmatrix} \right\}$.

29. It is clearly an additive abelian group and $\begin{bmatrix} n & x \\ 0 & m \end{bmatrix}\begin{bmatrix} n' & x' \\ 0 & m' \end{bmatrix} = \begin{bmatrix} nn' & nx' + m'x \\ 0 & mm' \end{bmatrix} \in R.$

The associative and distributive laws hold for all matrices. Units: $\left\{ \begin{bmatrix} \pm 1 & x \\ 0 & \pm 1 \end{bmatrix} \right\}$;

nilpotents: $\left\{ \begin{bmatrix} 0 & x \\ 0 & 0 \end{bmatrix} \right\}$; idempotents: $\left\{ \begin{bmatrix} 1 & 0 \\ 0 & 1 \end{bmatrix}, \begin{bmatrix} 0 & 0 \\ 0 & 0 \end{bmatrix}, \begin{bmatrix} 1 & x \\ 0 & 0 \end{bmatrix}, \begin{bmatrix} 0 & y \\ 0 & 1 \end{bmatrix} \right\}.$

31. If m is odd then $m^2 - m = (m-1)m$ is a multiple of $2m$. Thus $\bar{m}^2 = \bar{m}$ in $\mathbb{Z}_{2m}$.

33. If $r^2 = r$ for all $r \in R$, then
$$r + r = (r + r)^2 = (r + r)(r + r) = r^2 + r^2 + r^2 + r^2 = r + r + r + r.$$
Hence $0 = r + r = 2r$. Since $R \neq 0$, this shows that R has characteristic 2. In particular, $-r = r$ for all $r \in R$. If $r, s \in R$, then
$$r + s = (r + s)^2 = r^2 + rs + sr + s^2 = r + rs + sr + s.$$
Thus $rs + sr = 0$, so $sr = -rs = rs$. Thus R is commutative.

35. Define
$$\sigma : \begin{bmatrix} R & R \\ 0 & R \end{bmatrix} \to \begin{bmatrix} R & 0 \\ R & R \end{bmatrix} \text{ by } \sigma \begin{bmatrix} r & s \\ 0 & t \end{bmatrix} = \begin{bmatrix} t & 0 \\ s & r \end{bmatrix}.$$
It is clear that σ is a bijection and preserves addition. As to multiplication
$$\sigma \begin{bmatrix} r & s \\ 0 & t \end{bmatrix} \sigma \begin{bmatrix} r' & s' \\ 0 & t' \end{bmatrix} = \begin{bmatrix} t & 0 \\ s & r \end{bmatrix} \begin{bmatrix} t' & 0 \\ s' & r' \end{bmatrix} = \begin{bmatrix} tt' & 0 \\ st' + rs' & rr' \end{bmatrix}$$
$$= \sigma \begin{bmatrix} rr' & st' + rs' \\ 0 & tt' \end{bmatrix} = \sigma \left(\begin{bmatrix} r & s \\ 0 & t \end{bmatrix} \begin{bmatrix} r' & s' \\ 0 & t' \end{bmatrix} \right).$$
Hence σ is an isomorphism.

36. (a) If $\sigma : \mathbb{C} \to \mathbb{R}$ is an isomorphism let $\sigma(i) = a \in \mathbb{R}$. Then
$$a^2 = \sigma(i^2) = \sigma(-1) = -\sigma(1) = -1,$$
which is impossible for $a \in \mathbb{R}$. So no such isomorphism exists.

(c) If $\sigma : \mathbb{Q} \to \mathbb{Z}$ is an isomorphism and $\sigma(\frac{1}{2}) = n$, then
$$2n = n + n = \sigma \left(\tfrac{1}{2} + \tfrac{1}{2} \right) = \sigma(1) = 1,$$
a contradiction in $\mathbb{Z}$. So no such isomorphism exists.

37. Put $\sigma(1) = e$. Given $r' \in R'$, write $r' = \sigma(r)$, $r \in R$. Then
$$er' = \sigma(1) \cdot \sigma(r) = \sigma(1 \cdot r) = \sigma(r) = r'.$$
Similarly $r'e = r'$, so e is the unity of R'.

38. Let $R \overset{\sigma}{\to} S$ be an isomorphism.
(a) If $z \in Z(R)$ let $s \in S$, say $s = \sigma(r)$. Then
$$\sigma(z) \cdot s = \sigma(z) \cdot \sigma(r) = \sigma(zr) = \sigma(rz) = \sigma(r) \cdot \sigma(z) = s \cdot \sigma(z),$$
so $\sigma(z) \in Z(S)$. If $z' \in Z(S)$ and $z' = \sigma(w)$, then for any
$$r \in R : \quad \sigma(wr) = z' \cdot \sigma(r) = \sigma(r) \cdot z' = \sigma(rw).$$
Since σ is one-to-one, $wr = rw$. Thus $w \in Z(R)$. It follows that $\sigma : Z(R) \to Z(S)$ is onto. It is clearly one-to-one and so is an isomorphism of rings.

39. It is a routine matter to verify that $\alpha + \beta$ and $\alpha\beta$ are again endomorphisms. The distributive law $\alpha(\beta + \gamma) = \alpha\beta + \alpha\gamma$ follows because, for all $x \in X$:

$$[\alpha(\beta + \gamma)](x) = \alpha[\beta x + \gamma x] = \alpha(\beta x) + \alpha(\gamma x)$$
$$= (\alpha\beta)(x) + (\alpha\gamma)(x) = (\alpha\beta + \alpha\gamma)(x)$$

The other distributive law is similar, as are the rest of the axioms. The zero is the zero endomorphism $\theta : X \to X$ where $\theta(x) = 0$ for all $x \in X$, and 1_X is the unity.

41. If $a \in (eRe)^*$, let $ab = e$, $b \in (eRe)^*$. Write $f = 1 - e$, so $ef = 0 = fe$ and $e + f = 1$. Then

$$\sigma(a) \cdot \sigma(b) = (a + f)(b + f) = ab + af + fb + f^2$$
$$= ab + 0 + 0 + f = e + f = 1 \tag{*}$$

because $af = (ae)f = 0$ and $fb = f(eb) = 0$. Thus $\sigma : (eRe)^* \to R^*$ is a mapping, and (*) shows it is a group homomorphism. It is one-to-one because $\sigma(a) = \sigma(b)$ means $a + f = b + f$, so $a = b$.

42. (a) Let $n = p_1^{n_1} \cdots p_r^{n_r}$, p_i distinct primes, $n_i \geq 1$. If $\bar{k}$ is nilpotent in $\mathbb{Z}_n$ then $k^m \equiv 0 \pmod{n}$ so $n | k^m$. Hence $p_i | k^m$ for all i so $p_i | k$. Conversely, if $p_i | k$ for all i then $p_1 p_2 \cdots p_r | k$ because the p_i are relatively prime in pairs. Hence if $m = \max\{n_1, \ldots, n_r\}$ then $k^m = 0$.

    (c) Let $\bar{e}^2 = \bar{e}$ in $\mathbb{Z}_n$. Hence $n | e(1 - e)$. Then we can write $n = ab$, $a | e$, $b | (1 - e)$. [Exercise 35 §1.2], say $e = xa$, $1 - e = yb$. Thus $1 = xa + yb$ and $e = xa$.

43. Let $|R| = 4$. This is an additive group of order 4, so $o(1) = 4$ or $o(1) = 2$. If $o(1) = 4$ then $R = \{0, 1, 2, 3\}$ is isomorphic to $\mathbb{Z}_4 = \{\bar{0}, \bar{1}, \bar{2}, \bar{3}\}$ via the obvious map. So assume $o(1) = 2$; that is the characteristic of R is 2. Hence $r + r = 0$ for all r in R. If $a \neq 0, 1$ then $1 + a \neq 0, 1, a$. Hence $R = \{0, 1, a, 1 + a\}$. Thus

+	0	1	a	$1+a$
0	0	1	a	$1+a$
1	1	0	$1+a$	a
a	a	$1+a$	0	1
$1+a$	$1+a$	a	1	0

+	0	1	a	$1+a$
0	0	0	0	0
1	0	1	a	$1+a$
a	0	a		
$1+a$	0	$1+a$		

the addition table is as shown (it is the Klein group). All but four entries in the multiplication table are prescribed as shown. The rest of the table is determined by the choice of a^2.

(1) $a^2 = 1 + a$. The table is then as shown. This is clearly a field if it is associative. (See Section 4.3).

	0	1	a	$1+a$
0	0	0	0	0
1	0	1	a	$1+a$
a	0	a	$1+a$	1
$1+a$	0	$1+a$	1	a

$a^2 = 1 + a$

(2) $a^2 = a$. The table is as shown. This is isomorphic to $\mathbb{Z}_2 \times \mathbb{Z}_2$ with $a = (1, 0)$ and $1 + a = (0, 1)$.

	0	1	a	$1+a$
0	0	0	0	0
1	0	1	a	$1+a$
a	0	a	a	0
$1+a$	0	$1+a$	0	$1+a$

$a^2 = a$

(3) $a^2 = 1$. The table is as shown. This is isomorphic to

$$L = \left\{ \begin{bmatrix} a & b \\ 0 & a \end{bmatrix} \middle| a, b \in \mathbb{Z}_2 \right\}$$

with $a = \begin{bmatrix} 1 & 1 \\ 0 & 1 \end{bmatrix}$ and $1 + a = \begin{bmatrix} 0 & 1 \\ 0 & 0 \end{bmatrix}$.

	0	1	a	$1+a$
0	0	0	0	0
1	0	1	a	$1+a$
a	0	a	1	$1+a$
$1+a$	0	$1+a$	$1+a$	0

$a^2 = 1$

(4) $a^2 = 0$. If $b = 1 + a$ then $1 + b = a$ and $b^2 = 1 + a^2 = 1$. Hence this is the same as Case 3.

3.2 INTEGRAL DOMAINS AND FIELDS

1. (a) $1, -4$.

   (c) $0, 1$.

3. If $e^2 = e$ in a domain, then $e(1 - e) = 0$ so $e = 0$ or $e = 1$. If $a^n = 0$, $n \geq 1$, then $a = 0$. For if $a \neq 0$ then $aa^{n-1} = 0$ gives $a^{n-1} = 0, \ldots$, and eventually $a = 0$, a contradiction.

5. Let $A = \begin{bmatrix} 0 & 0 & \cdots & 0 & 1 \\ 0 & 0 & & 0 & 0 \\ \vdots & \vdots & & \vdots & \vdots \\ 0 & 0 & \cdots & 0 & 0 \end{bmatrix}$. Then $A^2 = 0$ but $A \neq 0$. So $M_n(R)$ is not a domain.

7. If $ab = 0$ then $(ba)^2 = b(ab)a = 0$, so $ba = 0$ by hypothesis.

9. In $\mathbb{Z}_5$, $1^2 + 2^2 = 0$; in $\mathbb{Z}_3$, let $a^2 + b^2 = 0$. If either $a = 0$ or $b = 0$, the other is 0 ($x^2 = 0 \Rightarrow x = 0$ in a field). If $a \neq 0$, $b \neq 0$ then $a, b \in \{1, 2\}$. But $1^2 + 1^2 \neq 0$, $1^2 + 2^2 = 2 \neq 0$, $2^2 + 2^2 = 3 \neq 0$.

11. The group $F^* = F \smallsetminus \{0\}$ has order $q - 1$ so $a^{q-1} = 1$ for all $a \neq 0$ (by Lagrange's theorem). Thus $a^q = a$ if $a \neq 0$; this also holds if $a = 0$.

13. Since $|F| = p$ is prime, $(F, +)$ is cyclic and is generated by 1 (or any nonzero element) by Lagrange's theorem. Hence the map $\mathbb{Z}_p \to F$ given by $\bar{k} \to k1$ is an isomorphism of additive groups. It is a ring isomorphism because $(km)1 = (k1)(m1)$ in F.

15. Let Z denote the center of a division ring D. If $0 \neq z \in Z$ we have $zd = 1 = dz$ for some $d \in D$; we must show that $d \in Z$. Given $r \in D$ we have $(rz)d = r(zd) = r$ and $(zr)d = d(zr) = (dz)r = r$. Hence $(rz)d = (zr)d$ so $rz = zr$.

16. (a) If K is a subfield, let $0 \neq a \in K$. If a' is the inverse of a in K then $aa' = 1$. But $aa^{-1} = 1$ in F so $a' = a^{-1}$ by cancellation. Hence $a^{-1} \in K$. Conversely, if the condition holds, then the inverse of a in F serves as its inverse in K.

   (c) Here $|K| = 1, 2, 4, 8, 16$, and $|K^*|$ divides 15 so $|K^*| = 1, 3, 5, 15$. Thus $|K| = 2, 4, 8, 16$. The common values are $|K| = 2, 4, 16$. So $K = \{0, 1\}$, $|K| = 4$, or $K = F$.

17. It is clearly a subring of $\mathbb{C}$. If $a = r + si \neq 0$ then $a\bar{a} = r^2 + s^2 \neq 0$ (one of $r \neq 0$ or $s \neq 0$). Since $r^2 + s^2 \in \mathbb{Q}$ we have $(r^2 + s^2)^{-1} \in \mathbb{Q}$ too, so
$$a^{-1} = (r^2 + s^2)^{-1}\bar{a} \in \mathbb{Q}(i).$$

18. (a) It is clearly a subring of $\mathbb{C}$. As in Example 4, if $a = r + s\sqrt{5}\,i \in \mathbb{Q}(\sqrt{5}\,i)$ define $a^* = r - s\sqrt{5}\,i$ and $N(a) = aa^* = r^2 + 5s^2$. If $a \neq 0$ then $N(a) \neq 0$ in $\mathbb{Q}$ so $N(a)^{-1} \in \mathbb{Q}$. Thus $a^{-1} = N(a)^{-1}a^* \in \mathbb{Q}(\sqrt{5}\,i)$.

19. $\mathbb{Q}(\sqrt{2})$ is a subfield of $\mathbb{R}$ by Example 4, and it contains $\sqrt{2}$. If F is any subfield of $\mathbb{R}$ then $\mathbb{Z} \subseteq F$ (because $1 \in \mathbb{R}$), and hence $\mathbb{Q} \subseteq F$ (because $\frac{n}{m} = nm^{-1} \in F$ for all $n, m \neq 0$ in $\mathbb{Z}$). If also $\sqrt{2} \in F$, this means $r + s\sqrt{2} \in F$ for all $r, s \in \mathbb{Q}$. Thus $\mathbb{Q}(\sqrt{2}) \subseteq F$.

21. (a) $\mathbb{Z}(w)$ is a subring of $\mathbb{C}$, and so is an integral domain by Example 3.

    (c) $r^{**} = r$ is obvious. If $r = n + mw$ and $s = n' + m'w$ then
$$(rs)^* = (nn' + mm'w^2) - (nm' + mn')w = (n - mw)(n' - m'w) = r^*s^*$$
$$(pr + qs)^* = (pn + qn') - (pm + qm')w = p(n - mw) + q(n' - m'w)$$
$$= pr^* + qs^*.$$

    (e) If r is a unit in $\mathbb{Z}(w)$ then $rr^{-1} = 1$ in $\mathbb{Z}(i)$ so, by (d) $N(r)N(r^{-1}) = N(1) = 1$ in $\mathbb{Z}$. It follows that $N(r) = \pm 1$. Conversely, if $N(r) = \pm 1$ then $rr^* = \pm 1$ so $r^{-1} = \pm r^*$.

23. Let $R = \{r_1, r_2, \ldots, r_n\}$ be a domain with n elements. If $0 \neq a \in R$, then the elements of $aR = \{ar_1, ar_2, \ldots, ar_n\}$ are distinct (because a can be cancelled) so $|aR| = n = |R|$. Hence $aR = R$ so $ab = 1$ for some $b \in R$. Similarly $Ra = \{r_1a, \ldots, r_na\} = R$ so $ca = 1$ for some $c \in R$. Thus $c = c(ab) = (ca)b = b$, and this element is the inverse of a.

24. (a)
$$\binom{n}{r} = \frac{n!}{r!(n-r)!} = \frac{n}{r} \cdot \frac{(n-1)!}{(r-1)!(n-r)!}$$
$$= \frac{n}{r} \cdot \frac{(n-1)!}{(r-1)![(n-1)-(r-1)]!} = \frac{n}{r}\binom{n-1}{r-1}.$$

    (c) If F is a field of characteristic p, the map $\sigma : F \to F$ with $\sigma(a) = a^p$ satisfies $\sigma(1) = 1$, $\sigma(ab) = (ab)^p = a^pb^p = \sigma(a) \cdot \sigma(b)$, and, using (b), $\sigma(a + b) = (a + b)^p = a^p + b^p = \sigma(a) + \sigma(b)$. Hence σ is a homomorphism. We claim σ is one-to-one. Let $a \in \ker \sigma$, that is $\sigma(a) = 0$. Then $a^p = 0$ so $a = 0$ because F is a field). Since F is finite, σ is also onto, and so is an automorphism of F.

25. Given $\sigma : R \to R$, we have $Q = \{ru^{-1} \mid r \in R, 0 \neq u \in R\}$. Since $u \neq 0$ implies $\sigma(u) \neq 0$, if $\bar{\sigma} : Q \to Q$ exists it must be given by
$$\bar{\sigma}(ru^{-1}) = \bar{\sigma}(r)[\bar{\sigma}(u)]^{-1} = \sigma(r)[\sigma(u)]^{-1}.$$
So define $\bar{\sigma}$ by this formula. If $ru^{-1} = sv^{-1}$ then $rv = su$ so
$$\sigma(r) \cdot \sigma(v) = \sigma(s)\sigma(u);$$
that is $\sigma(r)[\sigma(u)]^{-1} = \sigma(s)[\sigma(v))]^{-1}$. Hence $\bar{\sigma}$ is well defined. Now
$$\bar{\sigma}(ru^{-1} + xv^{-1}) = \bar{\sigma}[(rv + su)(uv)^{-1}] = [\sigma(r)\sigma(v) + \sigma(s)\sigma(u)][\sigma(u)]^{-1}[\sigma(v)]^{-1}$$
$$= \sigma(r)[\sigma(u)]^{-1} + \sigma(s)[\sigma(v)]^{-1}$$
$$= \bar{\sigma}(ru^{-1}) + \bar{\sigma}(sv^{-1}).$$

Similarly $\bar{\sigma}$ preserves multiplication. If $\bar{\sigma}(ru^{-1}) = 0$ then $\sigma(r)[\sigma(u)]^{-1} = 0$ so $\sigma(r) = 0$. Hence $r = 0$, $ru^{-1} = 0$; $\bar{\sigma}$ is one-to-one. Finally let $sv^{-1} \in Q$. Let $r = \sigma^{-1}(s)$ and $u = \sigma^{-1}(v)$. Then $u \neq 0$ and $\sigma(ru^{-1}) = \sigma(r) \cdot [\sigma(u)]^{-1} = sv^{-1}$. Thus $\bar{\sigma}$ is onto.

26. (a) If $\dfrac{r}{u} = \dfrac{r'}{u'}$ and $\dfrac{s}{v} = \dfrac{s'}{v'}$ then $ru' = ur'$ and $sv' = vs'$, and so

$$(rs)(u'v') = (ru')(sv') = (r'u)(s'r) = (r's')(uv).$$

This shows $\dfrac{rs}{uv} = \dfrac{r's'}{u'v'}$.

(c) $\dfrac{r}{u}\left(\dfrac{s}{v} + \dfrac{t}{w}\right) = \dfrac{r}{u} \cdot \dfrac{sw + tv}{vw} = \dfrac{r(sw + tv)}{uvw} = \dfrac{rsw + rtv}{uvw} = \dfrac{(rsw + rtv)u}{uvwu}$

$$= \dfrac{rs \cdot uw + rt \cdot uv}{uv \cdot uw} = \dfrac{rs}{uv} + \dfrac{rt}{uw} = \dfrac{r}{u} \cdot \dfrac{s}{v} + \dfrac{r}{u} \cdot \dfrac{t}{w}.$$

27. Let $R \subseteq F$ where F is a field, and let $Q = \left\{\dfrac{r}{u} \mid r, u \in R,\ u \neq 0\right\}$. Define $\sigma : Q \to F$ by $\sigma\left(\dfrac{r}{u}\right) = ru^{-1}$. Then $\dfrac{r}{u} = \dfrac{s}{v} \Leftrightarrow rv = su \Leftrightarrow ru^{-1} = sv^{-1}$, so σ is well defined and one-to-one. We have

$$\sigma\left(\dfrac{r}{u} + \dfrac{s}{v}\right) = (rv + su)(uv)^{-1} = (rv + su)u^{-1}v^{-1} = ru^{-1} + sv^{-1}$$

$$= \sigma\left(\dfrac{r}{u}\right) + \sigma\left(\dfrac{s}{v}\right).$$

$$\sigma\left(\dfrac{r}{u} \cdot \dfrac{s}{v}\right) = (rs)(uv)^{-1} = rsu^{-1}v^{-1} = ru^{-1} \cdot sv^{-1} = \sigma\left(\dfrac{r}{u}\right)\sigma\left(\dfrac{s}{v}\right).$$

Hence $R \cong \sigma(R) = \{ru^{-1} \mid r, u \in R,\ u \neq 0\}$ and $\sigma(R)$ is a subring of F. It is a subfield because, if $ru^{-1} \neq 0$ then $r \neq 0$ so $(ru^{-1})^{-1} = ur^{-1} \in \sigma(R)$.

29. (a) If $r = i$ and $s = 1$ in $\mathbb{C}$, consider $a = r + s\omega$ in $\mathbb{C}(\omega)$. Then $aa^* = r^2 + s^2 = 0$, but $a \neq 0$ and $a^* \neq 0$ in $\mathbb{C}(\omega)$. Thus $\mathbb{C}(\omega)$ is not a field. In $\mathbb{Z}_5(\omega)$ let $a = 1 + 2\omega$. Then $aa^* = 1^2 + 2^2 = 0$, and $a \neq 0 \neq a^*$. So $\mathbb{Z}_5(\omega)$ is not a field. However $\mathbb{Z}_7(\omega)$ is a field. If $a = r + si \neq 0$ in $\mathbb{Z}_7(\omega)$ then $aa^* = r^2 + s^2$ and it suffices to show $r^2 + s^2 \neq 0$ in $\mathbb{Z}_7$. Suppose $r^2 + s^2 = 0$. If $r = 0$ or $s = 0$ then $a = 0$, contrary to hypothesis. Thus $r \neq 0 \neq s$. Then $0 = s^{-1}(r^2 + s^2) = (s^{-1}r)^2 + 1$ so $(s^{-1}r)^2 = -1$ in $\mathbb{Z}_7$. This is not the case because $0^2 = 0$, $1^2 = 1 = 6^2$, $2^2 = 4 = 5^2$, $3^2 = 2 = 4^2$ in $\mathbb{Z}_7$.

(c) Let $a = r + si \neq 0$ in $\mathbb{Z}_p(\omega)$, $p \equiv 3 \pmod 4$. Then $aa^* = r^2 + s^2$ so it suffices to show $r^2 + s^2 \neq 0$ (than $a^{-1} = (r^2 + s^2)^{-1}a^*$). Suppose $r^2 + s^2 = 0$. Now $r \neq 0$ or $s \neq 0$ (because $a \neq 0$). If $s \neq 0$ then $0 = s^{-2}(r^2 + s^2) = (s^{-1}r)^2 + 1$. Thus $x = s^{-1}r$ satisfies $x^2 = -1$ in $\mathbb{Z}_p$, contrary to the Corollary to Theorem 8 §1.3. Similarly if $r \neq 0$.

(e) If $e = r + s\omega$ and $e^2 = e$ then $r^2 - s^2 = r$ and $2rs = s$. If $s = 0$ then $r^2 = r$; $r = 0, 1$; $e = 0, 1$. If $s \neq 0$ then $2r = 1$, $r = \frac{1}{2}$, $s^2 = r^2 - r = \frac{1}{4} - \frac{1}{2} = -\frac{1}{4}$, $(2s)^2 = -1$.

30. (a) and (c). These are routine calculations.

(e) $N(pq) = (pq)(pq)^* = pqq^*p^* = pN(a)p^* = pp^*N(a) = N(b)N(g)$.

31. (a) and (c) are routine verifications.

(e) If $H = \{a + bi + cj + dk \mid a, b, c, d \in \mathbb{R}\}$ then (b) shows that

$$H = \left\{ \begin{bmatrix} z & w \\ -\bar{w} & \bar{z} \end{bmatrix} \middle| z, w \in \mathbb{C} \right\},$$

where $\bar{z}$ denotes the conjugate of z. In this form H is easily verified to be a subring of $M_2(\mathbb{C})$. Since (a) and (d) determine the multiplication, we have $\mathbb{H} \cong H$.

32. Since R is commutative, Lemma 1 holds in $\mathbb{H}(R)$; the proof is the same.

(a) If q is a unit in $\mathbb{H}(R)$, then $1 = N(1) = N(qq^{-1}) = N(q)N(q^{-1})$, so $N(q)$ is a unit in R. Conversely $qq^* = N(q)$ shows $q^{-1} = N(q)^{-1}q^*$ if $N(q) \in R^*$.

(c) Let $q = a + bi + cj + dk$ in $\mathbb{H}(R)$. Then

$$qi = iq \Leftrightarrow -b + ai - dj + ck = -b + ai + dj - ck \Leftrightarrow 2c = 2d = 0.$$

Similarly $qj = jq \Leftrightarrow 2a = 2c = 0$ and $qk = kq \Leftrightarrow 2b = 2c = 0$. Since $q \in Z[\mathbb{H}(R)] \Leftrightarrow qi = iq$, $qj = jq$ and $qk = kq$, the result follows. If R has characteristic 2, then $2r = 0$ for all $r \in R$, so $Z[\mathbb{H}(R)] = \mathbb{H}(R)$, that is $\mathbb{H}(R)$ is commutative. Conversely, if $\mathbb{H}(R)$ is commutative, $(1 + ai)j = j(1 + ai)$ for all $a \in R$, so $ak = -ak$, $2a = 0$. Thus R has characteristic 2. Finally, $A_2(\mathbb{Z}_6) = \{0, 3\} = 3\mathbb{Z}_6$, so

$$Z[\mathbb{H}(\mathbb{Z}_6)] = \{a + 3ri + 3sj + 3tk \mid a, r, s, t \in \mathbb{Z}_6\}.$$

3.3 IDEALS AND FACTOR RINGS

1. (a) No. $1 \in A$, $A \neq R$.

   (c) Yes. $\begin{bmatrix} a & b \\ 0 & c \end{bmatrix} \begin{bmatrix} 0 & x \\ 0 & y \end{bmatrix} = \begin{bmatrix} 0 & x' \\ 0 & y' \end{bmatrix}$, $\begin{bmatrix} 0 & x \\ 0 & y \end{bmatrix} \begin{bmatrix} a & b \\ 0 & c \end{bmatrix} = \begin{bmatrix} 0 & x' \\ 0 & y' \end{bmatrix}$.

   (e) No. $1 \in A$.

3. (a) $(1 + A)(r + A) = (1 \cdot r + A) = r + A$ and $(r + A)(1 + A) = (r \cdot 1 + A) = r + A$.

   (c) If R is commutative then, for all $r + A$, $s + A$ in R/A:

   $$(r + A)(s + A) = rs + A = sr + A = (s + A)(r + A).$$

4. (a) $mr + ms = m(r + s) \in mR$, $-(mr) = m(-r) \in mR$; $s(mr) = m(sr) \in mR$; $(mr)s = m(rs) \in mR$. If $mr = 0$ and $mt = 0$ then $m(r + t) = mr + mt = 0$; $m(-r) = -mr = 0$; $m(rs) = (mr)s = 0$; and $m(sr) = s(mr) = 0$ for all $s \in R$.

5. (a) $A \times B$ is clearly an additive subgroup and $(r, s)(a, b) = (ra, sb) \in A \times B$ for all $(r, s) \in R \times S$ and $(a, b) \in A \times B$. Similarly $(a, b)(r, s) \in A \times B$.

   (c) By (b) let $A \times B$ be a maximal ideal of $R \times S$. Then either $A = R$ or $B = S$ (otherwise $A \times B \subset A \times S \subset R \times S$). If $B = S$ then A is maximal in R [if $A \subseteq B \subseteq R$ then $A \times S \subseteq B \times S \subseteq R \times S$ so $B = A$ or $B = R$]. Similarly, if $A = R$ then B is maximal in S. Conversely $A \times S$ is maximal in $R \times S$ if A is maximal in R, with a similar statement for $R \times B$.

7. Let $ab \in \mathbb{Z} \times 0$ where $a = (n, m)$, $b = (k, l)$. Then $(nk, ml) \in \mathbb{Z} \times 0$ so $ml = 0$. As $\mathbb{Z}$ is a domain either $m = 0$ or $n = 0$; that is $a \in \mathbb{Z} \times 0$ or $b \in \mathbb{Z} \times 0$. Hence $\mathbb{Z} \times 0$ is a prime ideal of $\mathbb{Z} \times \mathbb{Z}$; similarly for $0 \times \mathbb{Z}$.

9. (a) Since $i \in A = Ri$, and since i is a unit in R, $A = R$. So R/A is the zero ring.

   (c) $A = R(1 + 2i)$. By the Hint $3 + i \in A$ so $i + A = -3 + A$. Hence $(m + ni) + A = (m - 3n) + A$, so each coset in R/A has the form $k + A$, $k \in \mathbb{Z}$. Moreover, $5 = 2(3 + i) - (1 + 2i) \in A$ so (as in Example 6)
   $$R/A = \{0 + A, \ 1 + A, \ 2 + A, \ 3 + A, \ 4 + A\}.$$
   These are distinct since $r + A = s + A$, $0 \le r \le s \le 4$ means $s - r \in A$, $0 \le s - r \le 4$. Hence $s - r = (m + ni)(1 + 2i)$ for m, $n \in \mathbb{Z}$, so taking absolute values, $(s - r)^2 = (m^2 + n^2)(1^2 + 2^2)$. Thus $5 | (s - r)^2$ so $5 | (s - r)$. This forces $s = r$.

11. (a) Assume $nR \ne 0$ and let $A = \{r \in R \mid nr = 0\}$. Then A is an ideal of R and $A \ne R$ because $nR \ne 0$. So $A = 0$.

12. (a) If $a, b \in \text{ann } X$ then $(a \pm b)x = ax \pm bx = 0$ and $(ra)x = 0$ for all $r \in R$.

   (c) $a \in \text{ann}(X \cup Y) \iff at = 0$ for all $t \in X \cup Y \iff at = 0$ for all $t \in X$ and $at = 0$ for all $t \in Y$.

   (e) By (d) and (b), $\text{ann}(X) \supseteq \text{ann}\{\text{ann}[\text{ann}(X)]\}$. Let $b \in \text{ann}(X)$. If $y \in \text{ann}[\text{ann}(X)]$ then $by = 0$. Hence $b \in \text{ann}\{\text{ann}[\text{ann}(X)]\}$.

13. If S is not commutative and C is commutative, then $R = S \times C$ is not commutative, but, if $A = S \times 0$, $R/A \cong C$ is commutative.

   More interesting example: Let $R = \begin{bmatrix} F & F \\ 0 & F \end{bmatrix}$, F a field, and $A = \begin{bmatrix} 0 & F \\ 0 & 0 \end{bmatrix}$. Then R is not commutative but $\begin{bmatrix} a & b \\ 0 & c \end{bmatrix} + A = \begin{bmatrix} a & 0 \\ 0 & c \end{bmatrix} + A$ for all $\begin{bmatrix} a & b \\ 0 & c \end{bmatrix} \in R$, and
   $$\left[\begin{bmatrix} a & 0 \\ 0 & c \end{bmatrix} + A\right]\left[\begin{bmatrix} a' & 0 \\ 0 & c' \end{bmatrix} + A\right] = \begin{bmatrix} aa' & 0 \\ 0 & cc' \end{bmatrix} + A = \left[\begin{bmatrix} a' & 0 \\ 0 & c' \end{bmatrix} + A\right]\left[\begin{bmatrix} a & 0 \\ 0 & c \end{bmatrix} + A\right].$$
   So R/A is commutative.

14. (a) $X + Y$ is a subgroup because $0 = 0 + 0 \in X + Y$ and, if $r = x + y$ and $r' = x' + y'$ are in $X + Y$ then $r + r' = (x + x') + (y + y') \in X + Y$ and $-r = (-x) + (-y) \in X + Y$. We have $X \subseteq X + Y$ because $x = x + 0 \in X + Y$ for all $x \in X$. Similarly $Y \subseteq X + Y$.

   (c) $S + A$ is an additive subgroup because S and A are; and $1 = 1 + 0 \in S + A$. Finally $(s + a)(s' + a') = ss' + (sa' + as' + aa') \in S + A$ because $sa' + as' + aa' \in A$.

15. $A \cap S$ is an additive subgroup since A and S are. If $a \in A \cap S$ and $s \in S$ then $as \in S$ (because $a \in S$) and $as \in A$ because A is an ideal of R. Thus $as \in S \cap A$, and $sa \in S \cap A$ is similar.

17. (a) If $Z = Z(R)$ is an ideal then $1 \in Z$ forces $Z = R$, and R is commutative. Conversely, R commutative implies $Z = R$ is an ideal.

   (c) Write $Z = Z(R)$ and let $R/Z = \langle b + Z \rangle$ as an additive group. Let r, $s \in R$. Then $r + Z = m(b + Z)$ and $s + Z = n(b + Z)$ where m, $n \in \mathbb{Z}$, say

$r = mb + z$ and $s = nb + z'$ where $z, z' \in Z$. Hence

$$rs = (mb + z)(nb + z') = mnb^2 + mbz' + nzb + zz'$$
$$= nmb^2 + nbz + mz'b + z'z = (nb + z')(mb + z) = sr.$$

18. (a) We have $\frac{B \cap C}{A} \subseteq \frac{B}{A} \cap \frac{C}{A}$ by (3) of Theorem 4. If $x \in \frac{B}{A} \cap \frac{C}{A}$, let $x = b + A$ and $x = c + A$ where $b \in B$ and $c \in C$. Hence $b + A = c + A$, so $b - c \in A \subseteq C$. But then $b \in c + C = C$, so $b \in B \cap C$. Finally then, $x = b + A \in \frac{B \cap C}{A}$, proving that $\frac{B}{A} \cap \frac{C}{A} \subseteq \frac{B \cap C}{A}$.

19. A ring S has no nonzero nilpotents if and only if $s^2 = 0$ in S implies $s = 0$. (Exercise 11 §3.1). If $r^2 \in A$ then $(r + A)^2 = 0 + A$ in R/A so $r + A = 0 + A$, $r \in A$. Conversely, if $r^2 \in A \Rightarrow r \in A$ then $(r + A)^2 = 0 + A$ in R/A implies $r + A = 0 + A$. This shows R/A has no nonzero nilpotents.

20. (a) If M is a maximal ideal then R/M is a field, so R/M is an integral domain, so M is prime.

(c) No. 0 is a prime ideal of $\mathbb{Z}$ because $\mathbb{Z}/0 \cong \mathbb{Z}$ is a domain, but 0 is not maximal because $\mathbb{Z}$ is not a field.

21. (a) $(r + A)^2 = r^2 + A = r + A$ for all $r \in R$.

(c) Given $r + A$, if $rs = 1 = sr$ then $(r + A)(s + A) = 1 + A = (s + A)(r + A)$. If $r^n = 0$ then $(r + A)^n = r^n + A = 0 + A$.

22. (a) Suppose $e^2 = e$ in R. Then $(e + A)^2 = e + A$ in R/A so $e + A = 0 + A$ or $e + A = 1 + A$ by hypothesis. Thus $e \in A$ or $1 - e \in A$. But $e \in A$ means e is nilpotent (hypothesis) so $e^n = 0$, $n \geq 1$. But $e^2 = e \Rightarrow e^n = e$ for all $n \geq 1$, so $e = 0$. If $1 - e \in A$ then $1 - e = 0$ in the same way because $(1 - e)^2 = 1 - e$.

(c) Suppose the only unit in R/A is $1 + A$. If $u \in R^*$ then $u + A$ is a unit in R/A, so $u + A = 1 + A$ by hypothesis. Hence $u - 1 \in A$, that is $u \in 1 + A$. This shows that $R^* \subseteq 1 + A$; the other inclusion holds because A consists of nilpotents.

   Conversely, suppose that $R^* = 1 + A$. If $u + A$ is a unit in R/A then $u \in R^*$ by (b), so $u \in 1 + A$ by hypothesis. But then $u + A = 1 + A$. This shows that $1 + A$ is the only unit in R/A, as required.

23. (a) $\mathbb{Z}_5$ is a field so 0 is the only maximal ideal.

(c) The divisors of 10 are 1, 2, 5, 10. So the lattice of additive subgroups is as shown. They are clearly ideals and $\langle 2 \rangle$ and $\langle 5 \rangle$ are the maximal ones,

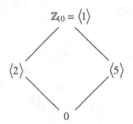

25. (a) Since R is commutative, $rab = (ra)b \in Rb$ and $rab = (rb)a \in Ra$ for all $r \in R$.

(c) If u is a unit then $u \in Ru$ implies $Ru = R$ by Theorem 2. Conversely, if $Ru = R$ then $1 \in Ru$, say $1 = vu$, $v \in R$. Hence u is a unit (R is commutative).

(e) If $a = ub$, $u \in R^*$, then $a \in Rb$ so $Ra \subseteq Rb$. But $b = u^{-1}a \in Ra$, so $Rb \subseteq Ra$ too. Conversely, if $Ra = Rb$ then $a \in Rb$, say $a = ub$, $u \in R$. Similarly $b = va$, $v \in R$, so $a = u(va) = (uv)a$. If $a = 0$ then $b = va = 0$ so $a = 1b$. If $a \neq 0$ then cancellation (R is a domain) gives $1 = uv$.

26. (a) AB is clearly an additive subgroup and $r(\sum_{i=1}^{n} a_ib_i) = \sum_{i=1}^{n}(ra_i)b_i \in AB$ for all r. Similarly $(\sum_{i=1}^{n} a_ib_i)r \in AB$. Now $AB \subseteq A$ because A is an ideal. Similarly $AB \subseteq B$, so $AB \subseteq A \cap B$.

(c) $AR \subseteq A$ because A is an ideal; $A \subseteq AR$ because $a = a \cdot 1$ for all $a \in A$. Thus $A = AR$; similarly $A = RA$.

27. RaR is clearly an additive subgroup of R, and $r(\sum_{i=1}^{n} r_ias_i) = \sum_{i=1}^{n}(rr_i)as_i$ and $(\sum_{i=1}^{n} r_ias_i)r = \sum_{i=1}^{n} r_ia(s_ir)$ show it is an ideal. Clearly $a = 1a1 \in RaR$. If $a \in A$, A an ideal, then $r_ias_i \in A$ for all r_i, s_i, so $\sum_{i=1}^{n} r_ias_i \in A$. Thus $RaR \subseteq A$.

29. We have $0, \begin{bmatrix} F & F \\ 0 & 0 \end{bmatrix}, \begin{bmatrix} 0 & F \\ 0 & F \end{bmatrix}, \begin{bmatrix} 0 & F \\ 0 & 0 \end{bmatrix}$ and R are all ideals of R. Let $A \neq 0$, R be an ideal, $0 \neq \begin{bmatrix} a & x \\ 0 & b \end{bmatrix} \in A$. If $a \neq 0$, $b \neq 0$ then $\begin{bmatrix} a & x \\ 0 & b \end{bmatrix}$ is a unit so $A = R$. If $a = 0$, $b \neq 0$. Then $\begin{bmatrix} 0 & 0 \\ 0 & 1 \end{bmatrix} = \begin{bmatrix} 0 & 0 \\ 0 & b^{-1} \end{bmatrix}\begin{bmatrix} 0 & x \\ 0 & b \end{bmatrix} \in A$, so $\begin{bmatrix} 0 & F \\ 0 & F \end{bmatrix} = R\begin{bmatrix} 0 & 0 \\ 0 & 1 \end{bmatrix} \subseteq A$. If $\begin{bmatrix} 0 & F \\ 0 & F \end{bmatrix} \subset A$ then A contains a unit so $A = R$. Thus $A = \begin{bmatrix} 0 & F \\ 0 & F \end{bmatrix}$ or $A = R$. Similarly, if $\begin{bmatrix} a & x \\ 0 & 0 \end{bmatrix} \in A$, $a \neq 0$, then $A = \begin{bmatrix} F & F \\ 0 & 0 \end{bmatrix}$ or $A = R$. So assume $A \subseteq \begin{bmatrix} 0 & F \\ 0 & 0 \end{bmatrix}$. If $\begin{bmatrix} 0 & x \\ 0 & 0 \end{bmatrix} \in A$, $x \neq 0$, then $\begin{bmatrix} 0 & 1 \\ 0 & 0 \end{bmatrix} = \begin{bmatrix} x^{-1} & 0 \\ 0 & 0 \end{bmatrix}\begin{bmatrix} 0 & x \\ 0 & 0 \end{bmatrix} \in A$. So $\begin{bmatrix} 0 & F \\ 0 & F \end{bmatrix} = R\begin{bmatrix} 0 & 1 \\ 0 & 0 \end{bmatrix} \subseteq A$, that is $A = \begin{bmatrix} 0 & F \\ 0 & 0 \end{bmatrix}$.

31. Put $R = \mathbb{Z}_2(i) = \{0, 1, i, 1 + i\}$. Then $1, i$ are units but $A = \{0, 1, +i\}$ is an ideal because $i(1 + i) = 1 + i$ and $(1 + i)(1 + i) = 0$. Clearly $A \neq 0$ is the only proper ideal.

32. (a) If $x = a + b\sqrt{2}$ in $\mathbb{Z}_3(\sqrt{2})$ let $x^* = a - b\sqrt{2}$. Then $xx^* = a^2 - 2b^2 = a^2 + b^2$ in $\mathbb{Z}_3$. Now $0^2 = 0$, $1^2 = 2^2 = 1$, in $\mathbb{Z}_3$ so $a^2 + b^2 \neq 0$ if $a \neq 0$ or $b \neq 0$. Thus $x \neq 0 \Rightarrow x^{-1} = (a^2 + b^2)^{-1}x^*$.

33. (a) If $a^n = 0$ then $(ra)^n = r^na^n = 0$ for all r. If also $b^m = 0$ consider

$$(a + b)^{n+m} = \sum_{k=0}^{n+m} \binom{n+m}{k}a^kb^{n+m-k}.$$

If $k \geq n$ then $a^n = 0$; if $n + m - k \geq m$ (i.e. $k \leq n$) then $b^{n+m-k} = 0$. So every term in the sum is zero; that is $(a + b)^{n+m} = 0$. Thus $N(R)$ is an ideal.

(c) Let $R = M_2(\mathbb{Z}_2)$. Then $a = \begin{bmatrix} 0 & 1 \\ 0 & 0 \end{bmatrix}$ and $b = \begin{bmatrix} 0 & 0 \\ 1 & 0 \end{bmatrix}$ are nilpotent, but $a + b = \begin{bmatrix} 0 & 1 \\ 1 & 0 \end{bmatrix}$ is not $\left((a + b)^2 = \begin{bmatrix} 1 & 0 \\ 0 & 1 \end{bmatrix}\right)$.

34. (a) Here $J(R) = \{0\}$ is an ideal.

(c) Write $\mathbb{Z}_{p^n} = \mathbb{Z}/p^n\mathbb{Z}$. If $A = \mathbb{Z}/k\mathbb{Z} \neq 0$ is an ideal of $\mathbb{Z}_{p^n}$, then $p^n\mathbb{Z} \subseteq k\mathbb{Z}$, so $k \mid p^n$. Hence $k = p^t$ for $t \leq n$, so $A \subseteq M$ where $M = \mathbb{Z}/p\mathbb{Z}$. It follows that M is the unique maximal ideal of $\mathbb{Z}_{p^n}$, so $\mathbb{Z}_{p^n}$ is local and $J(\mathbb{Z}_{p^n}) = M$.

(e) Write $J = J(R)$ and $J/A = \{r + A \mid r \in J\}$. If $r + A$ is a nonunit in R/A then r is a nonunit in R $[rs = 1 \Rightarrow (r + A)(s + A) = 1 + A]$. Hence $J(R/A) \subseteq J/A$. Conversely, let $r + A \in J/A$, so $r \in J$. We must show $r + A$ is a nonunit in R/A. Suppose not, and write $(r + A)^{-1} = s + A$. Then $rs - 1 \in A$ so $rs - 1 \in J$. But $r \in J$ and J is an ideal, so $rs \in J$. Thus $-1 \in J$, $1 \in J$, a contradiction.

35. (a) If $a, b \in M$ then $ab \in M$ because $ab \in P \Rightarrow a \in P$ or $b \in P$ (P is prime). Clearly $1 \in M$ because $P \neq R$.

    (c) Define $J = \{\frac{a}{u} \in R_P \mid a \in P\}$. This is an ideal of R_P because $\frac{a}{u} + \frac{b}{v} = \frac{av + bu}{uv} \in J$ since $av + bu + P$ if $a, b \in P$; and $\frac{r}{u} \cdot \frac{a}{v} = \frac{ra}{v} \in J$ because $ra \in P$ if $a \in P$. If $\frac{a}{u} \cdot \frac{r}{v} = 1$, $a \in P$, then $ar = uv$ so $uv \in P$. P is prime so $u \in P$ or $v \in P$, a contradiction. Thus $\frac{a}{u}$ is a nonunit if $a \in P$ so $J \subseteq J(R_P)$. Conversely, let $\frac{r}{u} \in J(R_P)$. Then we must show $r \in P$. But $r \notin P \Rightarrow r \in \dot{M}$ so $\frac{r}{u} \cdot \frac{uv}{r} = 1$ in R_P, a contradiction. So $r \in P$ and $\frac{r}{u} \in J$. Hence $J = J(R_P)$ so $J(R_P)$ is an ideal.

36. (a) We show that $J(R) = A$. Clearly $A \neq R$ so A consists of nonunits; that is $A \subseteq J(R)$. Let $r \in J(R)$. If $r \notin A$ then $r + A$ is a unit in R/A, say $(r + A)^{-1} = s + A$. Thus $rs - 1 \in A$, say $rs - 1 = a$. Hence $rs = 1 + a$ so rs is a unit in R because a is nilpotent by Example 17 §3.1. Similarly sr is a unit in R. If $(rs)u = 1 = v(sr)$, we see that r is a unit, contrary to $r \in J(R)$.

    (c) The power series for e^x is $1 + x + \frac{x^2}{2!} + \frac{x^3}{3!} + \cdots$. If $a \in A$ then a is nilpotent so $a + \frac{a^2}{2!} + \frac{a^3}{3!} + \cdots$ is an element of R which is nilpotent (R is commutative; see Exercise 34(a)). Thus $\sigma : A \to 1 + A$ given by $\sigma(a) = e^a = 1 + a + \frac{a^2}{2!} + \cdots$ is a map. Similarly $\ln(1 + x) = x - \frac{x^2}{2} + \frac{x^3}{3} - \frac{x^4}{4} + \cdots$, so if $a \in A$, $\ln(1 + a) = a - \frac{a^2}{2} + \frac{a^3}{3} - \frac{a^4}{4} + \cdots \in A$. Thus $\tau : 1 + A \to A$ is well defined by $\tau(u) = \ln u$. Now $\tau\sigma(a) = \ln[e^a] = a$ and $\sigma\tau(u) = e^{\ln u} = u$ hold, so σ and τ are inverses. Hence σ is a bijection. Finally $\sigma(a + b) = e^{a + b} = e^a e^b = \sigma(a)\sigma(b)$ for $ab \in A$. Thus σ is a group isomorphism.

3.4 HOMOMORPHISMS

1. (a) No. θ is a general ring homomorphism, because $4^2 = 4$ in $\mathbb{Z}_{12}$. But $\theta(1) = 4$, and $4 \neq 1$ in $\mathbb{Z}_{12}$.

    (c) No. $\theta[(r, s) \cdot (r', s')] = rr' + ss'$ need not equal
$$(r + s)(r' + s') = \theta(r, s) \cdot \theta(r', s').$$

    (e) Yes. $\theta(fg) = (fg)(1) = f(1)g(1) = \theta(f) \cdot \theta(g)$. Similarly for $f \pm g$. The unity of $F[\mathbb{R}, \mathbb{R}]$ is $\hat{1} : \mathbb{R} \to \mathbb{R}$ given by $\hat{1}(x) = 1$. Thus $\theta(\hat{1}) = \hat{1}(1) = 1$.

2. (a) Write $\theta(1) = e$, and let $s \in S$. Then $s = \theta(r)$ for some $r \in R$ (θ is onto) so
$$es = \theta(1)\theta(r) = \theta(1r) = \theta(r) = \theta(r1) = \theta(r)\theta(1) = se.$$

    Hence e is the unity of S.

3. If $\theta : \mathbb{Z} \to \mathbb{Z}$ is a general ring homomorphism, let $\theta(1) = e$. Then $e^2 = e$ so either $e = 1$ (θ a ring homomorphism) or $e = 0$. In the last case

$$\theta(k) = \theta(k \cdot 1) = \theta(k) \cdot \theta(1) = \theta(k) \cdot 0 = 0,$$

for all $k \in \mathbb{Z}$, so θ is trivial.

5. If $z \in Z(R)$ and $s \in R_1$, write $s = \theta(r)$, $r \in R$. Then

$$\theta(z) \cdot s = \theta(z) \cdot \theta(r) = \theta(zr) = \theta(rz) = \theta(r) \cdot \theta(z) = s \cdot \theta(z).$$

Thus $\theta(z) \in Z(R_1)$. Any non-onto ring homomorphism $\theta : R \to R_1$ where R_1 is commutative shows this need not be equality.

7. If $R \xrightarrow{\theta} S \xrightarrow{\varphi} T$ are ring homomorphisms then $\varphi\theta$ is a group homomorphism and $\varphi\theta(1_R) = \varphi[\theta(1_R)] = \varphi(1_S) = 1_T$,

$$\varphi\theta(rr_1) = \varphi[\theta(rr_1)] = \varphi[\theta(r) \cdot \theta(r_1)] = \varphi[\theta(r)] \cdot \varphi[\theta(r_1)] = \varphi\theta(r) \cdot \varphi\theta(r_1).$$

9. If R is a division ring and $\theta : R \to S$ is a ring homomorphism, then ker θ is an ideal of R so ker $\theta = 0$ or ker $\theta = R$. If ker $\theta = 0$ then θ is one-to-one so $\theta(R) \cong R$. If ker $\theta = R$ then $\theta(R) = 0$. Thus the images of R are R and 0 up to isomorphism.

10. Clearly $\theta(r^0) = \theta(1) = 1 = [\theta(r)]^0$. If $\theta(r^n) = [\theta(r)]^n$ then

$$\theta(r^{n+1}) = \theta(r) \cdot \theta(r^n) = \theta(r) \cdot [\theta(r)]^n = [\theta(r)]^{n+1}.$$

Hence (4) follows by induction.

11. In $\mathbb{Z}_4$ this is $x^3 + x - 1 = 0$. If $x = 0, 1, 2, -1$ then $x^3 + x - 1 = -1, 1, 1, 1$, so there is no solution in $\mathbb{Z}_4$, and hence none in $\mathbb{Z}$.

13. In $\mathbb{Z}_7$ this is $4n^2 = 2$ and this *has* a solution ($n = 2$) in $\mathbb{Z}_7$. However, in $\mathbb{Z}_{11}$ it is $7m^2 = 9$, or $m^2 = 8 \cdot 9 = 72 = 6$. But $m^2 = 0, 1, 3, 4, 5, 9$ in $\mathbb{Z}_{11}$, so there is no solution in $\mathbb{Z}_{11}$, and hence no solution in $\mathbb{Z}$.

15. The inverse map is a group isomorphism by group theory. Given s and s_1 in S then

$$\sigma[\sigma^{-1}(s) \cdot \sigma^{-1}(s_1)] = \sigma[\sigma^{-1}(s)] \cdot \sigma[\sigma^{-1}(s_1)] = ss_1 = \sigma[\sigma^{-1}(ss_1)].$$

Since σ is one-to-one, $\sigma^{-1}(s) \cdot \sigma^{-1}(s_1) = \sigma^{-1}(ss_1)$.

17. $R \cong R$ because $1_R : R \to R$ is an automorphism. If $R \cong S$, say $\sigma : R \to S$ in an isomorphism, then $\sigma^{-1} : S \to R$ is an isomorphism by Exercise 15, so $S \cong R$. Finally if $\tau : S \to T$ is an isomorphism, so is $\tau\sigma : R \to T$ so $R \cong T$.

19. (a) $\theta(A)$ is an additive subgroup of S by Theorem 1 §2.10. If $x \in \theta(A)$ and $y \in S = \theta(R)$, let $x = \theta(a)$, $y = \theta(r)$, $a \in A$, $r \in R$. Then

$$xy = \theta(a)\theta(r) = \theta(ar) \in \theta(A).$$

Similarly $yx \in \theta(A)$, so $\theta(A)$ is an ideal of S.

21. Let $\theta : \mathbb{C} \to \mathbb{R}$ be a ring homomorphism. Then ker θ is an ideal of the field $\mathbb{C}$ so ker $\theta = \mathbb{C}$ or ker $\theta = 0$. But ker $\theta = 0$ means $\mathbb{C} \cong \theta(\mathbb{C}) \subseteq \mathbb{R}$. Let $\theta(i) = a$. Then $a^2 = [\theta(i)]^2 = \theta(i^2) = \theta(-1) = -1$, a contradiction as a is real. Since ker $\theta \neq \mathbb{C}$ ($1\theta = 1 \neq 0$), there is no such θ.

23. (a) If $\bar{\theta}$ exists, then $\bar{\theta}(r + A) = \bar{\theta}\varphi(r) = \varphi'\theta(r) = \theta(r) + B$. Define $\bar{\theta}$ this way. Then $\bar{\theta}$ is well defined because

$$r + A = r_1 + A \Rightarrow (r - r_1) \in A \Rightarrow \theta(r) - \theta(r_1) \in \theta(A) \subseteq B \Rightarrow \theta(r) + B$$
$$= \theta(r_1) + B.$$

It is clearly a ring homomorphism and

$$\bar{\theta}\varphi(r) = \bar{\theta}(r + A) = \theta(r) + B = \varphi'\theta(r)$$

for all $r \in R$. Thus $\bar{\theta}\varphi = \varphi'\theta$.

24. (a) Define $\theta : R^* \to \mathrm{aut}\, R$ by $\theta(u) = \sigma_u$ for all $u \in R^*$. Then

$$\sigma_v\sigma_u(r) = v(uru^{-1})v^{-1} = (vu)r(vu)^{-1} = \sigma_{vu}(r)$$

for all r, so $\sigma_v\sigma_u = \sigma_{vu}$; that is, θ is a group homomorphism. Since $\theta(R^*) = \mathrm{inn}\ R$, this shows inn R is a subgroup of aut R. In fact $\mathrm{inn}\, R \lhd \mathrm{aut}\, R$. For if $\tau \in \mathrm{aut}\, R$ and $u \in R^*$, then $v = \tau(u) \in R^*$ and $(\tau\sigma_u\tau^{-1})(r) = \tau\{u\tau^{-1}(r)u^{-1}\} = vrv^{-1} = \sigma_v(r)$. Hence $\tau\sigma_u\tau^{-1} = \sigma_v$ is in inn R, as required.

25. (a) $e^2 = baba = b1a = ba = e$. Note that $e(bra) = babra = b1ra = bra$ and, similarly, $(bra)e = bra$. This σ is a mapping $R \to eRe$. Now $\sigma(r + s) = \sigma(r) + \sigma(s)$ is onto and $\sigma(r) \cdot \sigma(s) = bra \cdot bsa = br1sa = \sigma(rs)$. Finally $\sigma(1) = ba = e$ shows σ is a ring homomorphism. If $\sigma(r) = 0$ then $bra = 0$ so $r = 1r1 = (ab)r(ab) = a \cdot 0 \cdot b = 0$. Thus σ is one-to-one. Finally, if $r \in eRe$ then $r = ere = (ba)r(ba) = \sigma(arb)$. So σ is a ring isomorphism.

27. We use the isomorphism theorem by finding an onto ring homomorphism $\theta : R \to S \times S$ that has $A = \ker\theta$. If $\theta \begin{bmatrix} a & b \\ 0 & c \end{bmatrix} = (a, c)$, then θ is clearly onto, and the reader can verify that it is a ring homomorphism. Finally $\ker\theta = \left\{ \begin{bmatrix} a & b \\ 0 & c \end{bmatrix} \middle| (a, c) = (0, 0) \right\} = A$, as required.

29. (a) Define $\theta : R(\omega) \to \frac{R}{A}(\omega)$ by $\theta(a + b\omega) = \bar{a} + \bar{b}\omega$, $\bar{r} = r + A$. This is an onto R-homomorphism and $\ker\theta = A(\omega)$. Done by the isomorphism theorem.

31. Note that $e = (1, 0)$ is the unity of $\bar{S}$. Define $\theta : R = S \times T \longrightarrow T$ by $\theta(s, t) = t$. Then θ is an onto ring homomorphism and $\ker\theta = \bar{S}$. Done by the isomorphism theorem.

33. Define $\theta : R/A \to R/B$ by $\theta(r + A) = r + B$. This is well defined because $r + A = s + A \Rightarrow r - s \in A \subseteq B \Rightarrow r - s \in B \Rightarrow r + B = s + B$. It is clearly an onto ring homomorphism and $\ker\theta = \{r + A \mid r + B = 0\} = B/A$.

35. (a) Define $\theta : R(\eta) \to \frac{R}{A}(\eta)$ by $\theta(r + s\eta) = \bar{r} + \bar{s}\eta$ where $\bar{r} = r + A$. This is an onto ring homomorphism and $\ker\theta = \{r + s\eta \mid \bar{r} = \bar{s} = 1\} = A(\eta)$.

   (c) Observe first that $r + s\eta$ is a unit in $R(\eta)$ if and only if $r \neq 0$ (then the inverse is $r^{-1} - r^1sr^{-1}\eta$). Let $A \neq 0$ be an ideal of $R(\eta)$; we show that $A = R\eta = \{r\eta \mid r \in R\}$. As $A \neq 0$ let $0 \neq a + b\eta \in A$. Then $a = 0$ because $A \neq R(\eta)$, so $b \neq 0$. But then $\eta \in A$, whence $R\eta \subseteq A$. If $R\eta \subset A$ let $p + q\eta \in A$, $p \neq 0$, whence $A = R(\eta)$, as required.

36. (a) Define $\theta : R(\gamma) \to \frac{R}{A}(\gamma)$ by $\theta(r + s\gamma) = \bar{r} + \bar{s}\gamma$ where $\bar{r} = r + A$. This is an onto ring homomorphism and $\ker\theta = \{r + s\gamma \mid \bar{r} = \bar{s} = 1\} = A(\gamma)$.

(c) Let $A \neq 0, R(\gamma)$ be an ideal of $R(\gamma)$. Since $A \neq 0$ there exists $0 \neq r + s\gamma$ $\in A$, hence $1 + cg \in A$ for some c. Since $(1 + c\gamma)(1 - c\gamma) = 1 - c^2$, the fact that $A \neq R(\gamma)$ means that $c^2 = 1$. Hence $c = \pm 1$ because R is a division ring. So either $R(1 + \gamma) \subseteq A$ or $R(1 - \gamma) \subseteq A$.

Now consider the maps $\theta : R(\gamma) \to R$ and $\varphi : R(\gamma) \to R$ given by $\theta(r + s\gamma) = r + s$ and $\varphi(r + s\gamma) = r - s$ for all $r, s \in R$. These are onto ring homomorphisms, and ker $\theta = R(1 - \gamma)$ and ker $\varphi = R(1 + \gamma)$. It follows from the isomorphism theorem that $R(1 - \gamma)$ and $R(1 + \gamma)$ are both maximal ideals. Hence $A = R(1 - \gamma)$ or $A = R(1 + \gamma)$.

37. Define $\theta : \mathbb{Z} \to \mathbb{Z}_m \times \mathbb{Z}_n$ by $\theta(k) = (k + m\mathbb{Z}, k + n\mathbb{Z})$. This is a ring homomorphism and ker $\theta = \{k \mid k \in m\mathbb{Z} \text{ and } k \in n\mathbb{Z}\} = m\mathbb{Z} \cap n\mathbb{Z}$. Since $m|t$ and $n|t$, $m\mathbb{Z} \cap n\mathbb{Z} \supseteq t\mathbb{Z}$. Let $a \in m\mathbb{Z} \cap n\mathbb{Z}$. Then $m|a$ and $n|a$ so $t|a$ (because $t = lcm(m,n)$). Thus ker $\theta = t\mathbb{Z}$ and so $\mathbb{Z}_t = \mathbb{Z}/\text{ker } \theta \cong \theta(\mathbb{Z})$; a subring of $\mathbb{Z}_m \times \mathbb{Z}_n$.

39. By Example 10 and induction, all ideals of $R = R_1 \times \cdots \times R_n$ look like $A = A_1 \times \cdots \times A_n$, A_i an ideal of R. Hence $\frac{R}{A} \cong \frac{R_1}{A_1} \times \cdots \times \frac{R_n}{A_n}$. Thus A is maximal if and only if exactly one of the A_k is maximal in R_k, and $A_i = R_i$ if $i \neq k$.

41. Put $e = \frac{1}{2}(1 + u\sqrt{2})$. Thus $e^2 = \frac{1}{4}[(1 + 2u^2) + (2u\sqrt{2})] = \frac{1}{2}(1 + u\sqrt{2}) = e$. Then $R(\sqrt{2}) \cong R \times R$ by Theorem 8 if we can show $R \cong R(\sqrt{2})e$ and $R \cong R(\sqrt{2})(1 - e)$. Define $\sigma : R \to R(\sqrt{2})e$ by $\sigma(r) = re$. This satisfies $\sigma(r + s) = \sigma(r) + \sigma(s)$, and $\sigma(rs) = \sigma(r)\sigma(s)$, and σ is one-to-one ($\sigma(r) = 0$ means $0 = re = \frac{1}{2}(r + ru\sqrt{2})$, so $r = 0$). If $a, b \in R$ then

$$(a + b\sqrt{2})e = \tfrac{1}{2}(a + 2bu) + \tfrac{1}{2}(au + b)\sqrt{2} = \sigma(r)$$

where $r = a + 2bu$. Thus σ is onto so $R \cong R(\sqrt{2})e$ as rings. Similarly $R \cong R(\sqrt{2})(1 - e)$.

43. (a) Define $\theta : S \to R$ by $\theta(f) = f(x)$. Then

$$\theta(f + g) = (f + g)(x) = f(x) + g(x) = \theta(f) + \theta(g)$$

$$\theta(fg) = (fg)(x) = f(x)g(x) = \theta(f) \cdot \theta(g)$$

$$\theta(-f) = (-f)(x) = -(f(x)) = -\theta(f).$$

The unity $\bar{1}$ of S is the constant function onto 1, so

$$\theta(\bar{1}) = \bar{1}(x) = 1_R.$$

Thus θ is a ring homomorphism, onto because, if $r \in R$, $r = \theta(f)$ where $f(t) = r$ for all $t \in X$. Since ker $\theta = \{f \in S \mid f(x) = 0\}$ we are done by the isomorphism theorem because R is simple.

44. (a) The map $\theta : R \to R/A_1 \times \cdots \times R/A_n$ with $\theta(r) = (r + A_1, \ldots, r + A_n)$ is a ring homomorphism with kernel A.

3.5   ORDERED INTEGRAL DOMAINS

1. (a) $(b + c) - (a + c) = b - a \in R^+$.

   (c) $(-a) - (-b) = b - a \in R^+$.

(e) Since $a < b$ and $c > 0$, $ac < bc$ by Lemma 1. Similarly $bc < bd$ because $b > 0$, again by Lemma 1. Finally then $ac < bd$ by Lemma 1.

2. (a) $a \le a$ because $a = a$.

   (c) Let $a \le b$ and $b \le c$. If $a = b$ then $b \le c$ is $a \le c$; if $b = c$ then $a \le b$ is $a \le c$. Otherwise $a < b$ and $b < c$ so $a < c$ by Lemma 1.

3. (a) If $a \ge 0$ then $|a| = a \ge 0$. If $a < 0$ then $-a = 0 - a \in R^+$ so $|a| = -a > 0$.

   (c) If either $a = 0$ or $b = 0$ then $|ab| = 0 = |a||b|$.
   If $a > 0$, $b > 0$, then $ab > 0$ so $|ab| = ab = |a||b|$.
   If $a > 0$, $b < 0$, then $ab < 0$ so $|ab| = -ab = a(-b) = |a||b|$.
   If $a < 0$, $b > 0$, as above.
   If $a < 0$, $b < 0$ then $ab > 0$ so $|ab| = ab = (-a)(-b) = |a||b|$.

4. If a is such that $b \le a$ for all b then $a + 1 \le a$, whence $1 \le 0$, a contradiction.

5. (a) $-1 = i^2 \in \mathbb{Z}(i)^+$ and $1 \in \mathbb{Z}(i)^+$ would be a contradiction.

7. $\frac{r}{u}$ and $\frac{s}{v} \in Q^+$, so $ru \in R^+$ and $sv \in R^+$. Thus (using Lemma 1(5)):

$$\frac{r}{u} + \frac{s}{v} = \frac{rv + su}{uv} \in Q^+, \quad \text{because} \quad (rv + su)uv = (ru)v^2 + (sv)u^2 \in R^+$$

$$\frac{r}{u} \cdot \frac{s}{v} = \frac{rs}{uv} \in Q^+, \quad \text{because} \quad (rs)(uv) = (ru)(sv) \in R^+.$$

Hence Q^+ satisfies P1. Now let $\frac{r}{u} \in Q$. Then exactly one of $ru = 0$, $ru \in R^+$ and $-(ru) \in R^+$ is true. But $ru = 0 \iff \frac{r}{u} = 0$ (because $u \ne 0$); $ru \in R^+ \iff \frac{r}{u} \in Q^+$; and $-(ru) \in R^+ \iff -\frac{r}{u} = \frac{-r}{u} \in Q^+$. Thus Q^+ satisfies P2.

Chapter 4

Polynomials

4.1 POLYNOMIALS

1. (a) $f + g = 4 + 2x + 2x^2 + 5x^3$,
    $f g = 3 + 2x + 4x^2 + 4x^3 + 3x^4 + 4x^6$.

2. (a) $(1 + x)^5 = 1 + 5x + 10x^2 + 10x^3 + 5x^4 + x^5 = 1 + x^5$ in $\mathbb{Z}_5[x]$.

    (c) From the hint: $\binom{p}{k} = \frac{p!}{k!(p-k)!} = \frac{p}{k} \frac{(p-1)!}{(k-1)![(p-1)-(k-1)]!} = \frac{p}{k} \binom{p-1}{k-1}$. Hence p divides $k \binom{p}{k}$ for $1 \le k \le p - 1$. Since p is a prime and p does not divide k, this shows that p divides $\binom{p}{k}$. Hence $\binom{p}{k} = 0$ in $\mathbb{Z}_p$ for $1 \le k \le p - 1$. The binomial theorem gives
    $$(1 + x)^p = 1 + \binom{p}{1} x + \binom{p}{2} x^2 + \cdots + \binom{p}{p-1} x^{p-1} + x^p$$
    $$= 1 + x^p \quad \text{in} \quad \mathbb{Z}_p[x].$$

3. (a) The polynomials are $a_0 + a_1 x + a_2 x^2 + a_3 x^3$ where $a_i \in \mathbb{Z}_5$, for all i and $a_3 \neq 0$. Hence there are 5 choices for each of a_0, a_1 and a_2, and 4 choices for a_3, for $5^3 \cdot 4 = 500$ in all.

4. (a) If $f = (x - 4)(x - 5)$ then $f(4) = 0 = f(5)$ is clear. In $\mathbb{Z}_6$:
    $$f(0) = 20 = 2, \ f(1) = 12 = 0, \ f(2) = 6 = 0 \text{ and } f(3) = 2.$$
    Thus the roots in $\mathbb{Z}_6$ are 1, 2, 4 and 5. In $\mathbb{Z}_7 : f(a) = 0$ means $(a - 4)(a - 5) = 0$ so (since $\mathbb{Z}_7$ is a field) $a = 4$ or $a = 5$. These are the only roots in $\mathbb{Z}_7$ (or any integral domain!).

5. (a) In $\mathbb{Z}_4$, the roots are 0 and 1; in $\mathbb{Z}_2 \times \mathbb{Z}_2$, *every* element (a, b) is a root because $(a, b)^2 = (a^2, b^2) = (a, b)$; in any integral domain $a^2 - a = 0$ means $a(a - 1) = 0$ so $a = 0$ or $a = 1$; in $\mathbb{Z}_6 : 0, 1, 3, 4$ are the roots.

6. (a) Let $f = a_0 + a_1 x + \cdots + a_n x^n$, $a_n \neq 0$, and
    $$g = b_0 + b_1 x + \cdots + b_m x^m, \ b_m \neq 0.$$

Student Solution Manual to Accompany Introduction to Abstract Algebra, Fourth Edition.
W. Keith Nicholson.
© 2012 John Wiley & Sons, Inc. Published 2012 by John Wiley & Sons, Inc.

If $m < n$, the leading coefficient of $f + g$ is a_n, so $\deg(f + g) = n = \max(m, n)$. Similarly, $\deg(f + g) = \max(m, n)$ in case $n < m$. But if $m = n$, the coefficient of x^n is $a_n + b_m$. If this is nonzero, $\text{degree}(f + g) = n = \max(n, m)$. But if $a_n + b_n = 0$ then $\deg(f + g) < n = \max(m, n)$.

7. (a) Let ux^n and bx^m be the leading terms of f and g where u is a unit. The highest term appearing in fg is ubx^{n+m}. This is not zero ($ub = 0$ implies $b = u^{-1}(ub) = 0$), so certainly $fg \neq 0$ and

$$\deg(f g) = n + m = \deg f + \deg g.$$

9. Since $R \subseteq R[x]$, $o(1)$ in R is the same as $o(1)$ in $R[x]$.

11. $0 = f(a) = a_0 + a_1 a + \cdots a_{n-1}a^{n-1} + a_n a^n$. Multiply by a^{-n} to get

$$0 = a_0(a^{-1})^n + a_1(a^{-1})^{n-1} + \cdots + a_{n-1}a^{-1} + a_n = g(a^{-1}).$$

13. $x^3 - 4x + 5 = \left(\frac{1}{2}x^2 - \frac{1}{4}x - \frac{15}{8}\right)(2x + 1) + \frac{55}{8}$. In $\mathbb{Z}$, 2 is not a unit.

14. (a) $q = x^3 + 3x^2 - 3x + 5$, $r = -x - 3 = 5x + 3$.

   (c) $q = 3x^2 + 2x + 3$, $r = 7$.

   (e) $q = 3x + 2$, $r = -14x - 3$.

15. Only $x - 2$ because $f(1) \neq 0 \neq f(-1)$. $f = (x - 2)(x^3 - x + 1)$.

16. (a) $f(1) = 3 + 5 + 2 + 1 + 4 = 15$ is 0 in $\mathbb{Z}_p$ only if $p = 3$ or $p = 5$. We have $q = 2x^2 + x + 2$ in $\mathbb{Z}_3[x]$, and $q = 3x^3 + 3x^2 + 1$ in $\mathbb{Z}_5[x]$.

17. (a) $f = (x - 1)(x + 1)(x - 5)(x + 5)$.

   (c) $f = (x - 1)(x + 2)(x + 3)$.

19. Since $f(u) = 0$ we get $f = (x - u)g$. Then $g(v) = 0$ (as $v \neq u$) so $f = (x - u)(x - v)h$. Finally $f(w) = 0$ (because $u \neq w$ and $u \neq w$) so

$$f = (x - u)(x - v)(x - w)t, t \in F.$$

Since f is monic, $t = 1$. Thus

$$f = x^3 - (u + v + w)x^2 + (uv + uw + vw)x - uvw,$$

so the result follows from $f = x^3 + ax^2 + bx + c$.

21. (a) Clearly $2^2 = 0$ in $\mathbb{Z}_4$ so $(2x^n)^2 = 0$ in $\mathbb{Z}_4[x]$. Thus $2x^n$ is a nilpotent for all $n \geq 1$. This implies that $1 + 2x^n$ is a unit for all $n \geq 1$.

23. (a) Multiplicity 1. $f = (x - 3)(x^2 + x - 1)$.

   (c) Multiplicity 3. $f = (x - 1)^3(x^2 + x + 1)$.

24. (a) Fermat's theorem asserts that $a^{p-1} \equiv 1 \pmod{p}$ for all integers a not divisible by p. Hence $a^{p-1} = 1$ if $a \neq 0$ in $\mathbb{Z}_p$, so $a^p = a$ holds for all $a \in \mathbb{Z}_p$. Thus each $a \in \mathbb{Z}_p$ satisfies $x^p - x$.

   (c) The Chinese remainder theorem gives $\mathbb{Z}_{2p} \cong \mathbb{Z}_2 \times \mathbb{Z}_p$. Now $a^p = a$ for all $a \in \mathbb{Z}_p$ by (a) and $b^p = b$ if $b \in \mathbb{Z}_2$ because $\mathbb{Z}_2 = \{0, 1\}$. Hence $(b, a)^p = (b^p, a^p) = (b, a)$ in $\mathbb{Z}_2 \times \mathbb{Z}_p$.

   (e) $2^5 = 32 \neq 2$ in $\mathbb{Z}_{35}$; $2^7 = 128 = 23 \neq 2$ in $\mathbb{Z}_{35}$.

25. (a) $f = (4x - 3)(x^3 + x^2 + 1)$. The only rational root is $\frac{3}{4}$.

    (c) $f = (x - 2)(x + 1)(x^2 + 1)$. The rational roots are 2 and -1.

    (e) $f = (x^2 + x + 1)(x^2 + 2)$. There are no rational roots.

27. If f is monic, write $f = a_0 + a_1 x + \cdots + a_{a-1} x^{n-1} + x^n$. If $\frac{c}{d}$ is a rational root (in lowest terms) of f then $c | a_0$ and $d | 1$. Thus $d = \pm 1$ so $\frac{c}{d} = \pm c$ is an integer.

29. Write $h = f - g$; we wish to show $h = 0$. If not then $\deg h \leq n$. But h has $n + 1$ distinct roots in R by hypothesis, contradicting Theorem 8.

31. Let $\varphi_0 : R[x] \to R$ be evaluation at 0. This is an onto homomorphism for any ring R, and $\ker \varphi_0 = \langle x \rangle$. Hence $R[x]/\langle x \rangle \cong R$. If R is a field this shows that $\langle x \rangle$ is a maximal ideal of the (commutative) ring $R[x]$. If R is merely an integral domain, it shows $R[x]/\langle x \rangle$ is an integral domain, so $\langle x \rangle$ is a prime ideal of $R[x]$.

33. No, it need not preserve addition. For example, if $f = 1 + 2x$ and $g = 1 + 3x^2$ then $\theta(f) = 2$, $\theta(g) = 3$, but $\theta(f + g) = \theta(2 + 2x + 3x^2) = 3 \neq \theta(f) + \theta(g)$. Note that θ preserves multiplication if R is a domain (and $\theta(1) = 1$).

34. We have $\varphi_n(f) = f(n)$ for all f in $R[x]$.

    (a) If $f = r = r + 0x + 0x^2 + \cdots$, then
    $$\varphi_a(r) = f(a) = r + 0 \cdot a + 0 \cdot a^2 + \cdots = r.$$

    (c) If $u \in \mathbb{C}$, let $\bar{u}$ denote the conjugate of u. Define $\theta : \mathbb{C}[x] \to \mathbb{C}$ by $\theta[f(x)] = \overline{f(0)}$ This is a homomorphism (it is evaluation at 0 followed by conjugation) but it is not evaluation at a for any $a \in \mathbb{C}$. Indeed, if $\theta = \varphi_a$ then $\theta(i) = \bar{i} = -i$ while $\varphi_a(i) = i$.

35. Define $\theta : R[x] \longrightarrow R/A$ by $\theta(r_0 + r_1 x + \cdots) = r_0 + A$. It is easily verified that θ is an onto ring homomorphism, and
    $$\ker \theta = \{r_0 + r_1 x + \cdots \mid r_0 + A = 0\} = A.$$

    The Isomorphism Theorem does the rest.

37. (a) Clearly $\overline{f + g} = \overline{f} + \overline{g}$ and $\overline{1} = 1$. If $f = \sum_{i=1}^{n} a_i x^i$, $g = \sum_{j=1}^{m} b_j x^j$ then $f g = \sum_{k=1}^{p} c_k x^k$, $c_k = \sum_{i+j=k} a_i b_j$ Because θ is a ring homomorphism, so $\overline{fg} = \sum_{k=1}^{p} \bar{c}_k x^k = \overline{f} \cdot \overline{g}$. Clearly θ onto implies $\overline{\theta}$ onto.

    (c) If θ is an isomorphism then $\overline{\theta}$ is onto (θ is) and $\ker \overline{\theta} = A[x] = 0$ because $A = \ker \theta = 0$.

    (e) If $\overline{f}$ has no root in $S[x]$, assume $f(a) = 0$, $a \in R$. If
    $$f = \sum_{i=0}^{n} r_i x^i, \quad \overline{f(\bar{a})} = \sum_{i=0}^{n} \bar{r}_i \bar{a}^i = \sum_{i=0}^{n} \overline{r_i a^i} = \overline{f(a)} = \overline{0} = 0.$$

    Thus $\bar{a}$ is a root of $\overline{f}$, contrary to assumption.

38. (a) If P is prime ideal of R then R/P is an integral domain. By Exercise 37, this means $R[x]/P[x] \cong (R/P)[x]$ is an integral domain (Theorem 2); that is $P[x]$ is a prime ideal of $R[x]$.

39. If $fg = 1$ then $a_0 b_0 = 1$ where b_0 is the constant coefficient of g. In a commutative ring, if u is a unit and a is a nilpotent then $u^{-1}a$ is also nilpotent so $u + a = u(1 + u^{-1}a)$ is a unit (because $1 + u^{-1}a$ is a unit). If a_0 is a unit and

a_i is nilpotent for $i \geq 1$, then $a_i x^i$ is nilpotent for each i, so $a_0 + a_1 x$ is a unit; then $(a_0 + a_1 x) + a_2 x^2$ is a unit; etc.

40. (a) If $f = 1$ in $R[x]$, then $\tilde{f} : R \to R$ is given by $\tilde{f}(a) = 1$ for all $a \in R$. This is the unity of $F(R, R)$. Given f and g put $h = f + g$ so that $\theta(f + g) = \tilde{h}$. If $a \in R$, $\tilde{h}(a) = h(a) = f(a) + g(a) = \tilde{f}(a) + \tilde{g}(a) = (\tilde{f} + \tilde{g})(a)$. Thus $\tilde{h} = \tilde{f} + \tilde{g} = \theta(f) + \theta(g)$, so θ preserves addition. Similarly if $m = f g$ then $\tilde{m}(a) = f(a)g(a) = \tilde{f}(a)\tilde{g}(a) = \widetilde{(fg)}(a)$ for all a, so $\tilde{m} = \widetilde{fg}$. This shows θ preserves multiplication. Finally, $\theta(R[x]) = P(R, R)$ is clear.

    (c) If R is an infinite integral domain then $f \in \ker \theta$ means f has infinitely many roots (each element of R) and so $f = 0$ by Theorem 8.

41. Observe that $\deg c_k = n$ for each k and $c_k(a_i) = \begin{cases} 0 & \text{if } i \neq k \\ 1 & \text{if } i = k \end{cases}$. Now let

$$g = f(a_0)c_0 + f(a_1)c_1 + \cdots + f(a_n)c_n.$$

Then $g = 0$ or $\deg g \leq n$. Moreover, the choice of the c_k ensures that $g(a_i) = f(a_i)$ holds for each $i = 0, 1, 2, \ldots, n$. Hence Exercise 29 shows that $f = g$.

4.2 FACTORIZATION OF POLYNOMIALS OVER A FIELD

1. (a) $f = a(a^{-1}f)$.

2. (a) Yes, since $a \neq 0$, $f(b) = 0$ if and only if $af(b) = 0$.

3. (a) $f(1) = 0$; indeed $f = (x - 1)(x^2 - x + 2)$ over any field.

4. (a) Irreducible because it has no roots in $\mathbb{Z}_7$.

    (c) $x^2 + 11 = (x - \sqrt{11}i)(x + \sqrt{11}i)$ in $\mathbb{C}[x]$. Not irreducible.

    (e) Irreducible, because it has no root in $\mathbb{Z}_5$.

5.

		$\mathbb{Q}$	$\mathbb{R}$	$\mathbb{C}$	$\mathbb{Z}_2$	$\mathbb{Z}_3$	$\mathbb{Z}_5$	$\mathbb{Z}_7$
(a)	$x^2 - 3$	Yes	No	No	No	No	Yes	Yes
(c)	$x^3 + x + 1$	Yes	No*	No	Yes	No	Yes	Yes

*Every polynomial of odd degree in $\mathbb{R}[x]$ has a root in $\mathbb{R}$–see Exercise 9.

7. $f = [x - (1 - i)][x - (1 + i)][x - i][x + i] = (x^2 - 2x + 2)(x^2 + 1)$
   $= x^4 - 2x^3 + 3x^2 - 2x + 2$.
   The polynomial f^2 has the same roots, albeit of different multiplicities.

8. (a) As f is monic, we may assume that both factors are monic (Exercise 6). Hence $f = (x - u)(x - v) = x^2 - (u + v)x + uv$. Now equate coefficients.

9. Assume $f = a_n x^n + a_{n-1}x^{n-1} + \cdots + a_0$, $a_n \neq 0$, n odd. Then $g = a_n^{-1}f$ is monic and has the same roots as f. So assume

$$f = x^n + a_{n-1}x^{n-1} + \cdots + a_1 x \in a_0.$$

If $x \neq 0$ then

$$f = x^n \left(1 + \frac{a_{n-1}}{x} + \cdots + \frac{a_0}{x^n}\right).$$

Thus $\lim_{x \to \infty} f(x) = \infty$ and (because n is odd) $\lim_{x \to -\infty} f(x) = -\infty$. Thus there exist $a > 0$ and $b < 0$ such that $f(a) > 0$ and $f(b) < 0$. But then the intermediate value theorem of calculus gives c between a and b such that $f(c) = 0$.

11. If f is irreducible, then 0 is not a root so $f(0) \neq 0$ is the constant term. Hence f has the form $f = 1 + x^{n_1} + x^{n_2} + \cdots + x^{n_k}$. Then $0 \neq f(1) = k + 1 =$ the number of terms in f. Thus $k + 1$ is odd. The converse is false: $x^4 + x^2 + 1$ passes the test, but $x^4 + x^2 + 1 = (x^2 + x + 1)^2$ in $\mathbb{Z}_2[x]$.

13. If p satisfies (1) and (2), it has no linear factor by (1). If p fails to be irreducible it must factor as a quadratic and a cubic. Both are irreducible (by (1)), so by Example 5 and Exercise 10, either $p = (x^2 + x + 1)(x^3 + x + 1) = x^5 + x^4 + 1$ or $p = (x^2 + x + 1)(x^3 + x^2 + 1) = x^5 + x + 1$. These are ruled out by (2). Conversely, if p is irreducible then (1) comes from Theorem 1 and (2) from the above factorizations of $x^5 + x^4 + 1$ and $x^5 + x + 1$.

15. As in Exercise 1, this holds if the list is the set of all products of two (possibly equal) irreducible quadratics. By Exercise 14, they are

$(x^2 + 1)^2 = x^4 + 2x^2 + 1$ $(x^2 + 1)(x^2 + x + 2) = x^4 + x^3 + x + 2$

$(x^2 + x + 2)^2 = x^4 + 2x^3 + 2x^2 + x + 1$ $(x^2 + 1)(x^2 + 2x + 2) = x^4 + 2x^3 + 2x + 2$

$(x^2 + 2x + 2)^2 = x^4 + x^3 + 2x^2 + 2x + 1$ $(x^2 + x + 2)(x^2 + 2x + 2) = x^4 + 1.$

17. If $p \not\equiv 3 \pmod 4$, the only other possibility is $p \equiv 1 \pmod 4$. But then $x^2 + 1$ has a root in $\mathbb{Z}_p$ by the Corollary to Theorem 8 §1.3. This shows that $x^2 + 1$ is not irreducible.

18. (a) $3x^4 + 2 = 3(x - 1)(x + 1)(x - 3)(x + 3)$ in $\mathbb{Z}_5[x]$.

    (c) $x^3 + 2x^2 + 2x + 1 = (x + 1)(x + 3)(x + 5)$ in $\mathbb{Z}_7[x]$.

    (e) $x^4 - x^2 + x - 1 = (x - 1)(x - 2)(x^2 + 3x + 6)$ in $\mathbb{Z}_{13}[x]$.

19. $x^5 + x^4 + 1 = (x^2 + x + 1)(x^3 + x + 1)$ in $\mathbb{Z}_2[x]$.

20. $x^5 + x^2 - x + 1 = (x^2 + 1)(x^3 - x + 1)$ in $\mathbb{Z}_3[x]$.

21. (a) The possible rational roots are ± 1, ± 2, $\pm \frac{1}{3}$, $\pm \frac{2}{3}$. Direct checking shows none is a root, so it is irreducible by Theorem 1. Alternatively, modulo 5 the polynomial is $3x^3 + x + 2 = 3(x^3 + 2x - 1)$ and this has no root in $\mathbb{Z}_5$, again by a direct check.

    (c) Possible rational roots ± 1, ± 2, ± 3, ± 6. None work. Alternatively, modulo 5 it is $x^3 - x^2 + x + 1$, and this has no roots in $\mathbb{Z}_5$.

22. (a) The Eisenstein criterion applies with $p = 3$.

23. (a) $f(x + 1) = x^4 + 4x^3 + 6x^2 + 6x + 2$, so the Eisenstein criterion applies with $p = 2$.

    (c) $f(x + 1) = x^4 + 4x^3 + 6x^2 + 4x + m + 1$. Since $m + 1 = 4k - 2 = 2(2k - 1)$ the Eisenstein criterion applies with $p = 2$.

24. $f(x - 1) = (x - 1)^4 + 4(x - 1)^3 + 4(x - 1)^2 + 4(x - 1) + 5 = x^4 - 2x^2 + 4x + 2$. Use Eisenstein with $p = 2$.

25. $(1+x)f = 1 + x^p$. Replace x by $x - 1$:
$$xf(x-1) = 1 + (x-1)^p = x^p - \binom{p}{p-1}x^{p-1} + \cdots - \binom{p}{2}x^2 + px.$$
Then $f(x-1) = x^{p-1} - \binom{p}{p-1}x^{p-2} + \cdots - \binom{p}{2}x + p$. This is irreducible over $\mathbb{Q}$ by Eisenstein (with p), so f is also irreducible.

26. (a) $f_4(x) = x^3 + x^2 + x + 1 = (x+1)(x^2+1)$;
$f_6(x) = x^5 + x^4 + x^3 + x^2 + x + 1 = (x+1)(x^2+x+1)(x^2-x+1)$.

27. If $f = x^p + p^2 mx + (p-1)$ then
$$f(x+1) = (x+1)^p + p^2 m(x+1) + (p-1)$$
$$= x^p + \binom{p}{p-1}x^{p-1} + \cdots + \binom{p}{2}x^2 + p(pm+1)x + p(pm+1).$$
The Eisenstein criterion (using p) applies here because p does not divide $pm+1$.

29. The Eisenstein criterion applies using the prime p.

31. (a) If f is irreducible in $K[x]$ it cannot factor properly in $F[x]$ because $F[x] \subseteq K[x]$.

33. We are done by Theorem 1 as 1, 2, 3, and 7 are roots of x^2+x+1 in $\mathbb{Z}_p$ for $p = 3, 7, 13$ and 19 respectively, and x^2+x+1 has no roots in $\mathbb{Z}_p$ for $p = 2$, 5, 11 and 17.

34. Take $m = 7q$ where q is any prime except 7. Then the Eisenstein criterion (with $p = 7$) shows that f is irreducible over $\mathbb{Q}$.

35. (a) f has no rational roots. If $f = (x^2 + ax + b)(x^2 + cx + d)$, comparing coefficients gives $a + c = 3$, $b + ac + d = 1$, $ad + bc = 3$, $bd = 1$. Thus $b = d = \pm 1$, $c = 3 - a$, so $3 = b(a+c) = 3b$, $(b = 1)$. Then $b + ac + d = 1$ gives $2 + a(3 - a) = 1$, so $a^2 - 3a - 1 = a$. This has no integer roots, so f is already irreducible.

(c) f has no rational roots (the candidates are ± 1, ± 2). Suppose $f = (x^2 + ax + b)(x^2 + cx + d)$. Then, as in (a)
$$a + c = 2 \qquad b + ac + d = -2 \qquad ad + bc = 7 \qquad bd = -2.$$
Thus $(b, d) = (1, -2), (-1, 2), (2, -1)$ or $(-2, 1)$. By symmetry we consider only $(1, -2)$ and $(-1, 2)$.
Case 1. $(b, d) = (1, -2)$. Then
$$-2 = b + d + a(2 - a) = -1 + 2a - a^2; a^2 - 2a - 1 = 0;$$
no root in $\mathbb{Z}$.
Case 2. $(b, d) = (-1, 2)$. Now
$$7 = ad + bc = 2a - c = 2a - (2 - a) = 3a - 2,$$
so $a = 3$, $c = -1$. The factorization is $f = (x^2 + 3x - 1)(x^2 - x + 2)$. Both these quadratics are irreducible (no root in $\mathbb{Q}$) so this is the desired factorization.

(a) Not irreducible. $x^5 + x + 1 = (x^2 + x + 1)(x^3 - x^2 + 1)$. The factors are irreducible (no roots in $\mathbb{Q}$).

37. (a) Not irreducible. $x^5 + x + 1 = (x^2 + x + 1)(x^3 - x^2 + 1)$. The factors are irreducible (no roots in $\mathbb{Q}$).

39. (a) $g = (x+4)f + (4x+3)$ and $f = (4x+2)(4x+3) + 1$. Hence

$$\begin{aligned}
\gcd(f,g) = 1 &= f - (4x+2)[g - (x+4)f] \\
&= [1 + (4x+2)(x+4)]f - (4x+2)g \\
&= (4x^2 + 3x + 4)f - (4x+2)g.
\end{aligned}$$

(c) $g = (x^3 + x^2 - x - 1)f + (4x-8)$ and $f = \frac{1}{4}(x+1)(4x-8) + 0$. Hence

$$\begin{aligned}
\gcd(f,g) = x - 2 &= \frac{1}{4}(4x-8) \\
&= \frac{1}{4}[g - (x^3 + x^2 - x - 1)f] \\
&= \frac{1}{4}g - \frac{1}{4}(x^3 + x^2 - x - 1)f.
\end{aligned}$$

41. If $f = \gcd(f,g)$ then $f|g$ is clear. Conversely, if $g = qf$, let $d = \gcd(f,g)$. Write $d = hf + kg$ with h, k in $F[x]$. Then $d = hf + kqf$, so $f|d$. But also $d|f$ so $d = f$ by Theorem 9 because d and f are both monic.

42. Let $1 = mf + kg$ with m and k in $F[x]$.

(a) If $h = pf$ and $h = qg$, then

$$h = hmf + hkg = (qg)mf + (pf)hg = (qm + pk)fg$$

43. (a) Define σ and $\tau : F[x] \to F[x]$ by $\sigma(f) = f(x+b)$ and $\tau(f) = f(x-b)$. Then $\sigma\tau = 1_{F[x]}$ and $\tau\sigma = 1_{F[x]}$ so σ (and τ) is a bijection. Observe that $\sigma(f+g) = f(x+b) + g(x+b) = \sigma(f) + \sigma(g)$, when we use the evaluation theorem for $F[x]$. Similarly, $\sigma(fg) = \sigma(f) \cdot \sigma(g)$ so σ is a automorphism (and $\tau = \sigma^{-1}$).

(c) Let $\sigma : F[x] \to F[x]$ be an automorphism such that $\sigma(a) = a$ for all $a \in F$. Put $\sigma(x) = p$. If $f = \sum_{i=0}^{n} a_i x^i$, then

$$\sigma(f) = \sum_{i=0}^{n} \sigma(a_i)[\sigma(x)]^i = \sum_{i=0}^{n} a_i(p)^i = f(p).$$

Now write $\sigma^{-1}(x) = q$ so $\sigma^{-1}(f) = f(q)$ for all f in $F[x]$. Hence $x = \sigma[\sigma^{-1}(x)] = \sigma(q) = p(q)$. Now $p \neq 0$ because σ is one-to-one, and similarly $q \neq 0$. If $\deg p = m$ and $\deg q = n$ then $1 = \deg x = \deg\{p(q)\} = mn$. It follows that $m = n = 1$ so $p = ax + b$, $a \neq 0$. Thus $\sigma(f) = f(ax+b)$ for all f in $F[x]$, as required.

44. (a) This was done in Exercise 37 §4.1.

4.3 FACTOR RINGS OF POLYNOMIALS OVER A FIELD

1. (a) Here $A = \{f \mid f(0) = 0\}$ so $A = \langle x \rangle$ by Theorem 6 §4.1.

(c) Put $h = x(x-1)$. Clearly $h \in A$ so $\langle h \rangle \subseteq A$. If $f \in A$ then $f(0) = 0 = f(1)$ so x and $x-1$ divide f by Theorem 6 §4.1, say $f = xq = (x-1)p$ by Theorem 10, §4.2. But x and $x-1$ are relatively prime in $F[x]$ (indeed $1 = 1 \cdot x - 1(x-1)$) so

$$\begin{aligned}
f = xf - (x-1)f &= x[(x-1)p] - (x-1)[xq] \\
&= x(x-1)[p - q].
\end{aligned}$$

Hence $f \in \langle h \rangle$, proving that $A = \langle h \rangle$.

2. (a) $R = \{a + bt \mid a, b \in \mathbb{Z}_2; t^2 = 1\} = \{0, 1, t, 1+t\}$

+	0	1	t	$1+t$
0	0	1	t	$1+t$
1	1	0	$1+t$	t
t	t	$1+t$	0	1
$1+t$	$1+t$	t	1	0

$\times$	0	1	t	$1+t$
0	0	0	0	0
1	0	1	t	$1+t$
t	0	t	1	$1+t$
$1+t$	0	$1+t$	$1+t$	0

(c) $R = \{a + bt + ct^2 \mid a, b, c \in \mathbb{Z}_2; t^3 = 1\} = \{0, 1, t, t^2, 1+t, 1+t^2, t+t^2, 1+t+t^2\}$

+	0	1	t	t^2	$1+t$	$1+t^2$	$t+t^2$	$1+t+t^2$
0	0	1	t	t^2	$1+t$	$1+t^2$	$t+t^2$	$1+t+t^2$
1	1	0	$1+t$	$1+t^2$	t	t^2	$1+t+t^2$	$t+t^2$
t	t	$1+t$	0	$t+t^2$	1	$1+t+t^2$	t^2	$1+t^2$
t^2	t^2	$1+t^2$	$t+t^2$	0	$1+t+t^2$	1	t	$1+t$
$1+t$	$1+t$	t	1	$1+t+t^2$	0	$t+t^2$	$1+t^2$	t^2
$1+t^2$	$1+t^2$	t^2	$1+t+t^2$	1	$t+t^2$	0	$1+t$	t
$t+t^2$	$t+t^2$	$1+t+t^2$	t^2	t	$1+t^2$	$1+t$	0	1
$1+t+t^2$	$1+t+t^2$	$t+t^2$	$1+t^2$	$1+t$	t^2	t	1	0

$\times$	0	1	t	t^2	$1+t$	$1+t^2$	$t+t^2$	$1+t+t^2$
0	0	0	0	0	0	0	0	0
1	0	1	t	t^2	$1+t$	$1+t^2$	$t+t^2$	$1+t+t^2$
t	0	t	t^2	1	$t+t^2$	$1+t$	$1+t^2$	$1+t+t^2$
t^2	0	t^2	1	t	$1+t^2$	$t+t^2$	$1+t$	$1+t+t^2$
$1+t$	0	$1+t$	$t+t^2$	$1+t^2$	$1+t^2$	$t+t^2$	$1+t$	0
$1+t^2$	0	$1+t^2$	$1+t$	$t+t^2$	$t+t^2$	$1+t$	$1+t^2$	0
$t+t^2$	0	$t+t^2$	$1+t^2$	$1+t$	$1+t$	$1+t^2$	$t+t^2$	0
$1+t+t^2$	0	$1+t+t^2$	$1+t+t^2$	$1+t+t^2$	0	0	0	$1+t+t^2$

(e) $R = \{a + bt \mid a, b \in \mathbb{Z}_3; t^2 = 0\}$
   $= \{0, 1, -1, t, -t, 1+t, 1-t, -1+t, -1-t\}$

$\times$	0	1	-1	t	$-t$	$1+t$	$1-t$	$-1+t$	$-1-t$
0	0	0	0	0	0	0	0	0	0
1	0	1	-1	t	$-t$	$1+t$	$1-t$	$-1+t$	$-1-t$
-1	0	-1	1	$-t$	t	$-1-t$	$-1+t$	$1-t$	$1+t$
t	0	t	$-t$	0	0	t	t	$-t$	$-t$
$-t$	0	$-t$	t	0	0	$-t$	$-t$	t	t
$1+t$	0	$1+t$	$-1-t$	t	$-t$	$1-t$	1	-1	$-1+t$
$1-t$	0	$1-t$	$-1+t$	t	$-t$	1	$1+t$	$-1-t$	-1
$-1+t$	0	$-1+t$	$1-t$	$-t$	t	-1	$-1-t$	$1+t$	1
$-1-t$	0	$-1-t$	$1+t$	$-t$	t	$-1+t$	-1	1	$1-t$

3. The polynomial $x^3 + x + 1$ is irreducible over $\mathbb{Z}_2$ (no roots) so take

$$R = \mathbb{Z}_2[x]/\langle x^3 + x + 1 \rangle = \{a + bt + ct^2 \mid a, b, c \in \mathbb{Z}_2; t^3 = t + 1\}$$
$$= \{0, 1, t, t^2, 1+t, 1+t^2, t+t^2, 1+t+t^2\}$$

The multiplication table is:

×	0	1	t	t^2	$1+t$	$1+t^2$	$t+t^2$	$1+t+t^2$
0	0	0	0	0	0	0	0	0
1	0	1	t	t^2	$1+t$	$1+t^2$	$t+t^2$	$1+t+t^2$
t	0	t	t^2	$1+t$	$t+t^2$	1	$1+t+t^2$	$1+t^2$
t^2	0	t^2	$1+t$	$t+t^2$	$1+t+t^2$	t	$1+t^2$	1
$1+t$	0	$1+t$	$t+t^2$	$1+t+t^2$	$1+t^2$	t^2	1	t
$1+t^2$	0	$1+t^2$	1	t	t^2	$1+t+t^2$	$1+t$	$t+t^2$
$t+t^2$	0	$t+t^2$	$1+t+t^2$	$1+t^2$	1	$1+t$	t	t^2
$1+t+t^2$	0	$1+t+t^2$	$1+t^2$	1	t	$t+t^2$	t^2	$1+t$

5. (a) $27 = 3^3$ and $x^3 - x + 1$ is irreducible over $\mathbb{Z}_3$, so $\mathbb{Z}_3[x]/\langle x^3 - x + 1\rangle$ is such a field.

   (c) $121 = 11^2$ and $x^2 + x + 1$ is irreducible over $\mathbb{Z}_{11}$, so $\mathbb{Z}_{11}[x]/\langle x^2 + x + 1\rangle$ is such a field.

6. (a) $R = \{a + bt \mid t^2 = t\}$. Thus $(a+bt)^2 = a^2 + (2ab + b^2)t$. Write $r = a + bt$.
   Idempotents $r^2 = r$. Then $a^2 = a$, $b^2 + 2ab = b$. Either $a = 0$ ($b = 0, 1$) or $a = 1(b = 0, -1)$. Hence the idempotents are $0, t, 1, 1 - t$.
   Nilpotents $r^n = 0$. If $r^2 = 0$ then $a^2 = 0$, $2ab + b^2 = 0$; whence $a = b = 0$; $r = 0$. Thus there are no nilpotents except 0.
   Units $rs = 1$. If $s = x + yt$ then $rs = ax + (ay + bx + by)t = 1$ so $ax = 1$, $ay + bx + by = 0$. Thus a is a unit and $x = a^{-1}$; so $(a+b)y = -ba^{-1}$. If $a + b = 0$ this gives $b = 0$, $a = 0$, a contradiction. So $a + b \neq 0$, $y = -(a+b)^{-1}ba^{-1}$, and
   $$s = a^{-1} - (a+b)^{-1}ba^{-1} = a^{-1}(a+b)^{-1}[(a+b) + bn],$$
   where $a \neq 0$, $b \neq -a$.

7. (a) If $r^{-1} = a + bt + ct^2$ then $(a + bt + ct^2)(1 + t^2) = 1$. Since $t^3 = -1$ this gives $(a - b) + (b - c)t + (c + a)t^2 = 1$. Thus $a = b + 1$, $b = c$, $c = -a$. Thus $b = c$, $a = -c$ and $-c = c + 1$, $-2c = 1$. Since $2^{-1} = 6$ in $\mathbb{Z}_{11}$, $c = -2^{-1} = -6 = 5$. Thus $(1 + t^2)^{-1} = -5 + 5t + 5t^2 = 5(-1 + t + t^2)$.

9. $\mathbb{Q}[x]/\langle x^3 - 2\rangle = \{a + bt + ct^2 \mid a, b, c \in \mathbb{Q}, t^3 = 2\}$. Now write $w = \sqrt[3]{2}$ so that $w \in \mathbb{R}$, $w^3 = 2$. Define $R \subseteq \mathbb{R}$ by $R = \{a + bw + cw^2 \mid a, b, c, \in \mathbb{Q}\}$. This is clearly an additive subgroup of R containing 1. Moreover
   $$(a + bw + cw^2)(a_1 + b_1w + c_1w^2)$$
   $$= (aa_1 + 2bc_1 + 2cb_1) + (ab_1 + ba_1 + 2cc_1)w + (ac_1 + bb_1 + ca_1)w^2$$
   so R is a subring of R. Now the map
   $$\theta : \mathbb{Q}[x]/\langle x^3 - 2\rangle \to R \quad \text{where} \quad \theta(a + bt + ct^2) = a + bw + cw^2$$
   is clearly an onto ring homomorphism and it is one-to-one by Lemma 3.

10. (a) As in Theorem 2 with $R = F[x]/\langle x^2\rangle = \{a + bt \mid a, b \in F; t^2 = 0\}$. Define $\theta : R \to M_2[F]$ by $(a + bt)\theta = \begin{bmatrix} a & b \\ 0 & a \end{bmatrix}$. This is well defined by Lemma 3 and is clearly a one-to-one homomorphism of additive groups carrying 1 to 1. Finally
   $$[(a + bt)(c + dt)]\theta = \theta[ac + (ad + bc)t] = \begin{bmatrix} ac & ad + bc \\ 0 & ac \end{bmatrix}$$
   $$= \begin{bmatrix} a & b \\ 0 & a \end{bmatrix}\begin{bmatrix} c & d \\ 0 & c \end{bmatrix} = \theta(a + bt) \cdot \theta(c + dt)$$

so θ is a one-to-one ring homomorphism. Thus $R \cong \theta(R) = \left\{ \begin{bmatrix} a & b \\ 0 & a \end{bmatrix} \middle| a, b \in F \right\}$.

(c) Use the notation

$$d(a_0, a_1, \ldots, a_{m-1}) = \begin{bmatrix} a_0 & a_1 & a_2 & \cdots & a_{m-1} \\ 0 & a_0 & a_1 & \cdots & a_{m-2} \\ 0 & 0 & a_0 & \cdots & a_{m-3} \\ \vdots & \vdots & \vdots & & \vdots \\ 0 & 0 & 0 & \cdots & a_0 \end{bmatrix}$$

Then one checks

$$d(a_0, a_1, \ldots, a_{m-1}) + d(b_0, b_1, \ldots, b_{m-1})$$
$$= d(a_0 + b_0, a_1 + b_1, \ldots, a_{m-1} + b_{m-1})$$
$$d(a_0, a_1, \ldots, a_{m-1}) \cdot d(b_0, b_1, \ldots, b_{m-1})$$
$$= d(a_0 b_0, a_0 b_1 + a_1 b_0, a_0 b_2 + a_1 b_1 + a_2 b_0, \ldots, a_0 b_{m-1}$$
$$+ a_1 b_{m-2}, + \cdots + a_{m-1} b_0)$$

If $R = F[z]/\langle x^m \rangle = \{a_0 m_1 t + \cdots + a_{m-1} t^{m-1} \mid a_i \in F, \ t^m = 0\}$, define $\theta : R \to M_m(F)$ by

$$\theta(a_0, a_1 t + \cdots + a_{m-1} t^{m-n}) = d(a_0, a_1, \ldots, a_{m-1}).$$

This is well defined by Lemma 3, and is a ring homomorphism by the above formula. Since $\ker \theta = 0$ is clear,

$$R \cong \theta(R) = \{d(a_0, a_1, \ldots, a_{m-1}) \mid a_i \in F\}.$$

11. Write $R = F[z]/\langle x^2 - x \rangle = \{a + bt \mid a, b \in F; t^2 = t\}$. Define $\theta : R \to F \times F$ by $\theta(a + bt) = (a, a + b)$. This is well defined by Lemma 3, and so is clearly an isomorphism of additive groups satisfying $\theta(1) = (1, 1+0)$ — the unity of $F \times F$. Finally,

$$\theta[(a + bt)(c + dt)] = \theta[ac + (ad + bc + bd)t] = (ac, ac + ad + bc + bd)$$
$$= (ac, (a + b)(c + d))$$
$$= (a, a + b)(c, c + d)$$
$$= \theta(a + bt) \cdot \theta(c + dt).$$

13. (a) Write $R = F[x]/\langle h \rangle = \{a + bt \mid a, b \in f; t^2 = u + vt\}$. Define $\theta : R \to M_2(F)$ by $\theta(a + bt) = \begin{bmatrix} a & b \\ bu & a + bv \end{bmatrix}$. This is well defined by Lemma 3, $\theta(1) = \begin{bmatrix} 1 & 0 \\ 0 & 1 \end{bmatrix}$, and θ is a one-to-one additive group homomorphism. Finally:

$$\theta[(a + bt)(c + dt)] = \theta[ac + (ad + bc)t + bd(u + vt)]$$
$$= \theta[(ac + bdu) + (ad + bc + bdv)t]$$
$$= \begin{bmatrix} ac + bdu & ad + bc + bdv \\ adu + bcu + bduv & ac + bdu + adv + bcv + bdv^2 \end{bmatrix}$$
$$= \begin{bmatrix} a & b \\ bu & a + bv \end{bmatrix} \begin{bmatrix} c & d \\ du & c + dv \end{bmatrix}$$
$$= \theta(a + bt) \cdot \theta(c + dt).$$

Hence θ is a one-to-one ring homomorphism. Thus $S = \theta(R)$ is a subring of $M_2(F)$ and $R \cong S$.

(c) If $h = x^2 + 1$ then $u = -1$ and $v = 0$, so $\theta(a + bt) = \begin{bmatrix} a & b \\ -b & a \end{bmatrix}$. Since $\mathbb{C} \cong R[x]/\langle x^2 + 1 \rangle$ this gives $\mathbb{C} \cong \left\{ \begin{bmatrix} a & b \\ -b & a \end{bmatrix} \middle| a, b \in \mathbb{R} \right\}$.

14. (a) Here $E = \{a + bt + ct^2 \mid t^3 = 1 + t; a, b, c \in \mathbb{Z}_2\}$ and
$$p = [x + t][x + t^2][x + (t + t^2)] \text{ in } E[x].$$
(c) Here $E = \{a + bt + ct^2 \mid t^3 = t - 1; a, b, c \in \mathbb{Z}_3\}$ and
$$p = [x - t][x - (1 + t)][x + (1 - t)] \text{ in } E[x].$$

15. In E we have $p(t) = 0$ so $p = (x - t)q$ in $E[x]$. But $\deg(p) = 2$ so $q = ax + b$, $a \neq 0$. Moreover $a = 1$ as p is monic. $p = (x - t)(x + b)$ in $E[x]$.

16. (a) $E = \{a + bt + ct^2 \mid a, b, c \in F, t^3 = m\}$. In $E[x]$,
$$x^3 - m = x^3 - t^3 = (x - t)(x^2 + tx + t^2).$$
The discriminant of $x^2 + tx + t^2$ is $t^2 - 4t^2 = -3t^2$. Hence $x^3 - m$ factors into linear factors in $E[x] \Leftrightarrow -3t^2$ is a square in E; $\Leftrightarrow -3$ is a square in E. Clearly then, if -3 is a square in F it is a square in E.
Conversely, suppose $-3 = (a + bt + ct^2)^2$ in E. Then
$$a^2 + b^2 t^2 + c^2 t^4 + 2abt + 2act^2 + 2bct^3 = -3$$
$$a^2 + b^2 t^2 + c^2 mt + 2abt + 2act^2 + 2bcm = -3.$$
Thus $a^2 + 2bcm = -3$; $b^2 + 2ac = 0$; $c^2 m + 2ab = 0$. The last two give $b^3 = -2abc = c^3 m$. Thus if $c \neq 0$, $(bc^{-1})^3 = m$ in F, contrary to the irreducibility of $x^3 - m$ in F. So $c = 0$. Then $a^2 = -3$ in F by the first equation, as required.

17. First, A is an ideal of $F[x]$. For if f and $g \in A$ let $s = f + g$. Then $s(a_i) = f(a_i) + g(a_i) = 0$ for all i so $s \in A$. Similarly if $p = gf$, $g \in F[x]$ then $p(a_i) = 0$ for all i so $p \in A$. Now Theorem 1 gives $A = \langle d \rangle$ where d is monic. We show $d = m$. We have $m \in A$ so d divides m. Since both m and d are monic, it remains to show that m divides d (Theorem 9 §4.2). Now $d(a_1) = 0$ so $d = (x - a_1)q_1$. Next $0 = d(a_2) = (a_2 - a_1)q_1(a_2)$ so (since $a_2 \neq a_1$) $q_1(a_2) = 0$, and $d = (x - a_1)(x - a_2)q_2$. This continues (the a_i are distinct) to give $d = (x - a_1)(x - a_2) \cdots (x - a_n)q_n(x) = mq_n$. Hence m divides d, as required.

19. Let $a \neq 0$ in R. Let A be the set of polynomials in $R[x]$ with constant term of the form ra, $r \in R$. Then A is an ideal of $R[x]$ so let $A = \langle h \rangle$, $h \in R[x]$ by hypothesis. Then $a \in \langle h \rangle$ so $a = qh$, $q \in R[x]$. Hence $\deg h = 0 = \deg q$, say $h = b \in R$ and $q = c \in R$. Thus $a = cb$. Now $x \in A = \langle b \rangle$, so $x = q_1 b$, $q_1 \in R[x]$. Hence $\deg q_1 = 1$, say $q_1 = dx + e$. Thus $x = dbx + eb$, so $db = 1$. Thus $1 = db \in \langle b \rangle = A$, so $A = R$ because A is an ideal. Finally then
$$1 = ra + a_1 x + a_2 x^2 + m, \text{ for } r, a_1, a_2, \dots \text{ in } R.$$
This implies $ra = 1$, a is a unit.

21. (a) If $x^2 + ax + b$ is not irreducible over F, it must have a root $u \in F$. Thus $u^2 + au + b = 0$. Take $c = 2u + a$. Then $c^2 = 4(u^2 + ua) + a^2 = -4b + a^2$, contrary to hypothesis.

22. (a) A is an additive subgroup of $F[x]$ and $q(uf + vg) = (qu)f + (qv)g \in A$ for all $q \in F[x]$.

23. If $1 = q_1 f_1 + q_2 f_2 + \cdots + q_m f_m$ in $F[x]$ then any monic common divisor d of the f_i must divide 1. Hence $\deg d = 0$ so $d = a \in F$. But $a = 1$ because d is monic. Thus the f_i are relatively prime. Conversely, if $f_1, \ldots, f_m$ are relatively prime, consider $A = \{q_1 f_1 + \cdots + q_m f_m \mid q_i \in F[x]\}$. This is an ideal of $F[x]$ so, by Theorem 1, $A = \langle d \rangle$ where d is monic. Hence $d = q_1 f_1 + \cdots + q_m f_m$, and it remains to show $d = 1$. However $f_i \in A = \langle d \rangle$ for each i so d divides each f_i. Thus $d = 1$ by hypothesis.

24. (a) This is clear since the p_i are all monic.

    (c) The translation of the proof for $\mathbb{Z}$ works here. See Theorem 9 §1.2.

25. (a) Then $d \in \langle f \rangle + \langle g \rangle$ because $d = uf + vg$ for some $u, v \in F[x]$. On the other hand, $f \in \langle d \rangle$ and $g \in \langle d \rangle$ because d is a common divisor of f and g. Hence $\langle f \rangle \subseteq \langle d \rangle$ and $\langle g \rangle \subseteq \langle d \rangle$, so $\langle f \rangle + \langle g \rangle \subseteq \langle d \rangle$.

26. (a) Since $A \neq 0$, let $A = \langle h \rangle$ where h is a monic polynomial. Then $\deg h \geq 1$ because $A \neq F[x]$. Then A is a prime ideal if and only if $F[x]/A$ is an integral domain, and A is a maximal ideal if and only if $F[x]/A$ is a field. Hence the present result restates (1) $\Leftrightarrow$ (2) in Theorem 3.

27. We know (Theorem 12, §4.2) that $h = p_1^{n_1} p_2^{n_2} \cdots p_r^{n_r}$ where the $n_i \geq 1$ and the p_i are monic and irreducible. Write $R = F[x]/\langle h \rangle$. Assume first that R has no nonzero nilpotents. If $h = p^2 q$ where p is irreducible, let $r = pq + \langle h \rangle \in R$. Then $r^2 = 0$ in R but $r \neq 0$ because h does not divide pq. This contradiction shows that $h = p_1 p_2 \ldots p_r$.

    Conversely, assume $h = p_1 p_2 \ldots p_r$. It suffices to show that $r^2 = 0$ in R implies $r = 0$. Write $r = f + \langle h \rangle$. Then $r^2 = 0$ means $f^2 \in \langle h \rangle$, so $h \mid f^2$. Thus $p_1 \mid f^2$ so $p_1 \mid f$ by Theorem 11 §4.2. If $f = p_1 f_1$ then $p_2 \mid p_1 f_1$ and $\gcd(p_1, p_2) = 1$ because $p_1 \neq p_2$, so $p_2 \mid f_2$ by Exercise 47(b) §4.2. Thus $f = p_1 p_2 f_2$. Since p_3 is relatively prime to $p_1 p_2$, the process continues to show $f = p_1 p_2 \ldots p_r f_r = h f_r$. Thus $f \in \langle h \rangle$ so $r = f + \langle h \rangle = 0$.

29. Since p and q are relatively prime, $1 = up + vq$ for some $u, v \in F[x]$. Thus $\langle p \rangle + \langle q \rangle = F[x]$. Also, as for $\mathbb{Z}$, the fact that p and q are relatively prime gives $\text{lcm}(p, q) = pq = h$. This, with Exercise 25(b) above, shows that $\langle p \rangle \cap \langle q \rangle = \langle h \rangle$. Now the present result follows from the Chinese remainder theorem (Theorem 8 §3.4) with $R = F[x]$, $A = \langle p \rangle$ and $B = \langle q \rangle$.

31. (a) Let $\theta : F[x] \to R$ be the coset map $\theta(f) = f + A = f(x + A) = f(t)$. This is an onto ring homomorphism with $\ker \theta = A = \langle h \rangle$. Then $I = \theta(X)$ where $X = \{f \mid f(t) = \theta(f) \in I\}$ is an ideal of $F[x]$, and $X \supseteq A$. So $X = \langle d \rangle$ for a unique monic polynomial d. Since $X \supseteq A$, $h \in X$ and so d divides h. Then $I = \theta(X) = \{f(t) \mid f \in X\} = \{f(t) \mid d \text{ divides } f \text{ in } F[x]\}$.

    As to uniqueness, let $I = \langle d_1(t) \rangle$, d a monic divisor of h. It suffices to show that $X = \langle d_1 \rangle$. Clearly $d_1(t) \in I$ so $d_1 \in X$ and $\langle d_1 \rangle \subseteq X$. But $d(t) \in I$ so $d(t) = q(t)d_1(t)$; that is $d - qd_1 \in \ker \theta = A = \langle h \rangle$. But d_1 divides h so $d - qd_1 \in \langle d_1 \rangle$. Thus $X \subseteq \langle d_1 \rangle$.

(c) If $f(t) \in I$ then $f(t) = q(t)d(t)$. Hence $f(t)b(t) = q(t)d(t)b(t) = q(t)h(t) = 0$. Conversely, if $f(t)b(t) = 0$ then $fb \in \ker \theta = \langle h \rangle$, say $fb = qh$. But then $fb = qdb$, whence $f = qd$. Thus $f(t) = q(t)d(t) \in I$.

4.4 PARTIAL FRACTIONS

1. Suppose $r_0 + r_1 p + \cdots = s_0 + s_1 p + \cdots$ where p is monic in $R[x]$ and each r_i and s_i is either zero or has degree less than p. Then if $t_i = r_i - s_i$ we have $t_0 + t_1 p + t_2 p^2 + \cdots = 0$, so $t_0 + (t_1 + t_2 p + \cdots)p = 0$. The uniqueness in the division algorithm (Theorem 4 §4.1) shows $t_0 = 0$ and $t_1 + t_2 p + \cdots = 0$. Then do it again to get $t_1 = 0$ and $t_2 + t_3 p + \cdots = 0$. This continues to show $t_i = 0$ for all i.

2. (a) $\dfrac{x^2 - x + 1}{x(x^2 + x + 1)} = \dfrac{a}{x} + \dfrac{bx + c}{x^2 + x + 1}$, so

$$x^2 - x + 1 = a(x^2 + x + 1) + (bx + c)x.$$

Evaluating at 0, 1 and -1 gives $1 = a$, $1 = 3a + b + c$, $3 = a + b - c$. Hence $a = 1$, $b = 0$, $c = -2$.

(c) $\dfrac{x + 1}{x(x^2 + 1)^2} = \dfrac{a}{x} + \dfrac{bx + c}{x^2 - 1} + \dfrac{dx + e}{(x^2 + 1)^2}$, so

$$x + 1 = a(x^2 + 1)^2 + (bx + c)x(x^2 + 1) + (dx + e)x.$$

Evaluating at 0, gives $1 = a$; the coefficients x^4, x^3, x^2, x give

$$0 = a + b, \; 0 = c, \; 0 = 2a + b + d$$

and $1 = c + e$. Thus $a = e = 1$, $b = d = -1$ and $c = 0$.

3. $\dfrac{1}{(x - u_1) \cdots (x - u_n)} = \dfrac{a_1}{x - u_1} + \cdots + \dfrac{a_n}{x - u_n}$, so

$$1 = a_1 \prod_{i \neq 1} (x - u_i) + a_2 \prod_{i \neq 2} (x - u_i) + \cdots + a_n \prod_{i \neq n} (x - u_n).$$

Evaluation at u_k gives $1 = a_k \prod\limits_{i \neq k} (u_k - u_i)$, so $a_k = \left[\prod\limits_{i \neq i} (u_k - u_i) \right]^{-1}$ for each k.

4.5 SYMMETRIC POLYNOMIALS

1. The units in $R[x_1, \ldots, x_n]$ are just the units in R. If $n = 1$ this is Theorem 2 §4.1. In general, the units in $R[x_1, \ldots, x_n] = R[x_1, \ldots, x_{n-1}][x_n]$ are the units in $R[x_1, \ldots, x_{n-1}]$, so it follows by induction.

2. (a) $f(x, y, z) = x^3 + (x^2 + 2xyz + y^2 z^2) + (x^2 z + xz + 3x - xyz - yz - 3y)$
   $= (y^2 z^2) + (x^3 + xyz + x^2 z) + (x^2 + xz - yz) + (3x - 3y).$

3. $f(x, y) = x + y + xy$ is symmetric but not homogeneous.
   $f(x, y) = x^2 y$ is homogeneous but not symmetric.

5. Given $\theta : R \to S$, *any* homomorphism $\bar{\theta} : R[x_1 \ldots x_n] \to S$ with these properties must be given by

$$\bar{\theta}\left(\sum a_{i_1,\ldots,i_n} x_1^{i_1} \ldots x_n^{i_n}\right) = \sum \theta\left(a_{i_1,\ldots,i_n}\right) c_1^{i_1} \ldots c_n^{i_n}.$$

because the c_i are central in S, so $\bar{\theta}$ is unique if it exists. But this formula *defines* a map $R \to S$ because the coefficients $a_{i_1,\ldots,i_n}$ are uniquely determined by the polynomial. Then it is routine to verify that $\bar{\theta}$ is a homomorphism such that $\bar{\theta}(a) = \theta(a)$ for all $a \in R$ and $\bar{\theta}(x_i) = c_i$ for all i. This is what we wanted.

7. (a) $x_2^2 x_3 < x_1 x_3 < x_1 x_2^2 x_3 < x_1^2 x_2.$

8. (a) By Theorem 4:

$$f(x_1, x_2) = \sum_{i \neq j} x_i^2 x_j^3 = x_1^2 x_2^3 + x_1^3 x_2^2 = a s_1^5 + b s_1^3 s_2 + c s_1 s_2^2$$

for some a, b, c.

   If $(x_1, x_2) = (1, 0)$ we get $0 = a$.
   If $(x_1, x_2) = (1, 1)$ we get $2 = 32a + 8b + 2c$.
   If $(x_1, x_2) = (1, 2)$ we get $12 = 3^5 a + 27 \cdot 2b + 3 \cdot 2^2 c$.
   The solution is $a = b = 0$, $c = 1$, so $f = s_1 s_2^2$.

   (c) $f(x_1, x_2, x_3) = x_1^2 x_2^3 x_3 + x_2^2 x_1^3 x_3 + x_2^2 x_3^3 x_1 + x_3^2 x_2^3 x_1 + x_1^2 x_3^3 x_2 + x_3^2 x_1^3 x_2$
   $= (x_1 x_2 x_3)(x_1 x_2^2 + x_2 x_1^2 + x_2 x_3^2 + x_3 x_2^2 + x_1 x_3^2 + x_3 x_1^2).$

Now observe that

$$(x_1 + x_2 + x_3)(x_1 x_2 + x_1 x_3 + x_2 x_3)$$
$$= (x_1 x_2^2 + x_2 x_1^2 + x_2 x_3^2 + x_3 x_2^2 + x_1 x_3^2 + x_3 x_1^2) + 3 x_1 x_2 x_3.$$

Hence

$$f(x_1, x_2, x_3) = (x_1 x_2 x_3)[(x_1 + x_2 + x_3)(x_1 x_2 + x_1 x_3 + x_2 x_3) - 3 x_1 x_2 x_3]$$
$$= (x_1 x_2 x_3)[(x_1 + x_2 + x_3)(x_1 x_2 + x_1 x_3 + x_2 x_3) - 3(x_1 x_2 x_3)^2$$
$$= s_1 s_2 s_3 - 3 s_3^2.$$

9. $s_k(x_1, \ldots, x_n) = \sum_{\substack{i_1 < i_2 \\ i_1 < i_2 < \cdots < i_k}} x_{i_1} x_{i_2} \ldots x_{i_n}.$ The number of terms equals the number of k-subsets $\{i_1, i_2, \ldots, i_k\}$ in the n-set $\{1, 2, \cdots, n\}$. This is $\binom{n}{k}$. Thus for example $s_3(x_1, x_2, x_3, x_4, x_5)$ has $\binom{5}{3} = 10$ terms:

$$x_1 x_2 x_3, \ x_1 x_2 x_4, \ x_1 x_2 x_5, \ x_1 x_3 x_4, \ x_1 x_3 x_5,$$
$$x_1 x_4 x_5, \ x_2 x_3 x_4, \ x_2 x_3 x_5, \ x_2 x_4 x_5, \ x_3 x_4 x_5.$$

10. The monomial $x_1^{k_1} x_2^{k_2} \ldots x_n^{k_n}$ has degree $m = k_1 + k_2 + \cdots + k_n$, $k_i \geq 0$. So we must count the number of n-tuples $(k_1, k_2, \ldots, k_n)$ from $\mathbb{N}^n$ with

$$k_1 + k_2 + \cdots + k_n = m.$$

Take m circles 0 and $n - 1$ dividers | and line them up. For example

$$|\,0\,0\,0\,|\,0\,0\,|\,|\,\cdots\,|\,0\,0\,0\,|\,0\,|\,0\,0$$

The $n - 1$ dividers create n compartments (counting the ends). The number of circles in the ith compartment is k_i. Thus the above lineup yields

$$k_1 = 0, \ k_2 = 3, \ k_3 = 2, \ k_4 = 0, \ \cdots, \ k_{n-2} = 3, \ k_{n-1} = 1, \ k_n = 2.$$

So we must count lineups: there are $n + m - 1$ objects (circles and dividers). Label all the $n + m - 1$ positions and choose the m positions for circles in $\binom{m+n-1}{m}$ ways. Voilà!

11. $p_4 = p_3 s_1 - p_2 s_2 + p_1 s_3 - 4s_4$

$\quad = (s_1^3 - 3s_1 s_2 + 3s_3)s_1 - (s_1^2 - 2s_2)s_2 + s_1 s_3 - 4s_4$

$\quad = s_1^4 - 4s_1^2 s_2 + 4s_1 s_3 + 2s_2^2 - 4s_4$

$\quad p_5 = p_4 s_1 - p_3 s_2 + p_2 s_3 - p_1 s_4 + 5s_5$

$\quad = (s_1^4 - 4s_1^2 s_2 + 4s_1 s_3 + 2s_2^2 - 4s_4)s_1 - (s_1^3 - 3s_1 s_2 + 3s_3)s_2$

$\quad \quad + (s_1^2 - 2s_2)s_3 - s_1 s_4 + 5s_5$

$\quad = s_1^5 - 5s_1^3 s_2 + 5s_1^2 s_3 - 5s_1 s_4 - 5s_2 s_3 + 5s_1 s_2^2 + 5s_5$

(Conjecture: If q is a prime, $p_q = s_1^q + q \sum (-1)^{1 + \sum k_i} s_{i_1}^{k_1} s_{i_2}^{k_2} \cdots s_{i_r}^{k_r}$ summed over all i_j such that $1 \le i_1 < i_2 < \cdots < i_r \le q$ and $\sum_{j=1}^{r} k_j i_j = q$. Over $\mathbb{Z}_8$ this implies that $p_8 = s_1^8$, which is true).

$p_6 = p_5 s_1 - p_4 s_2 + p_3 s_3 - p_2 s_4 + p_1 s_5 - 6s_6$

$\quad = (s_1^6 - 5s_1^4 s_2 + 5s_1^3 s_3 + 5s_1^2 s_2^2 - 5s_1^2 s_4 - 5s_1 s_2 s_3 + 5s_1 s_5)$

$\quad \quad -(s_1^4 s_2 - 4s_1^2 s_2^2 + 4s_1 s_2 s_3 + 2s_2^3 - 4s_2 s_4) + (s_1^3 s_3 - 3s_1 s_2 s_3 + 3s_3^2)$

$\quad \quad -(s_1^2 s_4 - 2s_2 s_4) + s_1 s_5 - 6s_6$

$\quad = s_1^6 - 6s_1^4 s_2 + 6s_1^3 s_3 - 6s_1^2 s_4 + 9s_1^2 s_2^2 - 12s_1 s_2 s_3 + 6s_1 s_5 - 2s_2^3$

$\quad \quad + 6s_2 s_4 + 3s_3^2 - 6s_6.$

12. (a) $f(x_1, \ldots, x_n) = \sum_{i<j} (x_i - x_j)^2 = (n-1) \sum_{i=1}^{n} x_i^2 - 2 \sum_{i<j} x_i x_j$

$\quad \quad = (n-1)p_2 - 2s_2 = (n-1)[s_1^2 - 2s_2] - 2s_2$

$\quad \quad = (n-1)s_1^2 - 2ns_2.$

(c) $f(x_1, \ldots, x_n) = \sum_{i<j} x_i^3 x_j^3$. Now $\left[\sum_{i=1}^{n} x_i^3\right]^2 = \sum_{i=1}^{n} x_i^6 + 2f(x_i)$. Using the preceding exercise

$\quad f(x_1, \ldots, x_n) = \frac{1}{2}\left[\left(\sum_{i=1}^{n} x_i^3\right)^2 - \sum_{i=1}^{n} x_i^6\right] = \frac{1}{2}(p_3^2 - p_6)$

$\quad \quad = \frac{1}{2}[(s_1^3 - 3s_1 s_2 + 3s_3)^2 - p_6]$

$\quad \quad = \frac{1}{2}[(s_1^6 + 9s_1^2 s_2^2 + 9s_3^2 - 6s_1^4 s_2 + 6s_1^3 s_3 - 18s_1 s_2 s_3)$

$\quad \quad \quad - (s_1^6 - 6s_1^4 s_2 + 6s_1^3 s_3 - 6s_1^2 s_4 + 9s_1^2 s_2^2$

$\quad \quad \quad \quad - 12s_1 s_2 s_3 + 6s_1 s_5 - 2s_2^3 + 6s_2 s_4 + 3s_3^2 - 6s_6)]$

$\quad \quad = \frac{1}{2}[6s_3^2 - 6s_1 s_2 s_3 + 6s_1^2 s_4 - 6s_1 s_5 + 2s_2^3 - 6s_2 s_4 + 6s_6]$

$\quad \quad = 3s_3^2 - 3s_1 s_2 s_3 + 3s_1^2 s_4 - 3s_1 s_5 + s_2^3 - 3s_2 s_4 + 3s_6.$

13. (a) We have

$\quad x^3 - 5x^2 + 4x - 3 = (x - u)(x - v)(x - w)$

$\quad \quad \quad \quad = x^3 - (u + v + w)x^2 + (uv + uw + vw)x - uvw.$

Hence $u + v + w = 5$, $uv + uw + vw = 4$, and $uvw = 3$. The polynomial we want is

$\quad (x - u^2)(x - v^2)(x - w^2) = x^3 - (u^2 + v^2 + w^2)x^2$

$\quad \quad \quad \quad \quad + (u^2 v^2 + u^2 w^2 + v^2 w^2)x - u^2 v^2 w^2.$

Newton's formula $p_2 = s_1^2 - 2s_2$ gives
$$u^2 + v^2 + w^2 = p_2(u,v,w) = s_1^2(u,v,w) - 2s_2(u,v,w)$$
$$= (u+v+w)^2 - 2(uv+uw+vw) = 25 - 8 = 17.$$
$$u^2v^2 + u^2w^2 + v^2w^2 = p_2(uv, uw, vw)$$
$$= (uv+uw+vw)^2 - 2(u^2vw + uv^2w + uvw^2)$$
$$= 4^2 - 2(u+v+w)(uvw)$$
$$= 16 - 30$$
$$= -14$$
$$u^2v^2w^2 = (uvw)^2 = 3^2 = 9.$$
Hence $x^3 - 17x^2 - 14x - 9$ has roots u^2, v^2, w^2.

14. (a) Write $f(x_i)g(x_i) = h(x_i)$. Then
$$\theta_\sigma[h(x_i)] = f(x_{\sigma(i)})g(x_{\sigma(i)}) = \theta_\sigma[f(x_i)]\theta_\sigma[g(x_i)].$$
Similarly $\theta_\sigma(f+g) = \theta_\sigma(f) + \theta_\sigma(g)$ and $\theta_\sigma(1) = 1$. Now
$$\theta_\sigma\theta_\tau[f(x_i)] = \theta_\sigma[f(x_{\tau(i)})] = f(x_{\sigma\tau(i)}) = \theta_{\sigma\tau}[f(x_i)]$$
for all σ, τ and all f, so $\theta_{\sigma\tau} = \theta_\sigma\theta_\tau$. Similarly $\theta_\epsilon = 1$. In particular
$$\theta_\sigma\theta_{\sigma^{-1}} = \theta_\epsilon = 1 = \theta_{\sigma^{-1}}\theta_\sigma.$$
Thus $\theta_\sigma^{-1} = \theta_{\sigma^{-1}}$ and so θ_σ is a bijection.

(c) If f, $g \in S_G$ then $\theta(f) = f$ and $\theta(g) = g$ for all $\theta \in G$. Since θ is a ring homomorphism:
$$\theta(f+g) = \theta(f) + \theta(g) = f + g,$$
$$\theta(fg) = \theta(f) \cdot \theta(g) = fg, \ \theta(1) = 1,$$
$$\theta(-f) = -\theta(f) = -f$$
for all $\theta \in G$. Thus S_G is a subring. Note: S_G is a subring for any *subset* G of ring homomorphisms $R \to R$.

15. Prove this by induction on n. If $n = 1$ and $f(a) = 0$ for all $a \in \mathbb{Z}_p$ then $x, x-1, \ldots, x-p+1$ all divide $f(x)$. Thus $f(x) = x(x-1)\cdots(x-p+1)g(x)$, so deg $f(x) \geq p$, contrary to hypothesis. So $f(a) \neq 0$ for some $a \in \mathbb{Z}_p$. If $n \geq 2$ and $f(x_1, \ldots, x_n) \in \mathbb{Z}_p[x_1, \ldots, x_n]$ write
$$f = p_0 + p_1x_n + \cdots + p_mx_n^m, \ p_i \in \mathbb{Z}_p[x_1, \ldots, x_{n-1}].$$
Then $p_m \neq 0$ in $\mathbb{Z}_p[x_1, \ldots, x_{n-1}]$ so, by induction, let $p_m(a_1, \ldots, a_{n-1}) \neq 0$, $a_i \in \mathbb{Z}_p$. Write $g(x) = f(a_1, \ldots, a_{n-1}, x)$. Then $g(x) \neq 0$ so $g(a_n) \neq 0$ for some $a_0 \in \mathbb{Z}_p$ by the case $n = 1$. This is what we wanted.

17. If $\sigma \in S_n$,
$$f_1(x_{\sigma(1)}, \ldots, x_{\sigma(n)}) = p[f(x_{\sigma(1)}, \ldots, x_{\sigma(n)})] = p[sgn\sigma \cdot f(x_1, \ldots, x_n)]$$
$$= sgn\sigma \cdot p[f(x_1, \ldots, x_n)] = sgn\sigma \cdot f_1(x_1, \ldots, x_n).$$
The polynomial $p(x)$ has only odd exponents: $p(x) = a_1x + a_3x^3 + \cdots$. If $f = \Delta_n g$ this gives $f_1 = a_1\Delta_n g + a_3(\Delta_n g)^3 + \cdots = \Delta_n g(a_1 + a_3(\Delta_n g)^2 + \cdots)$. Since Δ_n^2 is symmetric, $g_1 = g[a_1 + a_3(\Delta_n g)^2 + a_5(\Delta_n g)^4 + \cdots]$ is symmetric.

19. (a) $a \leq a$ is obvious; if $a \leq b$ and $b \leq a$ then, if $a \neq b$ and k is the smallest integer with $a_k \neq b_k$, then $a_k < b_k$ and $b_k < a_k$. This is impossible. Now

suppose $a \leq b$, $b \leq c$. If $a = b$ or $b = c$, we are done. If $a < b$ and $b < c$ let $a_k < b_k$, $b_l < c_l$ where k is minimal with $a_k \neq b_k$ and l is minimal with $b_l \neq c_l$. Let $m = \min(k, l)$. If $i < m$ then $a_i = b_i = c_i$ by the choice of k and l.

Case 1. $k = l$. Then $m = k = l$ so $a_m < b_m < c_m$.
Case 2. $k < l$. Then $m = k < l$ so $a_m < b_m = c_m$.
Case 3. $l < k$. Then $m = l < k$ so $a_m = b_m < c_m$.

Thus $a_m < c_m$ in any case, so $a \leq c$.

(c) Let $X \neq \emptyset$, $X \subseteq \mathbb{N}^n$. Write $X_1 = \{x \mid (x, \ldots) \in X\}$. Then $\emptyset \neq X_1 \subseteq \mathbb{N}$ so let m_1 be the minimal element of X_1. Now let $X_2 = \{x \mid (m_1, x, \ldots) \in X\}$, and let m_2 be the minimal element of X_2. Then let m_3 be minimal in $X_3 = \{x \mid (m_1, m_2, x_1, \ldots) \in X\}$. Continue to get $m = (m_1, m_2, \ldots, m_n) \in \mathbb{N}^n$. If $a = (a_1, a_2, \ldots, a_n) \in \mathbb{N}^n$, $a \neq m$, consider two cases.

Case 1. $a_1 \neq m_1$. Then $m_1 \leq a_1$ by the choice of m_1 so $a \leq m$.
Case 2. $a_1 = m_1$. Let k be minimal such that $m_k \neq a_k$. Then $a_i = m_i$, $1 \leq i < k$ so $a = (m_1, m_2, \ldots, m_{k-1}, a_k, \ldots)$. Thus $m_k \leq a_k$ by the choice of m_k so $m \leq a$. Thus m is the least member of X.

20. (a) If $a, b \in G$ then $|a| < 1$ and $|b| < 1$, so $|ab| = |a| \, |b| < 1$. Thus $1 + ab > 0$. We have
$$(1 + ab) - (a + b) = (1 - a)(1 - b) > 0,$$
so $a * b < 1$. Similarly
$$(1 + ab) + (a + b) = (1 + a)(1 + b) > 0$$
implies $-1 < a * b$. Hence G is closed. The unity of G is 0 and the inverse of a is $-a$. Finally
$$(x_1 * x_2) * x_3 = \frac{(x_1 + x_2 + x_3) + x_1 x_2 x_3}{1 + (x_1 x_2 + x_1 x_3 + x_2 x_3)} = x_1 * (x_2 * x_3).$$

This proves (a).

Chapter 5

Factorization in Integral Domains

5.1 IRREDUCIBLES AND UNIQUE FACTORIZATION

1. Let $0 \neq a = bc$. If $a \sim b$, say $b = ua$, then $a = uac$ so $1 = uc$, c is a unit, $c \sim 1$. If $c \sim 1$ then c is a unit so $a = bc \sim b$.

3. (a) $2 + i = i(1 - 2i)$ in $\mathbb{Z}[i]$ and i is a unit.

5. Let $a \sim a'$, $b \sim b'$, say $a' = ua$, $b' = vb$. Then $a'b' = (uv)(ab)$ so $a'b' \sim ab$.

7. If $u = a + b\sqrt{-5}$ is a unit, $1 = N(u)N(u^{-1}) = (a^2 + 5b^2) \cdot N(u^{-1})$ shows $a^2 + 5b^2 = 1$. Thus $u = \pm 1$.

9. If $\langle p \rangle$ is maximal in $P = \{\langle a \rangle \mid a \in R, \ a \notin R^*\}$, let $p = ab$. Then $\langle p \rangle \subseteq \langle a \rangle$ so either $\langle a \rangle = \langle p \rangle$ or $a \in R^*$; that is $a \sim p$ or $a \sim 1$. Thus p is irreducible. Conversely, if p is irreducible and $\langle p \rangle \subseteq \langle a \rangle$, $a \notin R^*$, then $p = ab$ where b is a unit. Hence $a = pb^{-1} \in \langle p \rangle$, so $\langle p \rangle = \langle a \rangle$.

10. (a) 11 is irreducible. If $11 = xy$ then $11^2 = |x|^2|y|^2$ so $|x|^2$, $|y|^2$ each are $1, 11, 11^2$. If $|x|^2 = 1$ then x is a unit. So we rule out $|x|^2 = 11 = |y|^2$. If $x = a + bi$, this asks that $a^2 + b^2 = 11$. There are no such integers a, b.

    (c) 5 is not irreducible. We have $5 = (2 - i)(2 + i)$ and neither $2 - i$ nor $2 + i$ is a unit.

11. If $p \equiv 3 \pmod 4$, suppose $p = ab$ in $\mathbb{Z}(i)$. Then $p^2 = |a|^2|b|^2$. Since $|x| = 1$ in $\mathbb{Z}(i)$ means x is a unit, it suffices to show that $|a|^2 = p$ is impossible in $\mathbb{Z}(i)$ if $a = m + in$. This means $m^2 + n^2 = p$, whence $m^2 + n^2 = 0$ in $\mathbb{Z}_p$. If (say) $m \neq 0$ in $\mathbb{Z}_p$, this gives $1 + (m^{-1}n)^2 = 0$ in $\mathbb{Z}_p$. This cannot happen (Corollary to Theorem 8 §1.3) because $p \equiv 3 \pmod 4$. Hence $m = 0$ and $a = 0$ in $\mathbb{Z}_p$; that is $p|m$ and $p|n$. But then $m^2 + n^2 = p$ implies $p|1$, a contradiction.

Student Solution Manual to Accompany Introduction to Abstract Algebra, Fourth Edition. W. Keith Nicholson.
© 2012 John Wiley & Sons, Inc. Published 2012 by John Wiley & Sons, Inc.

12. (a) If $p = 6 + \sqrt{-5}$ then $N(p) = 6^2 + 5 \cdot 1^2 = 41$ is a prime. So $p = ab$ in $\mathbb{Z}(\sqrt{-5})$ means $41 = N(p) = N(a)N(b)$. Thus $N(a) = 1$ or $N(b) = 1$. This means a or b is a unit (if $a = m + n\sqrt{-5}$, then $1 = N(a) = m^2 + 5n^2$ implies $a = \pm 1$). So p is irreducible.

   (c) We have $29 = 9 + 20 = 3^2 + 5 \cdot 2^2 = (3 + 2\sqrt{-5})(3 - 2\sqrt{-5})$ in $\mathbb{Z}(\sqrt{-5})$. Since the only units in $\mathbb{Z}(\sqrt{-5})$ are ± 1, this shows 29 is not irreducible in $\mathbb{Z}(\sqrt{-5})$.

13. (a) If $p = 2 + \sqrt{-5}$, suppose $p = ab$ in $\mathbb{Z}(\sqrt{-5})$. Then $9 = N(p) = N(a)N(b)$. But $N(a) = 3$ is impossible in $\mathbb{Z}(\sqrt{-5})$ (because $m^2 + 5n^2 \neq 3$ for $m, n \in \mathbb{Z}$), so $N(a) = 1$ or $N(b) = 1$. Thus $a = \pm 1$ or $b = \pm 1$, and we have shown that p is irreducible in $\mathbb{Z}(\sqrt{-5})$. To see that it is not prime, the fact that $N(p) = 9$ gives $3 \cdot 3 = 9 = (2 + \sqrt{-5})(2 - \sqrt{-5})$. So if p is prime, then $p|3$ in $\mathbb{Z}(\sqrt{-5})$, say $3 = p \cdot q$. Then $9 = N(3) = N(p)N(q) = 9N(q)$ so $N(q) = 1$. This means $q = \pm 1$, whence $3 = \pm p = \pm(2 + \sqrt{-5})$, a contradiction. So p is not prime.

14. (a) $p = 3 + 2\sqrt{-3} = \sqrt{-3}(2 - \sqrt{-3})$ in $\mathbb{Z}(\sqrt{-3})$ and neither factor is a unit (that is ± 1). So p is not irreducible.

   (c) If $p = 5 = ab$ in $\mathbb{Z}(\sqrt{-3})$ then $25 = N(p) = N(a)N(b)$. Since $N(a) = 5$ is impossible in $\mathbb{Z}(\sqrt{-3})$ (because $m^2 + 3n^2 = 5$ is impossible with $m, n \in \mathbb{Z}$), either $N(a) = 1$ or $N(b) = 1$; that is $a = \pm 1$ or $b = \pm 1$. So 5 is irreducible.

15. If $p = 1 + \sqrt{-3}$ then $p = ab$ means $4 = N(p) = N(a)N(b)$. Since $m^2 + 3n^2 = 2$ is impossible for $m, n \in \mathbb{Z}$, we have $N(a) \neq 2$ and $N(b) \neq 2$. So $N(a) = 1$ or $N(b) = 1$; that is $a = \pm 1$ or $b = \pm 1$. Thus p is irreducible. If p is prime then $2 \cdot 2 = 4 = (1 + \sqrt{-3})(1 - \sqrt{-3})$ shows that $p|2$ in $\mathbb{Z}(\sqrt{-3})$, say $2 = pq$. Thus $4 = N(p)N(q) = 4N(q)$, whence $N(q) = 1$, $q = \pm 1$, $2 = \pm p$, a contradiction. So p is not prime.

16. (a) If $p \sim q$, suppose p is irreducible. If $q = ab$ in R then $p \sim ab$ so $p \sim a$ or $p \sim b$. Thus $q \sim a$ or $q \sim b$, so q is irreducible. The converse is proved the same way.

17. Assume $p \in \mathbb{Z}(\sqrt{-5})$ and $N(p)$ is as in Example 3. If $N(p)$ is a prime and $p = xy$ in R, then $N(p) = N(x)N(y)$. It follows that $N(x) = 1$ or $N(y) = 1$, say $N(x) = 1$. If $x = m + n\sqrt{-5}$, this means $m^2 + 5n^2 = 1$, whence $x = \pm 1$ is a unit. This shows that p is irreducible.

19. Suppose R is an integral domain with the DCCP. If $a \neq 0$ in R, we have $\langle a \rangle \supseteq \langle a^2 \rangle \supseteq \langle a^3 \rangle \supseteq \cdots$ so, by hypothesis, there exists $a \geq 1$ such that $\langle a^n \rangle = \langle a^{n+1} \rangle$. Thus $a^n \in \langle a^{n+1} \rangle = Ra^{n+1}$, say $a^n = ba^{n+1}$. Since $a \neq 0$ and R is a domain, this gives $1 = ba$, so a is a unit. This shows that R is a field. The converse is clear as a field has only two ideals.

21. If R has the ACCP, suppose F has no maximal member. Choose $\langle a_1 \rangle$ in F. Then $\langle a_1 \rangle$ is not maximal so $\langle a_1 \rangle \subset \langle a_2 \rangle$ for some $\langle a_2 \rangle$ in F. Now $\langle a_2 \rangle$ is not maximal, so $\langle a_2 \rangle \subset \langle a_3 \rangle$ for some $\langle a_3 \rangle$ in F. This continues to violate the ACCP.

   Conversely, suppose $\langle a_1 \rangle \subseteq \langle a_2 \rangle \subseteq \cdots$ and let F $= \{\langle a_i \rangle \mid i = 1, 2, \ldots\}$. By hypothesis, let $\langle a_n \rangle$ be maximal in F. If $m \geq n$ then $\langle a_n \rangle \subseteq \langle a_m \rangle$ so the maximality gives $\langle a_n \rangle = \langle a_m \rangle$. Thus R has the ACCP by Lemma 1.

23. No. $\mathbb{Z}(\sqrt{-5})$ is a subring of the UFD $\mathbb{C}$ which is itself not a UFD by Example 5 (and Theorem 7).

24. (a) Write $d = \gcd(a, b, \ldots)$. Then $d|0, d|a, d|b, \ldots$; if $r|0, r|a, r|b, \ldots$ then $r|d$ by definition. So $d \sim \gcd(0, a, b, \ldots)$. Clearly $0|0, a|0, b|0, \ldots$; if $0|r, a|r, b|r, \ldots$ then $r = 0$ (because $0|r$) so $0 \sim \text{lcm}(0, a, b, \ldots)$ by definition.

25. (a) Write $d = \gcd(a_1, \ldots, a_n)$ and $d' = \gcd(b_1, \ldots, b_n)$. Then $d|a_i$ for all i so $d|b_i$ for all i (Exercise 2), whence $d|d'$. Similarly, $d'|d$, whence $d' \sim d$.

27. Write $d = \gcd[a, \gcd(b, c)]$ and $d' = \gcd[\gcd(a, b), c]$. Then d divides a and $\gcd(b, c)$ so it divides all of a, b, c. Thus d divides $\gcd(a, b)$ and c, whence $d|d'$. Similarly $d'|d$ so $d \sim d'$. Moreover we have shown that $d|a$, $d|b$, $d|c$. If $r|a$, $r|b$ and $r|c$ then r divides a and $\gcd(b, c)$ so $r|d$. Hence $\gcd(a, b, c,)$ exists and $d \sim \gcd(a, b, c)$.

29. It is clear if $c = 0$. Let $a \sim p_1^{a_1} \cdots p_r^{a_r}$, $b \sim p_1^{b_1} \cdots p_r^{b_r}$ and $c \sim p_1^{c_1} \cdots p_r^{c_r}$. Then $a|bc$ means $a_i \leq b_i + c_i$ for all i, and $\gcd(a, b) \sim 1$ means $\min(a_i, b_i) = 0$ for all i. For each i, $\min(a_i, b_i) = 0$ means $a_i = 0$ or $b_i = 0$. Thus $a_i \leq c_i$ in either case (because $a_i \leq b_i + c_i$) so $a|c$.

31. If $m \sim \text{lcm}(a_1, \ldots, a_n)$ exists in R then $a_i|m$ for each i shows $\langle m \rangle \subseteq \langle a_i \rangle$ for each i, and hence that $\langle m \rangle \subseteq A$ where we write $A = \langle a_i \rangle \cap \cdots \cap \langle a_n \rangle$. But $r \in A$ means $a_i|r$ for each i, so $m|r$ by definition. Thus $r \in \langle m \rangle$ and we have $\langle m \rangle = A$. Conversely, if $A = \langle m \rangle$ then $a_i|m$ for each i (because $m \in \langle a_i \rangle$); and, if $a_i|r$ for each i, then $r \in A = \langle m \rangle$ so $m|r$. Thus m is a least common multiple of the a_i.

33. (1) If the nonzero coefficients of f are $a_1, a_2, \ldots, a_n$, then
$$c(f) = \gcd(a_1, \ldots, a_n).$$
Write $c = c(f)$ and $a_i = cb_i$ for each i. Then $f = c \cdot f_1$ where the nonzero coefficients of f_1 are $b_1, \ldots, b_n$. By Exercise 26, we have
$$c \cdot \gcd(b_1, \ldots b_n) \sim \gcd(a_1, \ldots, a_n) \sim c,$$
$$c \cdot \gcd(b_1, \ldots b_n) = \gcd(a_1, \ldots, a_n) = c,$$
so $\gcd(b_1, \ldots, b_n) \sim 1$ as required.

34. (a) If $p \in R$ is irreducible in R, let $p = ab$ in S. Then $a \in R$ and $b \in R$ by (2) so a or b is a unit in R. Hence a or b is a unit in S, and p is irreducible in S. Conversely, if p is irreducible in S, let $p = ab$ in R. Then one of a, b is a unit in S whence a or b is a unit in R by (1). So p is irreducible in R.

    (c) Take $S = R[x]$ above. Then (1) holds because units in $R[x]$ are units in R, and (2) holds because $fg \in R$ with f, g in $R[x]$ implies deg $f = 0 =$ deg g, that is $f \in R$ and $g \in R$. Now (b) gives (c).

35. Suppose $f = gh$, $h \in R[x]$. Then Gauss' lemma gives $c(g)c(h) \sim c(f) \sim 1$. Hence $c(g)c(h)$ is a unit, so $c(g)$ is a unit. This means g is primitive.

37. If $f \sim g$ in $R[x]$ then $f = ug$, u a unit in $R[x]$. Thus u is a unit in $F[x]$ so $f \sim g$ in $F[x]$. Conversely, let $f = ug$, u a unit in $F[x]$. Then $u \in F$, say $u = \frac{a}{b}$. Thus $bf = ag$ in R so, since f and g are primitive,
$$a \sim a \cdot c(g) \sim c(ag) \sim c(bf) \sim b \cdot c(f) \sim b.$$
If $a = bv$, $v \in R^*$, then $bf = ag$ gives $f = vg$, that is $f \sim g$ in $R[x]$.

39. (a) We have $R = \{a + bpx + x^2 f \,|\, a, b \in \mathbb{Z}, f \in \mathbb{Z}[x]\}$. This is a subring of $\mathbb{Z}[x]$ because

$$(a + pbx + x^2 f)(a' + pb'x + x^2 f') = aa' + p(ab' + ba')x$$
$$+ x^2[(a + pbx)f' + (a' + pb'x)f].$$

Thus R is an integral domain because $\mathbb{Z}[x]$ is.

40. (a) R is a subring of $\mathbb{Q}[x]$, hence a domain. If $f \in R$ is a unit in R, it is a unit in $\mathbb{Q}[x]$, hence a unit in $\mathbb{Q}$, hence a unit in $\mathbb{Z}$ (since the constant coefficient $cc(f)$ of f is in $\mathbb{Z}$). Now each of the following is a nontrivial factorization in R. $x = 2(\frac{1}{2}x)$, $\frac{1}{2}x = 2(\frac{1}{4}x)$, $\cdots$. Thus $\langle x \rangle \subset \langle \frac{1}{2}x \rangle \subset \langle \frac{1}{4}x \rangle \subset \cdots$, so the ACCP fails for R. Hence R is not a UFD.

(c) Let h be irreducible in R and suppose $h | fg$ in R, say

$$kh = fg = (m + xf_1)(n + xg_1)$$

where $m, n \in \mathbb{Z}$; $f_1, g_1 \in \mathbb{Q}[x]$. If $h = p$ is prime in $\mathbb{Z}$ then $p | m$, say $m = pm_1$. Thus $f = p(m_1 + \frac{x}{p}f_1)$, that is $h | f$. So assume $h = 1 + xh_1$, $h_1 \in \mathbb{Q}[x]$ and h is irreducible in $\mathbb{Q}[x]$. Then h is prime in $\mathbb{Q}[x]$ since $\mathbb{Q}$ is a field, so $h | f$ or $h | g$, say $f = kh$, $k \in \mathbb{Q}[x]$. But then the constant coefficient of k is m, the constant coefficient of f in $\mathbb{Z}$, so $h | f$ in R.

(e) Write $f = x^n f_1$, $n \geq 0$, $cc(f_1) \neq 0$. Since $f_1 \in \mathbb{Q}[x]$ write $f_1 = q_1 \ldots q_r$, q_i irreducible in $\mathbb{Q}[x]$. Let $t_i = cc(q_i)$. Then $t = t_1 \ldots t_n = cc(f_1) \neq 0$, so take $h_i = t_i^{-1} q_i$ for each i. Thus $h_i \sim g_i$ in $\mathbb{Q}[x]$ so h_i is irreducible in $\mathbb{Q}[x]$. Clearly $cc(h_i) = 1$ for all i. Finally $f = tx^n h_1 h_2 \cdots h_r$. If $f = t'x^{n'} h_1' \cdots h_s'$ is another such factorization, $x^n = x^{n'}$ so $n = n'$; whence

$$t' = t, \quad h_1 \ldots h_r = h_1' \ldots h_s'.$$

These are irreducible in $\mathbb{Q}[x]$ so $r = s$ and (after relabeling) $h_i \sim h_i'$. But $cc(h_i) = \pm 1 = cc(h_i')$ gives $h_i = \pm h_i'$ for all i.

5.2   PRINCIPAL IDEAL DOMAINS

1. No. $\mathbb{Z}[x]$ is a subring of the PID $\mathbb{Q}[x]$, but $\mathbb{Z}[x]$ is not a PID (Example 3).

3. The ideals are $0 = \langle 0 \rangle$ and $F = \langle 1 \rangle$, both principal.

4. No. If it were a PID it would be a UFD by Theorem 1, contrary to Example 5 §5.1.

5. Let $A = \langle a \rangle$, $a \neq 0$. If a is a unit then $|R/A| = 1$. Otherwise, by Theorem 4 §3.3, let B/A be any ideal of R/A, say $B = \langle b \rangle$. Then $\langle a \rangle \subseteq \langle b \rangle$ so $b | a$. Since a has a prime factorization, there are at most finitely many such divisors b of a up to associates, and hence only finitely many ideals $\langle b \rangle$.

6. (a) No. 0 is a prime ideal of $\mathbb{Z}$ ($\mathbb{Z}/0 \cong \mathbb{Z}$ is an integral domain) but 0 is not maximal.

7. (c) $\Rightarrow$ (a). Assume (c). If $a \neq 0$ in R consider $A = \{ra + gx \mid r \in R, g \in R[x]\}$. This is an ideal of $R[x]$ so let $A = \langle f \rangle$ by (c). Then $x \in A$ implies that $f | x$ so $f \sim 1$ or $f \sim x$. Thus $A = R[x]$ or $A = \langle x \rangle$. But $A = \langle x \rangle$ is impossible because

$a \in A$ would then imply $x|a$. So $1 \in A$, say $1 = ra + gx$. Clearly then $g = 0$, so a is a unit. This proves (c) $\Rightarrow$ (a).

8. (a) Write $R = \mathbb{Z}_{(p)}$. R is a subring of R because $-\left(\frac{m}{n}\right) = \frac{-m}{n}$, $\frac{m}{n} \cdot \frac{m'}{n'} = \frac{mm'}{nn'}$ and $\frac{m}{n} + \frac{m'}{n'} = \frac{mn'+m'n}{nn'}$ and p does not divide nn'. Thus R is an integral domain. Given $\frac{m}{n}$ in R, if p does not divide m then $\frac{m}{n}$ is a unit in R (with inverse $\frac{n}{m}$). Conversely, if $\frac{m}{n}$ is a unit, say $\frac{m}{n} \cdot \frac{m'}{n'} = 1$, then $mm' = nn'$ so p does not divide m (it does not divide n or n'), so $R^* = \{\frac{m}{n} \in R \mid p$ does not divide $m\}$.

    (c) R is a PID by (b) and the fact that $0 = \langle 0 \rangle$. By (b), the ideals of R are 0 and $R = \langle 1 \rangle \supset \langle p \rangle \supset \langle p^2 \rangle \supset \cdots$. Clearly $\langle p \rangle$ is the only maximal ideal.

9. If $a \neq 0$ in $R = \mathbb{Z}_{(p)}$, let $\langle a \rangle = \langle p^k \rangle$ where $a = \frac{m}{n}p^k$, p does not divide m or n (see (b) of the preceding exercise). Thus δ is well defined by $\delta(a) = k$. If $a, b \neq 1$ are given, we want $a = qb + r$ in R where $r = 0$ or $\delta(r) < \delta(b)$. Let $\delta(a) = k$ and $\delta(b) = m$ so $a = up^k$, $b = vp^m$ where $u, v \in R^*$. If $k < m$ then $a = 0b + a$ does it because $\delta(a) = k < m = \delta(b)$. If $k \geq m$ then
$$a = up^k = u \cdot p^{k-m} \cdot p^m = up^{k-m}v^{-1}b$$
so $a = qb + 0$ in this case. Hence δ is a division function on R.

Note: $a = up^k$, $u \in R^*$. If $b = vp^m$ then $ab = uvp^{m+k}$ so
$$\delta(ab) = k + m = \delta(a) + \delta(b).$$
Also, if $k \leq m$, $a + b = (u + vp^{m-k})p^k = wp^{k+t}$, $w \in R^*$. Thus
$$\delta(a + b) = k + t \geq k = \min\{\delta(a), \delta(b)\}.$$
In particular, if $a \neq 0 \neq b$ then $\delta(ab) = \delta(a) + \delta(b) \geq \delta(a)$. This proves that R is a euclidean domain.

11. (a) (1) If $a = m + n\omega$ then $aa^* = (m + n\omega)(m - n\omega) = m^2 - \omega^2 n^2 = N(a)$. Similarly $N(a^*) = aa^*$.

    (2) If $b = p + q\omega$ then $ab = (mp + nq\omega^2) + (mq + np)\omega$ and
$$a^*b^* = (m - n\omega)(p - q\omega) = (mp + nq\omega^2) - (mq + nq)\omega = (ab)^*.$$
    Clearly $(a^*)^* = (m - n\omega)^* = m + n\omega = a$.

13. (a) It suffices to verify Lemma 1. Given r, s in $\mathbb{Q}$ let m and n be the integers closest to r and s. Then $|r - m| \leq \frac{1}{2}$ and $|s - n| \leq \frac{1}{2}$ so
$$(r - m)^2 - \omega^2(s - n)^2 = (r - m)^2 + 2(s - m)^2 = \frac{1}{4} + 2 \cdot \frac{1}{4} = \frac{3}{4} < 1.$$

14. (a) Given r, $s \in \mathbb{Q}$, let m, n be the closest integers, so that
$$|r - m| \leq \tfrac{1}{2}, \ |s - n| \leq \tfrac{1}{2}.$$
    Case 1. $(r - m)^2 \leq 2(s - n)^2$. Then
$$|(r - m)^2 - 2(s - n)^2| = 2(s - m)^2 - (r - m)^2 \leq 2(s - n)^2 \leq \tfrac{1}{2}.$$
    Case 2. $(r - m)^2 > 2(s - n)^2$. Then
$$|(r - m)^2 - 2(s - n)^2| \leq (r - m)^2 \leq \tfrac{1}{4}.$$
This verifies Lemma 1.

15. (a) As in (a) of the preceding exercise, $|(r - m)^2 - 3(s - n)^2| \le 3(s - m)^2 \le \frac{3}{4}$ or $|(r - m)^2 - 3(s - m)^2| \le (r - m)^2 \le \frac{1}{4}$.

17. DA. If $a = qb + r$ where $r = 0$ or $\delta(r) < \delta(b)$, then
$$\delta'(r) = m\delta(r) + k < m\delta(b) + k = \delta'(b).$$

18. (a) Define $\delta(a) = 1$ for all $a \in F$. Then $\delta(ab) = 1 = \delta(a)$ proves E. As to DA, if a, $b \ne 0$ are in F then $a = (ab^{-1})b + 0$.

19. Let $a = ub$, a a unit. Then $\delta(a) = \delta(ub) \ge \delta(b)$ by E. Similarly $b = u^{-1}a$ implies $\delta(b) \ge \delta(a)$, so $\delta(b) = \delta(a)$.

21. We always have $\delta(ab) \ge \delta(a)$ by E. Assume b is a nonunit. If $\delta(ab) = \delta(a)$, write $c = ab$. Then $a|c$ and $\delta(a) = \delta(c)$ so, by the preceding exercise, $a \sim c$, say $c = ua$, u a unit. Thus $ab = ua$ so $b = u$ is a unit, contrary to assumption. Conversely, assume $\delta(ab) > \delta(a)$. If b is a unit then $ab \sim a$ so $\delta(ab) = \delta(a)$ by Exercise 19. So b is a nonunit.

23. (a) Let $P = \langle p \rangle$ be as given. If $a \in P$ then a is a nonunit, as $P \ne R$. If a is a nonunit, then $A = \langle a \rangle \ne R$ so (by Exercise 6), $A \subseteq P$. Then $a \in P$.

24. (a) Write $a = 1 + i$. Then $N(a) = 1^2 + 1 = 2$ is a prime in $\mathbb{Z}$, so a is irreducible in $\mathbb{Z}(i)$ by Theorem 5, and so R/A is a field by Theorems 7, 4 and 3. The preceding exercise shows that R/A is a finite field. Now $\delta(a) = N(a)$ so if $r = m + ni$ has $\delta(r) = m^2 + n^2 < 2$, then $r = 0, \pm 1, \pm i$, so
$$R/A = \{0 + A, 1 + A, -1 + A, i + A, -i + A\}.$$
However $i + A = -1 + A$ and $-i + A = 1 + A$, so
$$R/A = \{0 + A, 1 + A, -1 + A\}.$$
But $1 + A = -1 + A$ because $2 = (1 - i)(1 + i) \in A$. So
$$R/A = \{0 + A, 1 + A\}$$
is the field of two elements.

25. It is clear that $\mathbb{Q}(\omega)$ is a subring of $\mathbb{C}$ and that $\mathbb{Z}(\omega) \subseteq \mathbb{Q}(\omega)$. We show that $\mathbb{Q}(\omega)$ is a field, and that $\mathbb{Q}(\omega) = \{ab^{-1} \mid a, b \ne 0 \text{ in } \mathbb{Z}(\omega)\}$. If $0 \ne q = r + s\omega$ is in $\mathbb{Q}(\omega)$ then $N(q) = qq^* = r^2 - s^2\omega^2 \ne 0$ as $\omega \notin \mathbb{Q}$. Since
$$N(q) \in \mathbb{Q}, \quad q^{-1} = \frac{1}{N(q)}q^* \in \mathbb{Q}(\omega),$$
and we have shown that $\mathbb{Q}(\omega)$ is a field. Now let $x = \frac{m}{n} + \frac{m'}{n'}\omega$ in $\mathbb{Q}(\omega)$. Then $a = nn'q \in \mathbb{Z}(\omega)$ so $q = ab^{-1}$ where $b = nn' \in \mathbb{Z} \subseteq \mathbb{Q}(\omega)$.

26. (a) (1) $\Rightarrow$ (2). Given $a \ne 0$, $b \ne 0$, let $Ra + Rb = \langle d \rangle$. Then $d = ra + sb$ for some $r, s \in R$, so if $k|a$ and $k|b$ in R then $k|d$. But $d|a$ and $d|b$ because $a, b \in \langle d \rangle$. Thus $d = \gcd(a, b)$.

    (2) $\Rightarrow$ (1). Given $A = Ra + Rb$, clearly A is principal if $a = 0$ or $b = 0$. Otherwise, by (2) let $d = \gcd(a, b) \sim d$ where $d = ra + sb$ for $r, s \in R$. Then $d \in A$ so $\langle d \rangle \subseteq A$. On the other hand, $d|a$ and $d|b$ so $a \in \langle d \rangle$ and $b \in \langle d \rangle$. Hence $Ra + Rb \subseteq \langle d \rangle$.

27. If R is a PID, we must show that R satisfies the ACCP. This follows by the first part of the proof of Theorem 1. Conversely, if R satisfies the conditions,

every finitely generated ideal of R is principal by Exercise 26(b). If A is a non-finitely generated ideal of R, then $A \neq 0$ so let $a_1 \in A$. Then $A \neq \langle a_1 \rangle$ so let $a_2 \in A - \langle a_1 \rangle$ so that $Ra_1 \subset Ra_1 + Ra_2$. But $A \neq Ra_1 + Ra_2$ so let $a_3 \notin Ra_1 + Ra_2$. Thus $Ra_1 \subset Ra_1 + Ra_2 \subset Ra_1 + Ra_2 + Ra_3$. This process continues to give a strictly increasing chain of principal ideals, contrary to the ACCP.

29. Write $B = \langle b \rangle$ and $C = \langle c \rangle$. By Theorem 8 §3.4, it suffices to show that $B \cap C = \langle a \rangle$ and $B + C = R$. Now $\gcd(b, c) = 1$ means that $1 = rb + sc$ for some r, $s = R$. Thus $1 \in B + C$, so $B + C = R$. It is clear that

$$\langle a \rangle = \langle bc \rangle \subseteq B \cap C.$$

If $x \in B \cap C$ then $b|x$ and $c|x$, say $x = b'b = c'c$. Then $1 = rb + sc$ gives $x = rbx + scx = rb(c'c) + scb'b = (rc' + sb')bc = (rc' + sb')a \in \langle a \rangle$.

31. Let $v = m + n\sqrt{2}$ be a unit in $\mathbb{Z}(\sqrt{2})$, so that $m^2 - 2n^2 = N(v) = \pm 1$. Since $1 < u$ we have $0 < \cdots < u^{-3} < u^{-2} < u^{-1} < u^0 = 1 < u < u^2 < u^3 < \cdots$. Since u^{-k} approaches 0 for large k, either $v = u^k$ for $k \in \mathbb{Z}$ or $u^k < v < -u^{k+1}$. But the latter possibility implies $1 < vu^{-k} < u$ where vu^{-k} is a unit. We rule this out below, so $v = u^k$. If $v < 0$ then $-v = u^k$ in the same way.

Claim. $1 < v < u$ is impossible for a unit $v = m + n\sqrt{2}$ in $\mathbb{Z}(\sqrt{2})$.

Proof. $vv^* = \pm 1$ so $|v||v^*| = 1$. But $|v| = v > 1$ so $|v^*| < 1$, that is

$$-1 < m - n\sqrt{2} < 1.$$

Since also $1 < m + n\sqrt{2} < 1 + \sqrt{2}$, adding these inequalities gives

$$0 < 2m < 2 + \sqrt{2}.$$

Hence $m = 1$ and so $0 < n\sqrt{2} < \sqrt{2}$. This is impossible as n is an integer.

32. Write $a = m + n\omega$, $b = p + q\omega$ and $c = u + v\omega$ where m, n, p, q, u and v are in $\mathbb{Z}$.

   (a) $\langle a, b \rangle = mp - nq\omega^2 = \langle b, a \rangle$.

   (c) $\langle ka, b \rangle = (km)p - (kn)q\omega^2 = k(mp - nq\omega^2) = k\langle a, b \rangle$.

33. No. If so, write $w = \sqrt{-2}$. Then $w^2 = -2$ so $-2 > 0$ by Lemma 1 §3.5. But $1 > 0$ so $2 = 1 + 1 > 0$, whence $-2 < 0$ by Axiom P2. This is a contradiction.

34. (a) Clearly $\theta(a + b) = \theta(a) + \theta(b)$ and $\theta(1) = 1$. We have

$$\theta[(m + n\omega)(r + s\omega)] = \begin{bmatrix} mr + ns\omega^2 & (ms + nr)\omega^2 \\ ms + nr & mr + ns\omega^2 \end{bmatrix} = \begin{bmatrix} m & n\omega^2 \\ n & m \end{bmatrix} \begin{bmatrix} r & s\omega^2 \\ s & r \end{bmatrix}$$

$$= \theta(m + n\omega) \cdot \theta(r + s\omega).$$

   Clearly ker $\theta = 0$, so θ is one-to-one.

35. (a) We have $\tau(ab) = (ab)^* = a^*b^* = \tau(a) \cdot \tau(b)$ by Theorem 5. If $a = m + n\omega$ and $b = p + q\omega$. Then

$$\tau(a + b) = (m + p) - (n + q)\omega = (m - n\omega) + (p - q\omega)$$

$$= a^* + b^* = \tau(a) + \tau(b).$$

   Since $\tau(1) = 1^* = 1$ this shows τ is a ring homomorphism. But $a^{**} = a$ for all a shows $\tau^2 = 1_R$ so $\tau^{-1} = \tau$. Thus τ is an isomorphism.

Chapter 6

Fields

6.1 VECTOR SPACES

1. (a) No. $0 \notin U$.

   (c) No. $0 \notin U$ and not closed under addition.

2. (a) Yes. This is because $(2f)(x) = 2f(x)$ and $(f + g)(x) = f(x) + g(x)$ for all $f, g \in F[x]$.

   (c) No. $0 \notin U$.

3. $(a, b, c) = \frac{1}{2}(a + b - c)(1, 1, 0) + \frac{1}{2}(a - b + c)(1, 0, 1) + \frac{1}{2}(-a + b + c)(0, 1, 1)$.

4. (a) The inclusion $\supseteq$ is clear. Since
$$u = \tfrac{1}{2}[(u + v) + (u + w) - (v + w)\} \in \text{ span } \{u + v, \ u + w, \ v + w\}$$
and similarly for v and w, we have also proved the inclusion $\subseteq$.

5. We have $(-a)v + av = (-a + a)v = 0v = 0$, so adding $-(av)$ to both sides gives $(-a)v = -(av)$. Similarly $a(-v) + av = a(-v + v) = a \cdot 0 = 0$ yields $a(-v) = -(av)$.

7. (a) Dependent. $(1, 2, 3) + 2(4, 0, 1) + 3(2, 1, 0) = (0, 0, 0)$.

   (c) Independent. $a(x^2 + 1) + b(x + 1) + cx = 0$ gives $a=0$, $b + c=0$, $a + b = 0$; so $a = b = c = 0$.

9. (a) If $a + b\sqrt{2} + c\sqrt{3} = 0$ then $(a + b\sqrt{2})^2 = 3c^2$; $a^2 + 2ab\sqrt{2} + 2b^2 = 3c^2$. Thus $2ab = 0$ and $a^2 + 2b^2 = 3c^2$. If $a = 0$, $b \neq 0$ then $\frac{2}{3} = \left(\frac{c}{b}\right)^2$, a contradiction. If $a \neq 0$, $b = 0$ then $\frac{1}{3} = \left(\frac{c}{b}\right)^2$, a contradiction. So $a = b = 0 = c$.

11. $\{(1, -1, 0), (1, 1, 1), (a, 0, 0)\}$; $a \neq 0$ in $\mathbb{R}$.

13. If $rf + sg = 0$ then $rf(a) + sg(a) = 0$ so $s = 0$; and $rf(b) + sg(b) = 0$ implies that $a = 0$.

Student Solution Manual to Accompany Introduction to Abstract Algebra, Fourth Edition. W. Keith Nicholson. © 2012 John Wiley & Sons, Inc. Published 2012 by John Wiley & Sons, Inc.

15. Since dim $M_2(F) = 4$, the matrices I, A, A^2, A^3, A^4 cannot be independent by the fundamental theorem. The result follows.

17. If $M_n(F) = \text{span}\{A_1, \ldots, A_k\}$ then $I = \sum_{i=1}^{k} a_i A_i$, $a_i \in F$. Then for all

$V \in M_n(F)$ $V = IV = \sum_{i=1}^{k} a_i A_i v = \sum_{i=1}^{k} a_i 0 = 0$, a contradiction.

19. (a) $\{1, r, \ldots, r^n\}$ is not independent because $\dim_F R = n$. So
$$a_0 + a_1 r + \cdots + a_n r^n = 0$$
for some a_i not all zero. Thus $p(r) = 0$ when $p = \sum_{i=0}^{n} a_i x^i \neq 0$ in $F[x]$.

21. Clearly each of $u, v_2, \ldots, v_n \in \text{span}\{v_1, v_2, \ldots, v_n\}$ so it follows that $\text{span}\{u, v_2, \ldots, v_n\} \subseteq \text{span}\{v_1, v_2, \ldots, v_n\}$. If $a_1 \neq 0$ the other inclusion follows by the same argument because $v_1 = a_1^{-1} u - a_1^{-1} a_2 v_2 - \cdots - a_1^{-1} a_n v_n$.

22. (a) If $\{u_1, \ldots, u_m\} \subseteq \{u_1, \ldots, u_m, \ldots, u_n\}$ — independent, then $\sum_{i=1}^{m} a_i u_i = 0$
implies $\sum_{1}^{n} a_i u_i = 0$ where $a_{m+1} = \cdots = a_n = 0$. So $a_i = 0$ for all i by the independence of $\{u_1, \ldots, u_m, \ldots, u_n\}$.

23. (a) Given such a set $\{v_1, \ldots, v_n\}$, let $v \in V$. Then $v \notin \text{span}\{v_1, \ldots, v_n\}$ implies $\{v, v_1, \ldots, v_n\}$ is independent (Lemma 1), contrary to the choice of the v_i, so $v \in \text{span}\{v_i\}$ for all $v \in V$; that is $V = \text{span}\{v_1, \ldots, v_n\}$.

25. If $v_i \in \text{span}\{v_1, \ldots, v_{i-1}, v_{i+1}, \ldots, v_n\}$ then $v_i = \sum_{j \neq i} a_j v_j$, so
$\{v_1, \ldots, v_i, \ldots, v_n\}$ is dependent. If it is dependent, then $\sum_{i=1}^{n} a_i v_i = 0$, a_i not all
0. If $a_k \neq 0$ then $v_k = \sum_{i \neq k} -a_k^{-1} a_i v_i$.

26. (a) Clear by the subspace test.

(c) Pick a basis $\{x_1, \ldots, x_k\}$ of $U \cap V$. By Theorem 8, extend to a basis $\{x_1, \ldots, x_k, u_1, \ldots, u_m\}$ of U and a basis $\{x_1, \ldots, x_k, w_1, \ldots, w_n\}$ of W. It suffices to prove the following Claim.
    Claim: $\{u_i, x_j, w_u\}$ is a basis of $U + W$. It clearly spans. If
$$\sum_{i=1}^{m} a_i u_i + \sum_{j=1}^{k} b_j x_j + \sum_{h=1}^{n} c_h w_h = 0,$$
write $u = \sum_{i=1}^{m} a_i u_i$, $x = \sum_{j=1}^{k} b_j x_j$ and $w = \sum_{h=1}^{n} c_h w_h$; it suffices to prove $u = x = w = 0$. We have $u + x + w = 0$ so $u = -(x + w) \in U \cap W$. Thus $u = \sum_{j=1}^{k} d_j x_j$ that is $\sum_{i=1}^{m} a_i u_i + \sum_{j=1}^{k} (-d_j) x_j = 0$. Since $\{u_i, x_j\}$ is linearly independent, this forces $u = 0$. Then $x + w = 0$ so, similarly, $x = 0 = w$.

27. If $U = \text{span}\{u_i\}$ and $W = \text{span}\{w_j\}$ then $U + W = \text{span}\{u_i, w_j\}$.

28. (a) They are clearly subspaces. For example p, q even means
$$(p+q)(-x) = p(-x) + q(-x) = p(x) + q(x) = (p+q)(x), \text{ etc.}$$
If $p(x) \in U \cap W$ then $p(x) = -p(-x) = -p(x)$, so $p(x) = 0$ (because $2 \neq 0$ in F). Thus $U \cap W = 0$. If $f(x) \in V$ then $f(x) = p(x) + q(x)$ where
$$p(x) = \tfrac{1}{2}(f(x) + f(-x)) \in U$$
and $q(x) = \tfrac{1}{2}(f(x) - f(-x)) \in W$. So $V = U + W$.

29. W is clearly a subspace. Let $\{u_1, \ldots, u_m\}$ be a basis of U.
Case 1. $v \in U$. Then $W = U$ so dim $U = m$.
Case 2. $v \notin U$. Then $\{u_1, \ldots, u_m, v\}$ is independent (Lemma 1) and is clearly a basis of W (definition of W). So dim $W = m + 1$.

31. (a) They are additive subgroups by group theory. If $v \in \ker \varphi$ then
$$\varphi(av) = a\varphi(v) = a0 = 0 \text{ for all } a \in F. \text{ If } w \in \operatorname{im}\varphi, \text{ say } w = \varphi(v), \text{ then}$$
$$aw = a\varphi(v) = \varphi(av) \in \operatorname{im}\varphi.$$

(c) Let $\{u_1, \ldots, u_m\}$ be a basis of $\ker \varphi$; extend to a basis $\{u_1, \ldots, u_m, v_1, \ldots, v_k\}$ of V. It suffices to show that $\{\varphi(v_1), \ldots, \varphi(v_k)\}$ is a basis of $\operatorname{im}\varphi$.

Span. If $\varphi(v) \in \operatorname{im} \varphi$ and $v = \sum\limits_{i=1}^{m} a_i u_i + \sum\limits_{j=1}^{k} b_j v_j$ then

$$\varphi(v) = \sum_{j=1}^{k} b_j \varphi(v_j).$$

Independent. If $\sum\limits_{j=1}^{k} b_j \varphi(v_j) = 0$ then $\sum\limits_{j=1}^{k} b_j v_j \in \ker \varphi$, so

$$\sum_{j=1}^{k} b_j v_j = \sum_{i=1}^{m} a_j u_j.$$

Thus $\sum\limits_{j=1}^{k} b_j v_j + \sum\limits_{i=1}^{m}(-a_i)u_i = 0$, so $b_j = 0$ for all j.

6.2 ALGEBRAIC EXTENSIONS

1. (a) $u^2 = 8 + 2\sqrt{15}$, whence $(u^2 - 8)^2 = 60$. Hence $u^4 - 16u + 4 = 0$.
   (c) $u^2 = \sqrt{3} - 2i$ so $(u^2 - \sqrt{3})^2 = -4$, that is $u^4 - 2\sqrt{3}u^2 + 3 = -4$. Thus $(u^4 + 7)^2 = 12u^4$, so $u^8 + 2u^4 + 49 = 0$.

2. (a) $u^2 = 5 + 2\sqrt{6}$ so $(u^2 - 5)^2 = 24$, $u^4 - 10u^2 + 1 = 0$. We claim $m = x^4 - 10x^2 + 1$ is the minimal polynomial; it suffices to prove it is irreducible over $\mathbb{Q}$. It has no root in $\mathbb{Q}$ so suppose
$$x^4 - 10x^2 + 1 = (x^2 + ax + b)(x^2 + cx + d), a, b, c, d \text{ in } \mathbb{Q}.$$
Then
$$a + c = 0, \quad b + d + ac = -10, \quad ad + bc = 0, \quad bd = 1.$$

Then $c = -a$ and $b = d = \pm 1$. Hence $2b - a^2 = -10$, so $a^2 = 10 + 2b = 12$ or 8, a contradiction.

(c) $u^2 = 1 + \sqrt{3}$ so $(u^2 - 1)^2 = 3$; $u^4 - 2u^2 - 2 = 0$. The minimal polynomial is $x^4 - 2x^2 - 2$ because it is irreducible by Eisenstein's criterion.

3. (a) Algebraic; $u = \sqrt{\pi}$ is a root of $x^2 - \pi \in F[x]$ where $F = \mathbb{Q}(\pi)$.

(c) Transcendental. For if $f(\pi^2) = 0$ where $0 \neq f(x) \in \mathbb{Q}[x]$ then $g(\pi) = 0$ where $g(x) = f(x^2) \neq 0$. This would imply π itself is algebraic over $\mathbb{Q}$, a contradiction.

4. (a) $(u-1)^2 = i^2 = -1$, so $u^2 - 2u + 2 = 0$. We claim $m(x) = x^2 - 2x + 2$ in $F[x]$ is the minimal polynomial. The roots of m in $\mathbb{C}$ are $1 \pm i$ and neither is in $\mathbb{Q}(v) = \mathbb{Q}(\sqrt{2}) \subseteq \mathbb{R}$. Hence $m(x)$ is irreducible in $F[x]$.

5. Clearly $\mathbb{R}(u) \subseteq \mathbb{C}$ so $2 = [\mathbb{C} : \mathbb{R}] = [\mathbb{C} : R(u)][\mathbb{R}(u) : \mathbb{R}]$. Since $\mathbb{R}(u) \neq \mathbb{R}$ (because $u \notin \mathbb{R}$) this means $[\mathbb{R}(u) : \mathbb{R}] = 2$ so $[\mathbb{C} : R(u)] = 1$. Thus $\mathbb{R}(u) = \mathbb{C}$. Alternatively, let $u = a + bi$. Then $b \neq 0$ (as $u \notin \mathbb{R}$) so $i = b^{-1}(u - a) \in \mathbb{R}(u)$. Hence $\mathbb{C} = \mathbb{R}(i) \subseteq \mathbb{R}(u)$, so $\mathbb{C} = \mathbb{R}(u)$.

7. (a) $(u - \sqrt{3})^2 = (-i)^2 = -1$ so $u^2 - 2\sqrt{3}u + 4 = 0$. We claim

$$m(x) = x^2 - 2\sqrt{3}x + 4 \in \mathbb{R}[x]$$

is the minimal polynomial. Its roots in $\mathbb{C}$ are $\sqrt{3} \pm i$ and neither is in $\mathbb{R}$. So it is irreducible in $\mathbb{R}[x]$, as required.

9. If $v = f(u)$ with $f \in F[x]$, then $v \in F(u)$, whence $F(u, v) \subseteq F(u)$. Always $F(u) \subseteq F(u, v)$. Conversely, if $F(u) = F(u, v)$ then $v \in F(u)$ so

$$v = a_0 + a_1 u + \ldots + a_{n-1} u^{n-1}$$

with $a_i \in F$, $n \geq 1$, by Theorem 4. Take $f = a_0 + a_1 x + \cdots + a_{n-1} x^{n-1}$.

11. If $[F(u) : F]$ is finite then u is algebraic over F by Theorem 1 because $u \in F(u)$. The converse follows from Theorem 4.

12. (a) If $u = \sqrt[3]{2}$ then u is a root of $x^3 - 2$, which is irreducible over $\mathbb{Q}$ by the Eisenstein criterion. So the basis is $\{1, u, u^2\}$ by Theorem 4.

(c) If $u = \sqrt[3]{3}$ then u satisfies $x^3 - 3$, which is irreducible over $\mathbb{Q}$ by Eisenstein. Thus $\{1, u, u^2\}$ is a basis of $L = \mathbb{Q}(u)$ on $\mathbb{Q}$. We have $v = \sqrt{3}$ satisfies $x^2 - 3 \in L[x]$. This is irreducible because $\sqrt{3} \notin L$. [If $\sqrt{3} \in L = \mathbb{Q}(u)$ then $2 = \deg_{\mathbb{Q}}(\sqrt{3})$ divides $\deg_{\mathbb{Q}}(u) = 3$.] Thus $\{1, \sqrt{3}\}$ is a basis of $E = L(\sqrt{3})$ on L. By Theorem 5, $\{1, u, u^2, \sqrt{3}, u\sqrt{3}, u^2\sqrt{3}\}$ is a basis of E over $\mathbb{Q}$.

(e) We have $\sqrt{5} = (\sqrt{3})^{-1}\sqrt{15} \in E$, so $E = \mathbb{Q}(\sqrt{3}, \sqrt{5})$. Now $\{1, \sqrt{3}\}$ is a $\mathbb{Q}$-basis of $L = \mathbb{Q}(\sqrt{3})$. We have $\sqrt{5} \notin L$ [If $\sqrt{5} = a + b\sqrt{3}$ then $b \neq 0$, and hence $a \neq 0$]. Hence $x^2 - 5$ is irreducible in $L[x]$. Thus $\{1, \sqrt{5}\}$ is an L-basis of $L(\sqrt{5}) = E$. Finally $\{1, \sqrt{3}, \sqrt{5}, \sqrt{15}\}$ is a $\mathbb{Q}$-basis of E by Theorem 5.

13. (a) Put $u = \sqrt{3} + \sqrt{5}$. Then $(u - \sqrt{3})^2 = 5$ so u is a root of $m = x^2 - 2\sqrt{3}x - 2 \in F[x]$. The roots of m in $\mathbb{C}$ are $\sqrt{3} \pm \sqrt{5}$ and these are not in F because $\sqrt{5} \notin F$ [$\sqrt{5} \in F$ means $\sqrt{5} = a + b\sqrt{3}$, $a, b \in \mathbb{Q}$, and $b \neq 0$ and $a \neq 0$ ($\sqrt{5} \notin \mathbb{Q}$ and $\sqrt{5} \notin \mathbb{Q} \cdot \sqrt{3}$). Thus $5 = a^2 + 3b^2 + 2ab\sqrt{3}$ shows $\sqrt{3} \in \mathbb{Q}$, a contradiction.] Thus m is irreducible over F, so $[F(u) : F] = 2$. But $F(u) = E$.

(c) Put $u = \sqrt{3} + i$ so $(u-i)^2 = 3$ and u is a root of $m = x^2 - 2ix - 4 \in F[x]$. The roots of m in $\mathbb{C}$ are u and $2i - u = -\sqrt{3} + i$, and neither is in F because $\sqrt{3} \notin F$ [$\sqrt{3} = a + bi$ means $\sqrt{3} = a \in \mathbb{Q}$]. Thus m is F-irreducible so $E = F(u)$ has degree 2 over F; that is $[E : F] = 2$.

15. If $u \in E$, $u \notin F$, we have $E \supseteq F(u) \supseteq F$, so $[F(u) : F]$ divides $[E : F]$ by Theorem 5. Since $[F(u) : F] \neq 1$ because $u \notin F$, we have $[F(u) : F] = [E : F]$ because $[E : F]$ is prime. Thus $F(u) = E$ by Theorem 8 §6.1.

17. If $F(u) \supseteq L \supseteq F$ write $p = [F(u) : F] = [F(u) : L][L : F]$. Thus $[L : F] = 1$ or p; so $L = F$ or $[L : F] = [F(u) : F]$, whence $L = F(u)$ by Theorem 8 §6.1.

19. Let $u \in E \setminus \mathbb{Q}$. Then $[E : \mathbb{Q}] = 2$ means $f(u) = 0$ where $f \in \mathbb{Q}(u)$ has degree 1 or 2. Since $u \notin \mathbb{Q}$, deg $f = 2$, say $f = ax^2 + bx + c$. Thus $[\mathbb{Q}(u) : \mathbb{Q}] = 2$ and $\mathbb{Q}(u) \subseteq E$ so $E = \mathbb{Q}(u)$. Now $f(u) = 0$ so (clearing denominators) we may assume a, b and c are integers. Thus $u = \frac{1}{2}(-b \pm \sqrt{b^2 - 4ac}) = \frac{1}{2}(-b \pm \sqrt{d})$ where $d = b^2 - 4ac \in \mathbb{Z}$. Thus $E = \mathbb{Q}(u) = \mathbb{Q}(-b \pm \sqrt{d}) = \mathbb{Q}(\sqrt{d})$. If $d = p^2 e$, $e \in \mathbb{Z}$, p a prime, then $E = \mathbb{Q}(p\sqrt{e}) = \mathbb{Q}(\sqrt{e})$. Thus we can continue until $E = \mathbb{Q}(\sqrt{m})$ where m is square-free.

20. (a) $E(u) \supseteq E \supseteq F$ gives $E(u) \supseteq F$ finite because u is algebraic over E. Let m be the minimal polynomial of u over F, so $[F(u) : F] = \deg m$. Now u is also algebraic over E because $m(u) \in E[x]$, so let $p \in E(x)$ be the minimal polynomial of u over E. Since $m(u) = 0$, $m \in E[u]$, Theorem 3 gives $p|m$. Thus $[E(u) : E] = \deg p \leq \deg m = [F(u) : F]$.

21. (a) Write $L = F(u)$ so $F(u,v) = L(v) \supseteq L \supseteq F$. Now $[L : F] = m$ by hypothesis so, since $[F(u,v) : F] = [L(v) : L][L : F] = [L(v) : L] \cdot m$, it suffices to show $[L(v) : L] \leq n$. Let p and m be the minimum polynomials of v over L and F respectively. Then $p|m$ by Theorem 3, so

$$[L(v) : L] = \deg p \leq \deg m = [F(v) : F] = n,$$

as required.

(c) No. If $E = \mathbb{C}$, $F = \mathbb{Q}$, $u = \sqrt{2}$ and $v = \sqrt{5}$, then $m = 2$ and $n = 2$ are not relatively prime, but $[\mathbb{Q}(u,v) : \mathbb{Q}] = 4 = m \cdot n$ by Example 15.

23. If $\sqrt{2} \in \mathbb{Q}(\pi)$ then $\sqrt{2} = \frac{f(\pi)}{g(\pi)}$, $f, g \in \mathbb{Q}[x]$, $\gcd(f,g) = 1$. Then $h(\pi) = 0$ where $h(x) = 2g^2(x) - f^2(x)$, and this is a contradiction if $h \neq 0$ in $\mathbb{Q}[x]$. But $h = 0$ means $2g^2 = f^2$ so, since $\gcd(f,g) = 1$, $f|2$. Thus $f = \pm 1, \pm 2$, $g^2(x) = \pm 1, \pm\frac{1}{2}$. This forces $\deg g = 0$; $g \in \mathbb{Q}$, $g = \pm 1, \pm\frac{1}{\sqrt{2}}$. Thus $g = \pm 1$, $\sqrt{2} = \frac{f(\pi)}{g(\pi)} = \pm 1$, a contradiction. Thus $h \neq 0$ and $\sqrt{2} \in \mathbb{Q}(\pi)$ has led to a contradiction. So $\sqrt{2} \notin \mathbb{Q}(\pi)$.

25. Write $K = F(u_1, \ldots, u_k, \ldots, u_n)$ and $L = F(u_1, \ldots, u_k)$. Then $L \subseteq K$ becaue $u_i \in K$ for $1 \leq i \leq k$, and so $L(u_{k+1}, \ldots, u_n) \subseteq K$ because $u_i \in K$ for $k+1 \leq i \leq n$. On the other hand $F \subseteq L(u_{k+1}, \ldots, u_n)$ and every $u_i \in L(u_{k+1}, \ldots, u_n)$ because $u_i \in L$ if $1 \leq i \leq k$. So $K \subseteq L(u_{k+1}, \ldots, u_n)$.

26. (a) If u^2 is algebraic over F, let $f(u^2) = 0$, $f(x) \neq 0$ in $F[x]$. Hence $g(u) = 0$ where $g(x) = f(x^2) \neq 0$. Thus u is algebraic over F.

27. Let $\{F_i \mid i \in I\}$ be a family of subfields of E and write

$$F = \cap_{i \in I} F_i = \{a \in E \mid a \in F_i \text{ for all } i\}.$$

Then $1 \in F$ because $1 \in F_i$ for all i. If $a, b \in F$ then $a, b \in F_i$ for all i, so $a + b$, $a - b$, ab are in F_i. Thus $a + b$, $a - b$, ab are in F, so F is a subring of E. Finally, if $a \neq 0$ in F let a_i be its inverse in F_i. Thus $aa_i = 1 = aa^{-1}$ where a^{-1} is the inverse in E, so $a_i = a^{-1}$. Thus $a^{-1} \in F_i$ for all i, so $a^{-1} \in F$. Hence F is a field.

29. Yes. Take $F = \mathbb{Q}$, $u = \sqrt{2}$ and $v = \pi$. Then $\sqrt{2} \notin \mathbb{Q}(\pi)$ by Exercise 23 and $\sqrt{2}$ is algebraic over $\mathbb{Q}(\pi)$, being algebraic over $\mathbb{Q}$. On the other hand, π is transcendental over $\mathbb{Q}(\sqrt{2})$. For if $f(\pi) = 0$, $f(x) \in \mathbb{Q}(\sqrt{2})[x]$, write

$$f(x) = \sum_{i=0}^{n}(a_i + b_i\sqrt{2})x^i = \sum_{i=0}^{n} a_i x^i + \left(\sum_{i=0}^{n} a_i x^i\right)\sqrt{2} = g(x) + h(x)\sqrt{2}$$

where either $g(x) \neq 0$ or $h(x) \neq 0$. Thus $g(\pi) + h(\pi)\sqrt{2} = 0$. If $h(\pi) = 0$ then $g(\pi) = 0$, $g(x) \neq 0$, a contradiction. If $h(\pi) \neq 0$ then $\sqrt{2} = -\frac{g(\pi)}{h(\pi)} = \mathbb{Q}(\pi)$, contrary to Exercise 23.

31. (a) We show $F(u) = Q$ where

$$Q = \{f(u)g(u)^{-1} \mid f(x), \quad g(x) \in F[x], \quad g(u) \neq 0\}.$$

Since u is transcendental over F, $f(u) \neq 0$ whenever $f \neq 0$. Thus Q is a subfield of E containing F and u, so $F(u) \subseteq Q$. But any subfield of E containing F and u must contain Q, so $F(u) \supseteq Q$. Thus $F(u) = Q$.

  (c) Put $w = ab^{-1}$ where $a = f(u)$, $g = g(u)$ as in (a). If $h \neq 0$ in $F[x]$ satisfies $h(w) = 0$ then, clearing denominators leads to a polynomial $f(x, y) \neq 0$ in $F[x, y]$ such that $f(a, b) = 0$. Write $f(a, b) = \sum_i f_i(a)b^i$. Then $f_i(a) = 0$ for all i because b is transcendental; whence $f_i = 0$ for all i because a is transcendental. So $f(x, y) = 0$; contrary to assumption.

32. (a) Write $E = \mathbb{Q}(\sqrt{p}, \sqrt{q})$ and $u = \sqrt{p} + \sqrt{q}$. Clearly $\mathbb{Q}(u) \subseteq E$. We have $u^2 = (p + q) + 2\sqrt{pq}$, so

$$u^3 = (\sqrt{p} + \sqrt{q})(p + q) + 2(\sqrt{p} + \sqrt{q})\sqrt{pq} = (p + 3q)\sqrt{p} + (3p + q)\sqrt{q}.$$

Substituting $\sqrt{q} = u - \sqrt{p}$ leads to $u^3 = (3p + q)u + 2(q - p)\sqrt{p}$. Hence $2(q - p)\sqrt{p}$ is in $\mathbb{Q}(u)$ so, since $p \neq q$, $\sqrt{p} \in \mathbb{Q}(u)$. Then $\sqrt{q} = u - \sqrt{p} \in \mathbb{Q}(u)$, so $E \subseteq \mathbb{Q}(u)$.

  (c) We have $u^2 = (p + q) + 2\sqrt{pq}$, so $[u^2 - (p + q)]^2 = 4pq$. This gives $u^4 - 2(p + q)u^2 + (p + q)^2 = 4pq$, so $u^4 - 2(p + q)u^2 + (p - q)^2 = 0$. Thus $f(u) = 0$ where $f = x^4 - 2(p + q)x^2 + (p - q)^2$. If m is the minimal polynomial of u over $\mathbb{Q}$, this shows $m \mid f$. But $\deg m = [\mathbb{Q}(u) : \mathbb{Q}] = [E : \mathbb{Q}] = 4$ by (b), so we have $m = f$ (as required) because both are monic.

33. We show that if u is algebraic over A then u is algebraic over F, so $u \in A$ contrary to hypothesis. If $f(u) = 0$ where $f \neq 0$ in $A[x]$, let

$$f = w_0 + w_1 x + \cdots + w_n x^n, w_i \in A.$$

Then $L = F(w_1, \ldots, w_n)$ is a finite extension of F by Theorem 6. Moreover $f \in L[x]$ so $L(u) \supseteq L$ is a finite extension by Theorem 4. Thus $L(u) \supseteq F$ is finite by Theorem 5, so $u \in L(u)$ implies u is algebraic over F.

35 (a) Let $0 \neq u \in R$. Then u^{-1} exists in E and we must show it is in R. Now u is algebraic over F because $E \supseteq F$ is an algebraic extension, so let $f(u) = 0$, $0 \neq f \in F[x]$, say

$$a_n u^n + a_{n-1} u^{n-1} + \cdots + a_1 u + a_0 = 0, \quad a_i \in F, \quad a_n \neq 0.$$

If $a_0 = 0$ we can cancel u to reduce the degree. So we may assume $a_0 \neq 0$. Clearly $n \geq 1$. If $n = 1$, $a_1 u + a_0 = 0$ gives $u \in F$ so $u^{-1} \in F \subseteq R$. If $n > 1$ then $u(a_n u^{n-1} + \cdots + a_1) = -a_0$ so

$$u^{-1} = -a_0^{-1}(a_n u^{n-1} + \cdots + a_1) \in F(u) \subseteq R.$$

(c) If x and y are indeterminants over $\mathbb{Z}_2$, let $F = \mathbb{Z}_2(x, y)$ denote the field of fractions of the integral domain $\mathbb{Z}_2[x, y]$. Then define $E = F(u, v)$ where $u^2 = x$ and $v^2 = y$. If $R = \text{span}_F(1, u, v)$ then $E \supseteq R \supseteq F$, R is an F-space, but R is not closed under multiplication; in fact $uv \notin R$. For if $uv \in R$ then $uv = a + bu + cv$ where $a, b, c \in \mathbb{Z}_2$, and so (since the characteristic is 2)

$$xy = u^2 v^2 = (uv)^2 = (a + bu + cv)^2 = a^2 + b^2 x + c^2 y$$

a contradiction because x and y are indeterminates over F.

6.3 SPLITTING FIELDS

1. (a) $x^3 + 1 = (x + 1)(x^2 - x + 1)$, and the roots of $x^2 - x + 1$ are $u = \frac{1}{2}(1 + i\sqrt{3})$ and $v = \frac{1}{2}(1 - i\sqrt{3})$, so $E = \mathbb{Q}(1, u, v) = \mathbb{Q}(u, v)$. But $u - v = i\sqrt{3}$ so it follows that $E = \mathbb{Q}(i\sqrt{3})$. Since $i\sqrt{3}$ is a root of the polynomial $x^2 + 3$, and since $x^3 + 3$ is irreducible over $\mathbb{Q}$ (no root), then it is the minimal polynomial of $i\sqrt{3}$. Hence $[E : \mathbb{Q}] = 2$.

(c) $f = (x^2 - 7)(x^2 + 1)$ so $E = \mathbb{Q}(\sqrt{7}, i) = \mathbb{Q}(\sqrt{7})(i)$. Thus,

$$E = [\mathbb{Q}(\sqrt{2})](i) \supseteq \mathbb{Q}(\sqrt{2}) \supseteq \mathbb{Q}.$$

Moreover, $[\mathbb{Q}(\sqrt{2}) : \mathbb{Q}] = 2$ because $x^2 - 2$ is irreducible over $\mathbb{Q}$, and $[\mathbb{Q}(\sqrt{2})](i) : \mathbb{Q}(\sqrt{2})] = 2$ because $x^2 + 1$ has no root in $\mathbb{Q}(\sqrt{2})$. Hence $[E : \mathbb{Q}] = 2 \cdot 2 = 4$ by the multiplication theorem.

2. (a) $f = (x^2 - 2x - 2)(x^2 - 5)$ so the roots are $1 \pm \sqrt{3}$ and $\pm\sqrt{5}$. Hence $\mathbb{Q}(\sqrt{3}, \sqrt{5})$ is the splitting field.

4. (a) $f = (x + 1)(x^2 + x + 1)$ and $x^2 + x + 1$ is irreducible over $\mathbb{Z}_2$. If u is a root then the other is given by $u + v = 1$; $v = 1 + u$. Thus $E = \mathbb{Z}_2(u)$ and $f = (x + 1)(x + u)(x + 1 + u)$.

(c) f is irreducible over $\mathbb{Z}_2$. If u is a root then, by long division, $f = (x + u)g$ where $g = x^2 + (1 + u)x + (u + u^2)$. We claim that g also splits in $\mathbb{Z}_2(u)$, so $E = \mathbb{Z}_2(u)$. We try possibilities to get $g(u^2) = 0$. Then the other root v satisfies $u^2 + v = 1 + u$, so $v = 1 + u + u^2$. Thus

$$f = (x + u)(x + u^2)(x + 1 + u + u^2).$$

(e) $f = (x^2 - 2)(x^2 + 1) = (x^2 + 1)^2$ over $\mathbb{Z}_3$. Now $x^2 + 1$ is irreducible over $\mathbb{Z}_3$ (no root in $\mathbb{Z}_3$). If u is a root then $x^2 + 1 = x^2 - u^2 = (x - u)(x + u)$. Thus $E = \mathbb{Z}_3(u)$ and $f = (x - u)^2(x + u)^2$.

5. The roots are $\pm\sqrt{3}$ and $1 \pm \sqrt{3}$. So $\mathbb{Q}(\sqrt{3})$ is the splitting field for both.

6. (a) If $\mathbb{C}$ were the splitting field of $f(x) \in \mathbb{Q}[x]$ then $\mathbb{C} = \mathbb{Q}(u_1, \ldots, u_n)$. Thus $\mathbb{C} \supseteq \mathbb{Q}$ would be algebraic, contradicting the fact that π or e is transcendental.

7. We have $f = a(x - u_1) \cdots (x - u_m)$ in $E[x]$. If p is a monic, irreducible factor of g in $E[x]$ then $p|f$ so $p = x - u_i$ for some i. Thus
$$g = b(x - u_{i_1})(x - u_{i_2}) \cdots (x - u_{i_k})$$
by the unique factorization theorem, as required.

9. If $\gcd(f, g) = 1$ let $1 = fh + gk$; h, k in $F[x]$. If $E \supseteq F$ is an extension containing an element u such that $f(u) = 0 = g(u)$, substitution gives
$$1 = f(u)h(u) + g(u)k(u) = 0,$$
a contradiction. Conversely, let $d = \gcd(f, g)$. If $d \neq 1$ then deg $d \geq 1$ so let $E \supseteq F$ be a field containing a root u of d. Then $d|f$ and $d|g$ means $f(u) = 0 = g(u)$, contrary to hypothesis.

11. We have $E = F(u_1, \ldots, u_n)$ where $f = a(x - u_1) \cdots (x - u_n)$, $a \in F$, $u_i \in E$. Since $L \supseteq F$, $E = L(u_1, \ldots, u_n)$, and $a \in L$, $u_i \in E$, show f splits in E over L.

13. The roots of $x^p - 1$ are $1, w, w^2, \ldots, w^{p-1}$ (Theorem 6, Appendix A) so $\mathbb{Q}(w)$ is the splitting field. We have $x^p - 1 = (x - 1)\Phi_p$ where
$$\Phi_p = x^{p-1} + x^{p-2} + \cdots + x + 1$$
is the pth cyclotomic polynomial. Then Φ_p is irreducible over $\mathbb{Q}$ by Example 13 §4.2, and so is the minimal polynomial of w (w is a root of Φ_p). Hence $[\mathbb{Q}(w) : \mathbb{Q}] = p - 1$ by Theorem 4 §6.2.

15. Let $K \supseteq E$ be a field in which f has a root v. Write deg $f = m$ and deg $g = n$ so that $[F(u) : F] = n$ and $[F(v) : F] = m$ are relatively prime. By Exercise 21 §6.2, $[F(u, v) : F] = mn$. Now let p be the minimal polynomial of v over $F(u)$; we show $f = p$. We have $f(v) = 0$ so $p|f$ in $F(u)[x]$, and so it is enough to show that deg $p = m$. We have $[F(u, v) : F(u)] = $ deg p, so
$$mn = [F(u, v) : F] = [F(u, v) : F(u)][F(u) : F] = (\text{deg } p)n.$$
by the multiplication theorem. Thus deg $p = m$ as required.

17. The map $f \to f^\sigma$ is clearly onto (since σ is onto). If $f = \sum_{i=1}^{n} a_i x^i$ then
$$f^\sigma(x) = \sum_{i=1}^{n} \sigma(a_i) x^i$$
so $f^\sigma = 0$ implies $\sigma(a_i) = 0$ for all i, which in turn implies $a_i = 0$ for all i, whence $f = 0$. Thus f^σ is one-to-one if we can show it is a ring homomorphism. If $g = \sum_{j=1}^{n} b_j x^j$,
$$(f + g)^\sigma = \sum_{i=1}^{n} [\sigma(a_i + b_i)]x^i = \sum_{i=1}^{n} \sigma(a_i)x^\sigma + \sum_{i=1}^{n} \sigma(b_i)x^i = f^\sigma + g^\sigma.$$

Similarly

$$(fg)^\sigma = \sum_{k=1}^{n} \sigma \left(\sum_{i+j=k} a_i b_j \right) x^k = \sum_{k=1}^{n} \left(\sum_{i+j=k} \sigma(a_i) \cdot \sigma(b_j) \right) x^k$$

$$= \left(\sum_{i=1}^{n} \sigma(a_i) x^i \right) \left(\sum_{j=1}^{n} \sigma(b_i) x^i \right) = f^\sigma \cdot g^\sigma.$$

Hence the map is a ring isomorphism (clearly $1^\sigma = 1$). Finally, if $a \in F$ let $g = a$ be the constant polynomial. Then $g \mapsto g^\sigma$ means $a \mapsto \sigma(a)$. So the map extends σ.

19. If π were algebraic over $\mathbb{A}$ then π would satisfy a nonzero polynomial f in $\mathbb{A}[x]$. But f splits in $\mathbb{A}$ because $\mathbb{A}$ is algebraically closed, so this would imply that $\pi \in \mathbb{A}$, a contradiction.

20. (a) We show $A = \mathbb{Q}(i)$. Clearly $A \supseteq \mathbb{Q}$ is algebraic. We must show that if $u \in E$ is algebraic over $\mathbb{Q}$ then $u \in A$. Since u is algebraic over A, we show that $u \notin A$ implies u is transcendental over A. We have $E = A(\pi)$ so this follows from Exercise 31 §6.2 if we can show that π is transcendental over A. But if π were algebraic over A it would be algebraic over $\mathbb{A} \supseteq A$, contrary to the preceding exercise.

21. (1) $\Rightarrow$ (2). If $E \supseteq F$ is the splitting field of $f \in F[x]$, then $E = F(u_1, \ldots, u_m)$ where the u_i are the roots of f in E, so $[E : F]$ is finite by Theorem 6 §6.2. If $p \in F[x]$ is irreducible with a root $u \in E$, let v be a root in some field $K \supseteq E$. Then p is the minimal polynomial of both u and v so let $\sigma : F(u) \to F(v)$ be an isomorphism (Theorem 4 §6.2). Now E is a splitting field of f over $F(u)$, and $E(v)$ is a splitting field of f over $F(v)$, so Theorem 3 shows $E \cong E(v)$ via an isomorphism extending σ. Hence $[E : F(u)] = [E(v) : F(v)]$, so

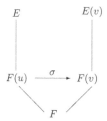

$$[E : F] = [E : F(u)][F(u) : F] = [E(v) : F(v)][F(v) : F] = [E(v) : F].$$

Since E is an F-subspace of $E(v)$, this shows $E = E(v)$, and so $v \in E$.

(2) $\Rightarrow$ (1). Since $E \supseteq F$ is finite, let $E = F(u_1, \ldots, u_n)$. Each u_i is algebraic over F, say with minimal polynomial $p_i \in F[x]$. Since p_i splits in E by (2), E is the splitting field of $f = p_1 p_2 \cdots p_n$.

6.4 FINITE FIELDS

1. (a) In $\mathbb{Z}_{11}$, $2^2 = 4$, $2^3 = 8$, $2^4 = 5$, $2^5 = -1$, so $o(2) = 10$. Thus 2 is a primitive element.

   (c) $GF(8) = \{a + bt + ct^2 \mid a, b, c \in \mathbb{Z}_2, t^3 = t + 1\}$ — since $x^3 + x + 1$ is irreducible in $\mathbb{Z}_2[x]$. In this case $GF(8)^*$ has order 7, so *every* nonzero element except 1 is primitive by Lagrange's theorem.

3. Both p and q have no root in $\mathbb{Z}_2$, so they are irreducible. Hence both rings are fields of order 2^3 and so are isomorphic by Theorem 4.

4. (a)

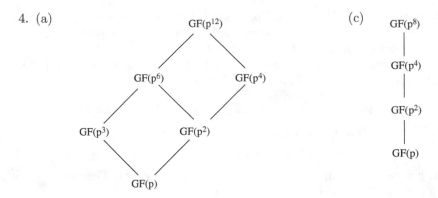

5. First $x^4 + x + 1$ is irreducible over $\mathbb{Z}_2$ (it has no root, $x^2 + x + 1$ is the only irreducible quadratic, and $(x^2 + x + 1)^2 = x^4 + x^2 + 1$). Hence

$$GF(16) = \{a + bt + ct^2 + dt^3 \mid a,\ b,\ c,\ d \in \mathbb{Z}_2,\ t^4 = t + 1\}.$$

Now $t^3 \neq 1$ and $t^5 = t(t+1) = t + t^2 \neq 1$, so $o(t) = 15$. Thus t is primitive. Since $16 = 2^4$ the subfields of $GF(2^4)$ are $GF(2)$, $GF(2^2)$ and $GF(2^4)$. Clearly $GF(2) = \mathbb{Z}_2$ and $GF(2^4) = GF(16)$. Finally $o(t^5) = 3$ so by the discussion following Corollary 2 of Theorem 7,

$$GF(2^2) = \{0\} \cup \langle t^5 \rangle = \{0, 1, t^5, t^{10}\} = \{0, 1, t + t^2, 1 + t + t^2\}.$$

7. E is finite so $E^* = \langle u \rangle, u \in E$. Thus $E = \{0\} \cup \langle u \rangle$, whence $E = F(u)$.

9. If $G \subseteq \mathbb{C}^*$ and $|G| = n$, then $G = \langle z \rangle$, $o(z) = n$. Thus $G = U_n$ — the group of all n^{th} roots of unity.

11. The Frobenius automorphism $\sigma : F \to F$ is given by $\sigma(b) = b^p$. This is onto, so each $a \in F$ has the form $a = \sigma(b) = b^p$ for some $b \in F$.

13. Write $f = a_0 + a_1 x + \cdots + a_n x^n$, $a_i \in \mathbb{Z}_p$. If $\sigma : E \to E$ is the Frobenius map defined by $\sigma(t) = t^p$, then $\sigma(a) = a$ for all $a \in \mathbb{Z}_p$ by Fermat's theorem. Since $0 = f(u) = a_0 + a_1 u + \cdots + a_n u^n$, we obtain

$$0 = \sigma(0) = a_0 + a_1[\sigma(u)] + \cdots + a_n[\sigma(u)]^n = f(\sigma(u)) = f(u^p).$$

15. If $u \in GF(2^n)$ is a root of $x^2 + x + 1$ then $u^2 = u + 1$. Hence $u^{2^2} = (u^2)^2 = u^2 + 1 = u$. Then

$$u^{2^3} = (u^{2^2})^2 = u^2 = u + 1, \quad u^{2^4} = (u^{2^3})^2 = u^2 + 1 = u.$$

In general $u^{2^k} = \begin{cases} u & \text{if } k \text{ is even} \\ u+1 & \text{if } k \text{ is odd} \end{cases}$. But $u^{2^n} = u$ here because $|GF(2^n)| = 2^n$, so we have a contradiction because n is odd.

17. Let $d = \gcd(f, f')$ and write $d = fg + f'h$ where g and h are in $F[x]$. If $d = 1$, suppose f has a repeated root a in $E \supseteq F$. Then $x - a$ divides both f and f' in $E[x]$, and so divides $d = 1$, a contradiction. Conversely, if $d \neq 1$, let E be a splitting field of f over F. Then $d \mid f$ implies d has a root a in E, so $x - a$ divides f and f', a contradiction by Theorem 3.

18. (a) Let f have no repeated root in a splitting field $E \supseteq F$, and suppose that $f' = 0$. If $f(a) = 0$, $a \in E$, then $f' = 0$ implies $(x - a)$ divides f', so a is a

repeated root by Theorem 3. Conversely, if $f' \neq 0$ let $d = \gcd(f, f')$. Then $d|f$ so $d = 1$ or $d = $ because f is irreducible. But $d = f$ is impossible because f does not divide f'. So $d = 1$, say $1 = fg + f'h$ with $g, h \in F[x]$. Now let a be a repeated root of f in a field $E \supseteq F$. Then $x - a$ divides f and f' by Theorem 3, so $x - a$ divides 1 in $E[x]$, a contradiction.

19. Use Exercise 18(a). If $f = \sum\limits_{i=0}^{n} a_i x^i$ then $f' = \sum\limits_{i=0}^{n} i a_i x^{i-1} = 0$ means $i a_i = 0$ for $i \geq 1$. Thus $a_i = 0$ if p does not divide i, so

$$f = a_0 + a_p x^p + a_{2p} x^{2p} + \cdots = g(x^p)$$

where $g = a_0 + a_p x + a_{2p} x^2 + \cdots$. Conversely, if $f = g(x^p)$, Theorem 2 gives $f' = g'(x^p) \cdot p x^{p-1} = 0$.

21. Let $E \supseteq \mathbb{Z}_p$ be a splitting field of f. Suppose that $u \in E$ satisfies $f(u) = 0$. If $a \in \mathbb{Z}_p$, then

$$f(u + a) = (u + a)^p - (u + a) - 1 = u^p + a^p - (u + a) - 1$$
$$= f(u) + a^p - a = 0 + 0 = 0.$$

Thus $\{u + a \mid a \in \mathbb{Z}_p\}$ consists of p distinct roots of f, and so is the set of *all* roots of f (since $\deg f = p$). Since E is generated over $\mathbb{Z}_p$ by these roots, we have $E = \mathbb{Z}_p(u)$.

22. Write $h = x^{p^n} - x$.

   (a) Let $K \supseteq \mathbb{Z}_p$ be a field containing a root u of f. Since f is irreducible, it is the minimal polynomial of u over $\mathbb{Z}_p$. If we write $E = \mathbb{Z}_p(u)$ then $[E : \mathbb{Z}_p] = n$ and so $|E| = p^n$. Then u is a root of h so $f|h$ in $E[x]$, say $h = qf$. But $h = q_0 f + r$ in $\mathbb{Z}_p[x]$ by the division algorithm, so this holds in $E[x]$. By the uniqueness in $E[x]$, we get $q = q_0 \in \mathbb{Z}_p[x]$ and $r = 0 \in \mathbb{Z}_p[x]$.

   (c) Here $h = x^8 - x = x^{2^3} - x$, so the irreducible divisors are of degree 1 or 3 by (b). Then

$$x^8 - x = x(x - 1)(x^6 + x^5 + x^4 + x^3 + x^2 + x + 1)$$
$$= x(x - 1)(x^3 + x + 1)(x^3 + x^2 + 1).$$

23. If char $F = 2$ then $|F| = 2^n$ so $a^{2^n} = a$ for all a. Thus $a = (a^{2^{n-1}})^2 + 0^2$. So assume char $F = p \neq 2$. Write $X = \{u^2 \mid u \in F\}$. If $F = \{0, u_1, -u_1, u_2, -u_2, \ldots\}$ then $X = \{0, u_1^2, u_2^2, \ldots\}$. If we write $|F| = q$, then $|X| = 1 + \frac{q-1}{2} = \frac{1}{2}(q + 1)$. Now let $u \in F$ and write $Y = \{a - u^2 \mid u \in F\}$. Then $|X| = |Y|$ so both sets have more than $\frac{1}{2}q = \frac{1}{2}|F|$ elements. Thus $X \cap Y \neq \emptyset$, say $a - u^2 = v^2$, $u, v \in F$, so $a = u^2 + v^2$.

6.5 GEOMETRIC CONSTRUCTIONS

1. (a) Given the line through A and B and the point C, locate D on AB such that $|AD| = |AC|$. Then locate E such that $|CE| = |DE|$. The line through C and E is parallel to AB ($ACED$ is a parallelogram).

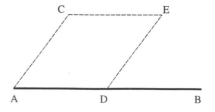

2. (a) The line through (a,b) and (c,d) has equation $(b-d)x + (c-a)y = \frac{bc-ad}{c-a}$.

   (c) Given $ax + by = c$ and $dx + ey = f$ in $F[x]$, the solution (if any) involves only field operations in F, and so both x and y are in F (if they exist).

3. The question asks whether $\pi/12 = 15°$ can be constructed. This is the result of bisecting $\pi/6 = 30°$. But the angle $\pi/6$ is constructible because $\sqrt{3}$ is constructible (see diagram) so angle $\pi/12$ is constructible.

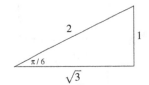

5. No. A sphere of radius 1 has volume $\frac{4}{3}\pi$ and, if this is the volume of a cube with side a, then $a = \left(\frac{4\pi}{3}\right)^{1/3}$. But a is not constructible since it is not even algebraic over $\mathbb{Q}$. For if a is a root of $f(x) \in \mathbb{Q}[x]$. Then π is a root of $g = f[\frac{3}{4}x^3]$. This is impossible as π is transcendental over $\mathbb{Q}$.

7. (a) If $a = \sin\theta$ is constructible so is $\cos\theta = \sqrt{1-a^2}$ by the Lemma. The converse is similar.

9. If a heptagon could be constructed so could $a = \cos(2\pi/7)$. Now De Moivre's Theorem gives $\cos 7\theta + i \cdot \sin 7\theta = (\cos\theta + i \cdot \sin\theta)^7$. Writing $c = \cos\theta$ and $s = \sin\theta$, the real parts are

$$\cos 7\theta = c^7 + \binom{7}{5}c^5(is)^2 + \binom{7}{3}c^3(is)^4 + \binom{7}{1}c(is)^6$$
$$= c^7 - 21c^5(1-c^2) + 35c^3(1-c^2)^2 - 7c(1-c^2)^3$$
$$= 64c^7 - 112c^5 + 56c^3 - 7c.$$

Taking $\theta = \frac{2\pi}{7}$, this shows that a is a root of $f = 64x^7 - 112x^5 + 56x^3 - 7x - 1$. By the hint $f = pq$, where $p = 8x^3 + 4x^2 - 4x - 1$, and

$$q = 8x^4 - 4x^3 - 8x^2 + 3x + 1.$$

With a calculator, show that $q(a) = -.7$ approximately. Hence $p(a) = 0$. But p is irreducible over $\mathbb{Q}$ (no roots by Theorem 9 §4.1) so it is the minimal polynomial of a. Since deg $p = 3$, a is not constructible.

6.7 AN APPLICATION TO CYCLIC AND BCH CODES

1. (a) If $n = 1$ it is clear, if $n = 2$, $(f_1 + f_2)^2 = f_1^2 + 2f_1f_2 + f_4^2 = f_1^2 + f_2^2$. In general, $(f_1 + \cdots + f_{n+1})^2 = (f_1 + \cdots + f_n)^2 + f_{n+1}^2 = f_1^2 + \cdots + f_n^2 + f_{n+1}^2$.

   (c) If $f = a_0 + a_1x + \cdots$ then $f' = a_1 + a_3x^2 + a_5x^4 + \cdots = g(x^2)$ where $g = a_1 + a_3x + \cdots$. Conversely, $f = g(x^2) = b_0 + b_1x^2 + b_2x^4 + \cdots$ clearly implies that $f' = 0$ (as $2 = 0$ in $\mathbb{Z}_2$).

3. If $n = 2^k$ then

$$1 - x^4 = (1-x^2)(1+x^2) = (1+x)^4, \quad 1 - x^8 = (1-x^4)(1+x^4) = (1+x^8), \cdots$$

and, by induction, $1 - x^{2^k} = (1+x)^{2^k}$; that is $1 - x^n = (1+x)^n$. So the divisors of $1 - x^n$ are $1, (1+x), (1+x)^2, \ldots, (1+x)^n$, and these are a chain under divisibility.

   Conversely, if $1 - x^n = (1+x)^k p^m \cdots$ where p is irreducible and $p \neq (1+x)$, then neither $\langle 1+t \rangle$ nor $\langle p(t) \rangle$ contains the other in B_n. If $n = 2^k m$,

m odd, then $1 + x^n = (1 + x^{2^k})^m = 1 + mx^{2^k} + \cdots + x^n$ a contradiction as $m = 1$ in $\mathbb{Z}_2$.

5. (a) In $B_4 : 1 + t, \; t + t^2 = t(1 + t), \; t^2 + t^3 = t^2(1 + t)$ and $1 + t^3 = t^3(1 + t)$. The other members $0, \; 1 + t^2, \; t + t^2$ and $1 + t + t^2 + t^3$ all lie in smaller ideals. (See Example 3.)

6. (a) We have $|D| = 2$ so $C \cap D = 0$ or D. If n is odd then $1 + t + \cdots + t^{n-1}$ has odd parity, so $C \cap D = 0$ in this case. Also, if $n - 1 = 2m$ then

$$1 + t + \cdots + t^{n-2} = (1 + t)(1 + t^2 + t^4 + \cdots + t^{2(m-1)})$$

is in C. Thus $t^{n-1} \in (C + D) - C$. Since $|C| = 2^{n-1}$ implies C is maximal, $C + D = B_n$. Thus $B_n \cong C \times D$ when n is odd by Theorem 7 §3.4.

7. (a) $1 + x^7 = (1 + x)(1 + x + x^3)(1 + x^2 + x^3)$ so there are $2 \cdot 2 \cdot 2 = 8$ divisors in all. Thus there are 7 codes (excluding $\langle 1 + x^7 \rangle = 0$).

   (c) $1 + x^8 = (1 + x)^8$ so there are $9 - 1 = 8$ codes in B_8.

   (e) $1 + x^{10} = (1 + x^2)(1 + x^2 + x^4 + x^6 + x^8) = (1 + x)^2(1 + x + x^2 + x^3 + x^4)^2$ (using Exercise 10 below). Hence there are $3 \cdot 3 - 1 = 8$ codes in B_{10}.

9. Since $u^i \in E^*$ and $|E^*| = n$, we have $(u^i)^n = 1$ by Lagrange's theorem. Then u^i is a root of $x^n - 1$, whence m_i divides $x^n - 1$.

10. (a) Since $g = 1 + x + x^2 + x^3 + x^4$ has no root in $\mathbb{Z}_2$, if it factors, it does so as

$$g = (a + bx + cx^2)(a' + b'x + c'x^2).$$

Hence $aa' = 1 = cc'$ so $a = a' = 1 = c = c'$. Hence

$$g = (1 + bx + x^2)(1 - b'x + x^2)$$

so $1 + bb' + 1 = 1$ (coefficient of x^2) whence $b = b' = 1$. But then the coefficient of x is $1 = b + b' = 0$, a contradiction.

11. Since $g = 1 + x + x^4$ has no root in $\mathbb{Z}_2$, if it factorizes, it does so as

$$g = (a + bx + cx^2)(a' + b'x + c'x^2).$$

Thus $aa' = 1 = cc'$ so $a = a' = 1 = c = c'$. Thus $g = (1 + bx + x^2)(1 + b'x + x^2)$ so (coefficient of x^3) $b + b' = 0$ and (coefficient of x) $b + b' = 1$. This is impossible.

13. $(1 + x + x^4)(1 + x^3 + x^4)(1 + x + x^2 + x^3 + x^4)$
$$= (1 + x + x^3 + x^4 + x^5 + x^7 + x^8)(1 + x + x^2 + x^3 + x^4)$$
$$= (1 + x^3 + x^6 + x^9 + x^{12}).$$
Since $(1 + x)(1 + x + x^2) = 1 + x^3 = 1 - x^3$, this verifies the factorization. We have checked that $1 + x + x^4$ and $1 + x + x^2 + x^3 + x^4$ are irreducible. If $1 + x^3 + x^4 = (a + bx + cx^2)(a' + b'x + c'x^2)$, then $aa' = 1 = cc'$ so $a = a' = 1 = c = c'$. Thus $1 + x^3 + x^4 = (1 + ax + x^2)(1 + a'x + x^2)$ so (coefficients of x, x^3) $a + a' = 0$ and $a + a' = 1$, a contradiction.

15. Write $m = 1 + x + \cdots + x^{n-1}$. Let $C = \langle g(t) \rangle$, where g divides $(x^n - 1)$. If $f(t) \in C$ has odd parity, then $f(t) = q(t)g(t)$ so $1 = f(1) = q(1)g(1)$. Thus $g(t)$ has odd parity. We must show that $m(t) \in C$. Let $gh = x^n - 1$ so

$$gh = x^n + 1 = (x + 1)m \quad \text{in} \quad \mathbb{Z}_2[x]. \tag{*}$$

Since $x + 1$ is prime, either $(x + 1) \mid g$ or $(x + 1) \mid h$. But $g = \lambda(x + 1)$ gives $g(1) = \lambda(1)(1 + 1) = 0$, a contradiction. So $h = k(x + 1)$, whence (*) gives $gk(x + 1) = (x + 1)m$; $m = hg$; $m(t) \in \langle g(t) \rangle$.

17. (a) By Theorem 10 §4.2, let $1 = qg + ph$ in $\mathbb{Z}_2[x]$ and define $e = qg$. Then $e(t) \in C$ so $\langle e(t) \rangle \subseteq C$. On the other hand $1 = e + qg$ so $g = eg + p(1 - x^n)$. This means $g(t) = e(t)g(t) \in \langle e(t) \rangle$, so $C = \langle e(t) \rangle$. Since $e(t) = q(t)g(t)$, this gives $e(t)^2 = q(t)e(t)g(t) = q(t)g(t) = e(t)$.

    (c) We have
    $$1 + x^7 = (1 + x)(1 + x^2 + x^3)(1 + x + x^3) = (1 + x + x^2 + x^4)(1 + x + x^3).$$
    We have $1 + x + x^2 + x^4 = x(1 + x + x^3) + 1$ so
    $$1 = x(1 + x + x^3) + 1(1 + x + x^2 + x^4).$$
    So take $e = x(1 + x + x^3) = x + x^2 + x^4$. Note that
    $$e(t)^2 = e(t^2) = t^2 + t^4 + t^8 = t^2 + t^4 + t = e(t).$$
    So $e(t) = t + t^2 + t^4$ is the idempotent generator.

    (e) Given $e(t)$ in B_n, $e(t)^2 = e(t^2)$; $e(t)^4 = e(t^2)^2 = e(t^4), \ldots, e(t)^{2^n} = e(t^{2^n})$. If $n = 2^k$, and $e(t)^2 = e(t)$, this gives $e(t) = e(t)^n = e(t^n) = e(1) = 0$ or 1. In B_6, $e(t) = 1 + t^2 + t^4$ is idempotent because
    $$e(t)^2 = 1 + t^4 + t^8 = 1 + t^4 + t^2 = e(t).$$

18. (a) Given $f(t) = a_0 + a_1 t + \cdots + a_{n-1} t^{n-1}$ in B_n, write $\bar{f} = a_0 a_1 \cdots a_{n-1}$ for the corresponding word. Then $\bar{f}H = [f(u_1), f(u_2), \ldots, f(u_n)]$ so $\bar{f}H = 0$ if and only if $f(u_i) = 0$ for all i. This is certainly true if $f(t) \in C = \langle g(t) \rangle$; the converse holds because the roots are distinct.

19. The roots of G are linearly independent over $\mathbb{Z}_2$ by choice, so G has rank k. Since rank $R = $ rank G, it follows that $R = [I_n \ A]$ by the definition of a row echelon matrix. To see that R generates C, it suffices to show that the rows of R span C. Since the rows of G span C, it suffices to show that the rows of G are all linear combinations (over $\mathbb{Z}_2$) of the rows of R. But $R = UG$, U invertible, so $G = U^{-1}R$ and the result follows.

Chapter 7

Modules over Principal Ideal Domains

7.1 MODULES

1. (a) $0x = (0+0)x = 0x + 0x$, so $0x = 0$.

   (c) Using (a), $x + (-1)x = (1 + (-1))x = 0x = 0 = x + (-x)$. Now "subtract" x from both sides by adding $-x$ to both sides.

2. (a) If $\alpha : M \to N$ is onto and R-linear, and if $M = Rx_1 + \cdots + Rx_n$, then $N = R\alpha(x_1) + \cdots + R\alpha(x_n)$. Since some of the $\alpha(x_i)$ may be zero, the result follows.

   (c) Let $K = Rx_1 + \cdots + Rx_m$ and let $M/K = R(y_1 + K) + \cdots + R(y_n + K)$ where the x_i and y_j are in M. If $x \in M$ let
   $$x + K = r_1(y_1 + K) + \cdots + r_n(y_n + K)$$
   with $r_i \in R$ for each i. Then $x - (r_1y_1 + \cdots + r_ny_n)$ is in K, so
   $$x - (r_1y_1 + \cdots + r_ny_n) = s_1x_1 + \cdots + s_mx_m$$
   where $s_j \in R$ for each j. Hence $\{x_1, \ldots, x_m, y_1, \ldots, y_n\}$ generates M.

3. If Re is an ideal then $er \in (Re)r \subseteq Re$ for any $r \in R$, say $er = se$, $s \in R$. Hence $ere = se^2 = se = er$, that is $er(1 - e) = 0$. As r was arbitrary, this proves $eR(1 - e) = 0$. Conversely, if $eR(1 - e) = 0$ then $er = ere$ for each $r \in R$. Hence $er \in Re$ for all r, so $Rer \subseteq Re$, proving that Re is a right ideal. Hence it is an ideal.

5. Recall that $AK = \{\Sigma_{i=1}^{n} a_i k_i \mid a_i \in A,\ k_i \in K,\ n \geq 1\}$.

   (a). Given $x = \Sigma a_i k_i$ in AK and $r \in R$, then $rx = \Sigma r(a_i k_i) = \Sigma(ra_i)k_i \in AK$ because A is a left ideal.

Student Solution Manual to Accompany Introduction to Abstract Algebra, Fourth Edition.
W. Keith Nicholson.
© 2012 John Wiley & Sons, Inc. Published 2012 by John Wiley & Sons, Inc.

(c). If $x \in (A + B)K$ then

$$rx = \Sigma(ra_i + rb_i)k_i = \Sigma(ra_i)k_i + \Sigma(rb_i)k_i \in AK + BK.,$$

This proves that $(A + B)K \subseteq AK + BK$. The reverse inclusion is similar.

7. Let K and N be submodules of M. We use the module isomorphism theorem.

   (a) Define $\alpha : N \to (K + N)/K$ by $\alpha(n) = n + K$ for all $n \in N$. Then α is R-linear and $\ker \alpha = \{n \mid n + K = K\} = K \cap N$. So by Theorem 1 it suffices to show that α is onto. But each element of $(K + N)/K$ has the form $(k + n) + K = n + K = \alpha(n)$, as required.

8. Let R be an integral domain. Given $_RM$ let $T(M) = \{x \in M \mid x \text{ is torsion}\}$.

   (a). We must show $T(M)$ is a submodule of M. Let $x \in T(M)$, say $ax = 0$ for $0 \neq a \in R$. If also $y \in T(M)$, say $by = 0$ where $0 \neq b \in R$, then (since R is commutative) $ab(x + y) = b(ax) + a(by) = 0 + 0 = 0$. Since $ab \neq 0$ (R is a domain), this shows that $s + t \in T(M)$. Similarly, if $r \in R$ then

   $$a(rx) = r(ax) = r0 = 0,$$

   so $rx \in T(M)$.

9. If $M = P \oplus Q$ are modules, we must show that $M/P \cong Q$ and $M/Q \cong P$. Define $\pi : M = P \oplus Q \to Q$ by $\pi(p + q) = q$ for all $p \in P$ and $q \in Q$. This is well defined because if $p + q = p_1 + q_1$ then $p - p_1 = q - q_1 \in P \cap Q = 0$, whence $q - q_1$. It is easy to see that π is an onto R-morphism, and $\ker \pi = P$. Hence the isomorphism theorem gives $M/P \cong Q$. A similar argument shows $M/Q \cong P$.

11. (a) Yes. $(m, n) = (n, n) + (m - n, 0)$ shows that $M = K + X$; clearly $K \cap X = 0$.

    (c) Yes. $(m, n) = (3m - 2n, 3m - 2n) + (2(n - m), 2(n - m))$ shows that $M = K + X$. If $(m, n) \in K \cap X$, then $(m, n) = (k, k)$ and $(m, n) = (2l, 3l)$ so $k = 2l$ and $k = 3l$. Hence $k = 0$, proving that $K \cap X = 0$.

13. Let $(k_1 + \cdots + k_s) + m_2 + \cdots + m_r = 0$ in $K_1 \oplus \cdots \oplus K_s \oplus M_2 \oplus \cdots \oplus M_r$. Then $(k_1 + \cdots + k_s) = 0$ and $m_2 = \cdots = m_r = 0$ because

    $$M_1 + M_2 + \cdots + M_r$$

    is direct. Then $k_1 = \cdots = k_s = 0$ because $K_1 + \cdots + K_s$ is direct.

15. As in the Hint, let $1 = e + f$ where $e \in A$ and $f \in B$. If $a \in A$ then

    $$a - ae = af \in A \cap B = 0$$

    because $a - ae \in A$ and $af \in B$ (using the fact that A and B are left ideals). Hence $a = ae$ for all $a \in A$ so, since $e \in A$, we obtain $e^2 = e$ and $A \subseteq Re$. But $Re \subseteq A$ because A is a left ideal and $e \in A$, so $A = Re$. Now observe that $f = 1 - e$ satisfies $f^2 = f$, so $B = Rf = R(1 - e)$ follows in the same way.

16. (a) We have $M = \pi(M) + \ker \pi$ because $m = \pi(m) + (m - \pi(m))$ for each $m \in M$ and $\pi[m - \pi(m)] = \pi(m) - \pi^2(m) = 0$. If $m \in \pi(M) \cap \ker \pi$, let $m = \pi(m_1)$ with $m_1 \in M$. Then $0 = \pi(m) = \pi^2(m_1) = \pi(m_1) = m$, so $\pi(M) \cap \ker \pi = 0$.

17. Suppose $M \xrightarrow{\beta} N \xrightarrow{\alpha} M$ with $\alpha\beta = 1_M$. If $m \in M$ observe that

$$\alpha[m - \beta\alpha(m)] = \alpha(m) - (\alpha\beta)[\alpha(m)] = 0$$

because $\alpha\beta = 1_M$. Hence the fact that $m = \beta\alpha(m) + [m - \beta\alpha(m)]$ shows that $M = \beta(M) + \ker(\alpha)$. But if $x \in \beta(M) \cap \ker(\alpha)$ and we write $x = \beta(m)$, then $0 = \alpha(x) = \alpha[\beta(m)] = m$, whence $x = \beta(m) = \beta(0) = 0$. Hence

$$\beta(M) \cap \ker(\alpha) = 0,$$

and we have proved that $N = \beta(M) \oplus \ker(\alpha)$.

18. We use Corollary 3 of Theorem 3 several times.

    (a) We have $K \cong X$ as both are isomorphic to $\mathbb{Z}_2$. However $\frac{M}{K} \not\cong \frac{M}{X}$ because

    $$\frac{M}{K} = \frac{\mathbb{Z}_2 \oplus \mathbb{Z}_4}{\mathbb{Z}_2 \oplus 0} \cong \mathbb{Z}_4, \text{ while } \frac{M}{X} = \frac{\mathbb{Z}_2 \oplus \mathbb{Z}_4}{0 \oplus \{0,2\}} \cong \mathbb{Z}_2 \oplus \mathbb{Z}_2 \text{ because } \frac{\mathbb{Z}_4}{\{0,2\}} \cong \mathbb{Z}_2.$$

19. Since $\gcd(m,n) = 1$ let $1 = xn + ym$, $x, y \in \mathbb{Z}$. Then if

    $$g \in G, \ g = xn \cdot g + ym \cdot g \in G_m + G_n.$$

    Hence $G = G_m + G_n$. If $g \in G_m \cap G_n$, then $mg = 0 = ng$ so

    $$g = 1 \cdot g = xng + ymg = 0.$$

    Thus $G = G_m \oplus G_n$.

21. Here A is an ideal of a ring R, and $_R W$ is a module.

    (a) To see that $\bar{\alpha}$ is well defined, let $w + AW = w_1 + AW$; we must show that $\alpha(w) + AV = \alpha(w_1) + AV$. We have $w - w_1 \in AW$, say $w - w_1 = \Sigma_i a_i w_i$ where each $a_i \in A$. Because α is R-linear, we obtain

    $$\alpha(w) - \alpha(w_1) = \alpha(w - w_1) = \Sigma_i \alpha(a_i w_i) = \Sigma_i a_i \left(\alpha w_i\right) \in AV.$$

    Hence $\alpha(w) + AV = \alpha(w_1) + AV$, as required.

23. Let $\alpha : M \to N$ where M and N are simple. If $\alpha \neq 0$, then $\alpha(M)$ is a nonzero submodule of N. By hypothesis, $\alpha(M) = N$, that is α is onto. Again, $\alpha \neq 0$ implies that $\ker(\alpha) \neq M$. But then $\ker(\alpha) = 0$, again because M is simple, and we have shown that α is one to one. Hence α is an isomorphism.

24. $(2) \Rightarrow (3)$. By (2), we can identify P with a summand of a finitely generated free module F, say $F = P \oplus Q$, an internal direct sum. Define $\pi : F \to P$ by $\pi(p + q) = p$ for all $p \in P$ and $q \in Q$. Let $\{x_1, \cdots x_n\}$ be a basis of F. Given the diagram we have $\beta\pi : F \to N$, so since α is onto, choose $m_i \in M$ such that $\alpha(m_i) = \beta\pi(x_i)$ for each i. By Theorem 6, there exists an R-homomorphism $\theta : F \to M$ such that $\theta(x_i) = m_i$ for each i. Then $\alpha\theta(x_i) = \alpha(m_i) = \beta\pi(x_i)$ for each i, so $\alpha\theta = \beta\pi$ because the x_i span F. But then, if $p \in P$, we obtain $\alpha\theta(p) = \beta\pi(p) = \beta(p)$ because $\pi(p) = p$. Hence the restriction $\gamma : P \to M$ given by $\gamma(p) = \theta(p)$ for $p \in P$, satisfies all our requirements.

25. As in the Hint:

    (i) $\mathbb{Q}$ is divisible. (If $0 < n \in \mathbb{Z}$ and $q = \frac{a}{b} \in \mathbb{Q}$ then $nx = q$ where $n = \frac{a}{nb}$).

    (ii) If $Q = K \oplus N$ is divisible, then K is divisible. (If $0 < n \in \mathbb{Z}$ and $k \in K$, let $nx = k$, $x \in G$. If $x = y + z$, $y \in K$, $z \in N$, then $nx = ny + nz$ so $nx = ny$ because $K + N$ is direct. Hence $ny = k$, $y \in K$.)

(iii) $\mathbb{Z}$ is not divisible. $(0 < 2 \in \mathbb{Z}$ and $3 \in \mathbb{Z}$ but $2x = 3$ has no solution in $\mathbb{Z}$.)

Now assume that $_{\mathbb{Z}}\mathbb{Q}$ is free and let $\{b_i \mid i \in I\}$ be a basis. Then $\mathbb{Z} \cong \mathbb{Z}b_1$ and $\mathbb{Q} = \mathbb{Z}b_1 \oplus (\Sigma_{i \neq 1}\mathbb{Z}b_i)$. It follows that $\mathbb{Z}$ is divisible, a contradiction.

7.2 MODULES OVER A PRINCIPAL IDEAL DOMAIN

1. (a) $\mathbb{Z}_9, \mathbb{Z}_3 \oplus \mathbb{Z}_3$.

    (c) $\mathbb{Z}_4 \oplus \mathbb{Z}_3, \mathbb{Z}_2 \oplus \mathbb{Z}_2 \oplus \mathbb{Z}_3$.

    (e) $\mathbb{Z}_2 \oplus \mathbb{Z}_3 \oplus \mathbb{Z}_5$.

    (g) $\mathbb{Z}_4 \oplus \mathbb{Z}_{27}, \mathbb{Z}_4 \oplus \mathbb{Z}_9 \oplus \mathbb{Z}_3, \mathbb{Z}_4 \oplus \mathbb{Z}_3 \oplus \mathbb{Z}_3 \oplus \mathbb{Z}_3; \mathbb{Z}_2 \oplus \mathbb{Z}_2 \oplus \mathbb{Z}_{27}$,
    $\mathbb{Z}_2 \oplus \mathbb{Z}_2 \oplus \mathbb{Z}_9 \oplus \mathbb{Z}_3, \mathbb{Z}_2 \oplus \mathbb{Z}_2 \oplus \mathbb{Z}_3 \oplus \mathbb{Z}_3 \oplus \mathbb{Z}_3$.

2. (a) The types are: (4), $(3,1)$, $(2,2)$, $(2,1,1)$ and $(1,1,1,1)$. Hence representative groups are $\mathbb{Z}_{p^4}$, $\mathbb{Z}_{p^3} \oplus \mathbb{Z}_p$, $\mathbb{Z}_{p^2} \oplus \mathbb{Z}_{p^2}$, $\mathbb{Z}_{p^2} \oplus \mathbb{Z}_p \oplus \mathbb{Z}_p$ and $\mathbb{Z}_p \oplus \mathbb{Z}_p \oplus \mathbb{Z}_p \oplus \mathbb{Z}_p$.

3. (a) $\mathbb{Z}_p \oplus \mathbb{Z}_{q^2}$, $\mathbb{Z}_p \oplus \mathbb{Z}_q \oplus \mathbb{Z}_q$.

4. (a) The types are: p-component (2), $(1,1)$; the q-component (3), $(2,1)$, $(1,1,1)$; and the r-component (4), $(3,1)$, $(2,2)$, $(2,1,1)$, $(1,1,1,1)$. Hence $2 \cdot 3 \cdot 5 = 30$ in all.

5. The types smaller than $(3,2,1)$ are:

$(3,2,1)$	$(3,1,1)$	$(2,2,1)$	$(2,1,1)$	$(1,1,1)$
$(3,2)$	$(3,1)$	$(2,2)$	$(2,1)$	$(1,1)$
(3)	(2)	(1)		

7. (a) Using the primary decomposition, we get
$$G = \mathbb{Z}_{12} \oplus \mathbb{Z}_{60} \oplus \mathbb{Z}_{75} = (\mathbb{Z}_4 \oplus \mathbb{Z}_3) \oplus (\mathbb{Z}_4 \oplus \mathbb{Z}_3 \oplus \mathbb{Z}_5) \oplus (\mathbb{Z}_3 \oplus \mathbb{Z}_{25})$$
$$= (\mathbb{Z}_4 \oplus \mathbb{Z}_4) \oplus (\mathbb{Z}_3 \oplus \mathbb{Z}_3 \oplus \mathbb{Z}_3) \oplus (\mathbb{Z}_{25} \oplus \mathbb{Z}_5).$$

    Thus $G(2) \cong \mathbb{Z}_4 \oplus \mathbb{Z}_4$ has type $(2,2)$; $G(3) \cong \mathbb{Z}_3 \oplus \mathbb{Z}_3 \oplus \mathbb{Z}_3$ has type $(1,1,1)$ and $G(5) \cong \mathbb{Z}_{25} \oplus \mathbb{Z}_5$ has type $(2,1)$.

    Thus $G(2) \cong \mathbb{Z}_4 \oplus \mathbb{Z}_2 \oplus \mathbb{Z}_2$ has type $(2,1,1)$; while $G(3) \cong \mathbb{Z}_9 \oplus \mathbb{Z}_3$ and $G(7) \cong \mathbb{Z}_{49} \oplus \mathbb{Z}_7$ each have type $(2,1)$.

8. (a) Let $|G| = p^n$ be of type $(n_1, n_2, \ldots, n_r)$. Since G has an element of order p^{n-1}, we must have $n_i \geq n - 1$ for some i. The possibilities are (n) and $(n-1,1)$.

9. (a) The types are $(2,2,2)$, $(2,2,1,1)$, $(2,1,1,1,1)$, $(1,1,1,1,1,1)$.

11. We have $\operatorname{ann}(x) = Rd$. Since $b(ax) = dx = 0$ we have $b \in \operatorname{ann}(ax)$. Hence $Rb \subseteq \operatorname{ann}(ax)$; we must prove equality. If $c \in \operatorname{ann}(ax)$ then $cax = 0$, so $ca \in \operatorname{ann}(x) = Rd$, say $ca = rd$, $r \in R$. Thus $ca = rba$ so, as $a \neq 0$ and R is a domain, $c = rb \in Rb$. Hence $\operatorname{ann}(ax) \subseteq Rb$, as required.

12. (a) $T(K) = \{k \in K \mid o(k) \neq 0\} = K \cap \{m \in M \mid o(m) \neq 0\} = K \cap T(M)$.

13. If M is torsion, it is clear that K is torsion; if $m + K \in M/K$ and $dm = 0$, $d \neq 0$, then $d(m + K) = 0$. Thus M/K is torsion. Conversely: Given $m \in M$,

we have $m + K \in M/K$, so let $b(m + K) = 0$, $b \neq 0$. Then $bm \in K$ so (since K is torsion) $a(bm) = 0$, $a \neq 0$. Thus $(ab)m = 0$ and $ab \neq 0$.

15. We may assume that the sum $M = M_1 \oplus \cdots \oplus M_n$ is internal. Clearly $T(M_i) \subseteq T(M)$ for all i, so $T(M_1) + \cdots + T(M_n) \subseteq T(M)$. This sum is direct because $M = M_1 \oplus \cdots \oplus M_n$. If $x \in T(M)$ write $x = x_1 + \cdots + x_n$, $x_i \in M_i$. If $dx = 0$, $d \neq 0$, then $0 = dx_1 + \cdots + dx_n$, so $dx_i = 0$ for each i because $M = M_1 \oplus \cdots \oplus M_n$ is direct. But then $x_i \in T(M_i)$ for all i, so $x \in T(M_1) \oplus \cdots \oplus T(M_n)$.

16. (a) Define $\sigma : K \to M/T(M)$ by $\sigma(k) = k + T(M)$. This is a group homomorphism and $\ker \sigma = \{k \in K \mid k \in T(M)\} = K \cap T(M) = T(K)$. Use the isomorphism theorem.

17. We are given abelian groups $M = H \oplus W$ where H is torsion and W is torsion-free. Then $H \subseteq T(M)$ because H is torsion. If $t \in T(M)$, write $t = h + f$, $h \in H$, $f \in W$. Then $t - h = f \in T(M) \cap W$, and $T(M) \cap W = 0$ because W is torsion free. So $H = T(M)$, whence $M = T(M) \oplus F$. Thus $W \cong M/T(M)$.

19. The torsion elements in $\mathbb{C}^\circ$ are the complex numbers of finite order, that is the roots of unity. Hence $T(\mathbb{C}^\circ) = \{z \in \mathbb{C} \mid z^n = 1 \text{ for some } n \geq 1\}$. Turning to $\mathbb{Q}^*$, the torsion elements are 1 and -1 so $T(\mathbb{Q}^*) = \{1, -1\}$.

20. We have $L_d(N) = \{x \in N \mid dx = 0\}$.
    (a) If $x, y \in L_d(N)$ then $d(x - y) = dx - dy = 0$. Clearly
    $$d(rx) = r(dx) = r0 = 0.$$
    (c) Let $m = \Sigma_i x_i$, $x_i \in M_i$. If $m \in L_d(M)$ then $0 = dm = \Sigma_i dx_i$ so $dx_i = 0$ for each i. Hence $L_d(M) \subseteq \Sigma_i L_d(M_i)$, and the other inclusion is obvious. Finally $\Sigma_i L_d(M_i)$ is a direct sum because $L_d(M_i) \subseteq M_i$ for each i.

21. We have $dN = \{dx \mid x \in N\}$.
    (a) If $x, y \in dN$, say $x = dw$ and $y = dz$, then $x - y = d(w - z)$, and $d(rx) = (r)dx$.

22. (a) Here $M = Rx$. We have $R(p^{m-1}x) \subseteq L_p(M)$ because $p(p^{m-1}x) = 0$. Conversely, if $rx \in L_p(M)$ then $0 = prx$ so $pr \in \text{ann}(x) = Rp^m$, say $pr = sp^m$. Since $m \geq 1$ and R is a domain, this gives $r = sp^{m-1}$, so
    $$rx = sp^{m-1}x \in R(p^{m-1}x).$$
    Hence $L_p(M) = R(p^{m-1}x)$. Finally, $p(Rx) = Rpx$ is a routine verification, and $px = 0$ if $m = 1$ because $\text{ann } x = p^m$.

23. Each nonzero element of G has order p. Hence G is a p-group, say of type $(n_1, n_2, \ldots, n_m)$. But this means that G has an element of order p^{n_i} for each i, so $n_i = 1$ for all i. Thus $G = G_1 \oplus \cdots \oplus G_m$ where $|G_i| = p$ for each i, that is $G \cong \mathbb{Z}_p \oplus \mathbb{Z}_p \oplus \cdots \oplus \mathbb{Z}_p$, m copies. The type is $(1, 1, \ldots, 1)$.

25. Let $|G| = p_1^{m_1} p_2^{m_2} \cdots p_r^{m_r}$. If n divides $|G|$, let $n = p_1^{n_1} p_2^{n_2} \cdots p_r^{n_r}$. We have $G = G(p_1) \oplus G(p_2) \oplus \cdots \oplus G(p_r)$ so it suffices to find a subgroup $H_i \subseteq G(p_i)$ with $|H_i| = p_i^{n_i}$ (then take $H = H_1 \oplus \cdots \oplus H_r$).
    So assume $|G| = p^n$, say of type $(n_1, n_2, \ldots, n_r)$. Thus
    $$G = G_1 \oplus \cdots \oplus G_r$$

where G_i is cyclic of order p^{n_i}. We must find a subgroup of order p^m for all $m \leq n$. It suffices to show that $m = m_1 + m_2 + \cdots + m_r$, where $m_i \leq n_i$ for each i (then G_i has a subgroup H_i of order p^{m_i}, so take

$$H = H_1 \oplus H_2 \oplus \cdots \oplus H_r).$$

If $m \leq n_1$ this is clear. Otherwise $m = n_1 + m_1$. If $m_1 \leq n_2$ we are done. Otherwise $m = n_1 + n_2 + n_3$. Continue. The process ends because $m \leq n$.

27. We have $G = G_1 \oplus \cdots \oplus G_r$ where G_i is cyclic and $|G_i| = p^{n_i}$ for each i.

    (a) The subgroup $L_p(G) = \{g \mid pg = 0\}$ consists of 0 and the elements of order p. Then $L_p(G) = L_p(G_1) \oplus \cdots \oplus L_p(G_r)$ by Exercise 20, and $|L_p(G_i)| = p$ for each i by the preceding exercise. Hence $|L_p(G)| = p^r$ and every nonzero element has order p.

29. Since G is a direct sum of cyclic subgroups, it suffices to prove it when G is cyclic (then the cyclic summands of G can be cancelled one by one, starting with $G \oplus H \cong G \oplus K$). So let $|G| = p^n$, G cyclic. Let H and K have types $(n_1, \ldots, n_r)$ and $(m_1, \ldots, m_s)$ respectively. Then $G \oplus H$ has type $(n_1, \ldots, n_k, n, n_{k+1}, \ldots, n_r)$ where $n_k \geq n \geq n_{k+1}$ (possibly n is at either end) and similarly, $G \oplus K$ has type $(m_1, \ldots, m_t, n, m_{t+1}, \ldots, m_s)$ where $m_t \geq n \geq m_{t+1}$. Since $G \oplus H \cong G \oplus K$, these types are identical. Thus $r + 1 = s + 1$ so $r = s$, and so

$$(n_1, \ldots, n_k, n, n_{k+1}, \ldots, n_r) = (m_1, \ldots, m_t, n, m_{t+1}, \ldots, m_r).$$

We may assume $k \leq t$ without loss of generality. Then $n_i = m_i$ for $1 \leq i \leq k$ and $t + 1 \leq i \leq r$. If $k = t$ we are done. If $k < t$ then (since these are types)

$$n \geq n_{k+1} \geq \cdots \geq n_t = n = m_{k+1} \geq \cdots \geq m_t \geq n.$$

Hence these are all equal so $n_i = m_i = n$ for $k + 1 \leq i \leq t$.

Chapter 8

p-Groups and the Sylow Theorems

8.1 PRODUCTS AND FACTORS

1. (a) $XY = \{\tau, \tau\sigma\}\{\tau, \tau\sigma^2\} = \{\tau^2, \tau^2\sigma^2, \tau\sigma\tau, \tau\sigma\tau\sigma^2\}$
$= \{\varepsilon, \sigma^2, \tau^2\sigma^2, \tau^2\sigma^4\} = \{\varepsilon, \sigma^2, \sigma\}.$
$YX = \{\tau, \tau\sigma^2\}\{\tau, \tau\sigma\} = \{\tau^2, \tau^2\sigma, \tau\sigma^2\tau, \tau\sigma^2\tau\sigma\}$
$= \{\varepsilon, \sigma, \tau^2\sigma, \tau^2\sigma^2\} = \{\varepsilon, \sigma, \sigma^2\}.$

3. If $G' \subseteq H$, then $H/G' \lhd G/G'$ because G/G' is abelian. Hence $H \lhd G$ by the correspondence theorem.

4. (a) $G = D_6 = \{1, a, \dots, a^5, b, ba, \dots, ba^5\}$, $o(a) = 6$, $o(b) = 2$, $aba = b$. We have $K = Z(D_6) = \{1, a^3\}$. Write $\bar{a} = Ka$, $\bar{b} = Kb$. Then $G/K = \langle \bar{a}, \bar{b}\rangle$, $o(\bar{a}) = 3$, $o(\bar{b}) = 2$, $\bar{a} \cdot \bar{b} \cdot \bar{a} = \bar{b}$. Hence $G/K \cong D_3$ and the only subgroups of G/K are $\{1\}, \langle\bar{a}\rangle, \langle\bar{b}\rangle, \langle\bar{b}\cdot\bar{a}\rangle, \langle\bar{b}\cdot\bar{a}^2\rangle$ and G/K. So the subgroups of G containing K are G, K and

$$H = \{x \in G \mid Kx \in \langle\bar{a}\rangle\} = \langle a\rangle$$
$$H_1 = \{x \in G \mid Kx \in \langle\bar{b}\rangle\} = \{1, a^3, b, ba^3\}$$
$$H_2 = \{x \in G \mid Kx \in \langle\bar{b}\cdot\bar{a}\rangle\} = \{1, a^3, ba, ba^4\}$$
$$H_3 = \{x \in Q \mid Kx \in \langle\bar{b}\cdot\bar{a}^2\rangle\} = \{1, a^3, ba^2, ba^5\}.$$

Note that $H_1 \cong H_2 \cong H_3 \cong K_4$ — the Klein group (in contrast with (b)).

(c) $G = A_4$ and $K = \{\varepsilon, (1\ 2)(3\ 4), (1\ 3)(2\ 4), (1\ 4)(2\ 3)\}$. Then $|G| = 12$, so $|G/K| = 3$ is cyclic of order 3. Hence the only subgroups of A_4 containing K are K and A_4. In particular, K is maximal normal in A_4.

Student Solution Manual to Accompany Introduction to Abstract Algebra, Fourth Edition.
W. Keith Nicholson.
© 2012 John Wiley & Sons, Inc. Published 2012 by John Wiley & Sons, Inc.

5. (a) Every subgroup of $\mathbb{Z}$ has the form $n\mathbb{Z}$, $n \in \mathbb{Z}$, and $\mathbb{Z}/n\mathbb{Z} = \mathbb{Z}_n$. This is simple (abelian) if and only if n is a prime. Thus $\{p\mathbb{Z} \mid p \text{ a prime}\}$ are the maximal normal subgroups of $\mathbb{Z}$.

(c) $G = D_{10} = \{1, a, \ldots, a^9, b, ba, \ldots, ba^9\}$, $o(a) = 10$, $o(b) = 2$, $aba = b$. If H is maximal normal in G, then G/H is abelian (it's order is ≤ 5) and so has order c, a prime dividing 20. Thus $|G/H| = 2, 5$, so $|H| = 10, 4$. The subgroups of order 10 are

$$H_1 = \langle a \rangle \qquad \text{and}$$
$$H_2 = \langle a^2, b \rangle = \langle a^6, b \rangle = \{1, a^2, a^4, a^6, a^8, b, ba^2, ba^4, ba^6, ba^8\}.$$

There is no element of order 4, so the subgroups of order 4 are

$$K_0 = \langle a^5, b \rangle = \{1, a^5, b, ba^5\}$$
$$K_1 = \langle a^5, ba \rangle = \{1, a^5, ba, ba^6\}$$
$$K_2 = \langle a^5, ba^2 \rangle = \{1, a^5, ba^2 ba^7\}$$
$$K_3 = \langle a^5, ba^3 \rangle = \{1, a^5, ba^3, ba^8\}$$
$$K_4 = \langle a^5, ba^4 \rangle = \{1, a^5, ba^4, ba^9\}.$$

7. If $K \lhd G$, G/K cyclic, $|K| = k$, $|G| = n$, let $k \mid m$ and $m \mid n$. Then $|G/K| = n/k$ and $\frac{m}{k}$ divides $\frac{n}{k}$. Hence G/K cyclic implies it has a unique subgroup X with $|X| = \frac{m}{k}$. Write $X = H/K$. Then $|H| / |K| = m/k$, so $|H| = m$. If $K \subseteq H_1 \subseteq G$ and $|H_1| = m$, then $|H'/K| = \frac{m}{k}$, so $H'/K = H/K$ by the uniqueness. Thus $H_1 = H$.

9. We must prove that $(H_1/K) \cap (H_2/K) = (H_1 \cap H_2)/K$. We have

$$(H_1 \cap H_2)/K \subseteq (H_1/K) \cap (H_2/K)$$

by Lemma 2. If $Kg \in (H_1/K) \cap (H_2/K)$, let $Kg = Kh_i$ for $h_i \in H_i$, $i = 1, 2$. Then $g \in Kh_i \subseteq H_i$ for each i, so $g \in H_1 \cap H_2$. This shows that $Kg \in (H_1 \cap H_2)/K$, and so proves that

$$(H_1/K) \cap (H_2/K) \subseteq (H_1 \cap H_2)/K.$$

10. (a) If $HH_1 = H_1H$, then $Kh \cdot Kh_1 = Khh_1 = Kh_1'h' = Kh_1' \cdot Kh' \in \frac{H_1}{K} \cdot \frac{H}{K}$. Hence $\frac{H}{K} \cdot \frac{H_1}{K} \subseteq \frac{H_1}{K} \cdot \frac{H}{K}$, and the other inclusion is similar. Conversely, if $\frac{H}{K} \cdot \frac{H_1}{K} = \frac{H_1}{K} \cdot \frac{H}{K}$ then $Khh_1 = Kh \cdot Kh_1 = Kh_1' \cdot Kh' = Kh_1'h'$; so $hh_1 = kh_1'h' \in H_1H$ because $KH_1 \subseteq H_1$. Thus $HH_1 \subseteq H_1H$ and the other inclusion is similar.

11. $\langle X \cup Y \rangle$ is a subgroup containing both X and Y, so $\langle X \rangle \subseteq \langle X \cup Y \rangle$ and $\langle Y \rangle \subseteq \langle X \cup Y \rangle$. Then $\langle X \rangle \langle Y \rangle \subseteq \langle X \cup Y \rangle$ because $\langle X \cup Y \rangle$ is closed. If $\langle X \rangle \langle Y \rangle = \langle X \cup Y \rangle$, then $\langle X \rangle \langle Y \rangle$ is a subgroup, so $\langle X \rangle \langle Y \rangle = \langle Y \rangle \langle X \rangle$ Lemma 2 §2.8. Conversely, if $\langle X \rangle \langle Y \rangle = \langle Y \rangle \langle X \rangle$ then $\langle X \rangle \langle Y \rangle$ is a subgroup (again by Lemma 2 §2.8). Since $\langle X \rangle \langle Y \rangle$ contains both X and Y, it contains $X \cup Y$, and so $\langle X \cup Y \rangle \subseteq \langle X \rangle \langle Y \rangle$ by Theorem 8 §2.4.

12. (a) $H^2 \subseteq H$ because H is closed; $H \subseteq H^2$ because $1 \in H$.

13. Write $m = |H \cap K|$. Since $H \cap K$ is a subgroup of H we have $m \in \{1, p, q, pq\}$ by Lagrange's theorem. Similarly $m \in \{1, q, r, qr\}$, so $m \in \{1, q\}$. But if $m = 1$ then $|HK| = |H| |K| = pq^2 r > |G|$, a contradiction. So $m = q$ as required.

15. Note that KA and KB are subgroups by Theorem 5 §2.8, and $KA = AK$, $KB = BK$. If $kb \in KB$, we have $Ab = bA$ and $Kb = bK$. Hence

$$KAkb = AKkb = AKb = AbK = bAK = bKA = KbA = kKbA = kbKA.$$

Thus $KA \lhd KB$.

17. Given $K \lhd G$, $|K| = p^n$, assume H is any subgroup of G with $|H| = p^n$. Then HK is a subgroup and $\frac{HK}{K} \cong \frac{H}{H\cap K}$ by the second isomorphism theorem. Since $|H| = p^n$, we have $\left|\frac{H}{H\cap K}\right| = \frac{|H|}{|H\cap K|} = p^k$ for some k such that $0 \leq k \leq n$. Hence $\left|\frac{HK}{K}\right| = \left|\frac{H}{H\cap K}\right| = p^k$. But $\frac{HK}{K}$ is a subgroup of $\frac{G}{K}$, and $\left|\frac{G}{K}\right| = \frac{|G|}{|K|} = m$. It follows that p^k divides m. Since $p \nmid m$ by hypothesis, we have $k = 0$. Thus $H = H \cap K$ and $HK = K$; so certainly $K \subseteq H$. Hence $H = K$ because $|H| = |K|$.

18. (a) We have $M \subset KM$ because $M = KM$ implies $K \subseteq M$. Since KM is normal, we have $M = KM$ because M is maximal normal.

    (c) By (a) and Theorem 3, $\frac{G}{K\cap M} = \frac{KM}{K\cap M} \cong \frac{K}{K\cap M} \times \frac{M}{K\cap M}$, so (c) follows from (b).

19. (a) The argument in Example 3 goes through with "cyclic" replaced by "abelian". All that is needed is that subgroups and factor groups of abelian groups are again abelian. However, we give a slick argument using (b) below. Let G' be metabelian, that is G' is abelian by (b). If H is a subgroup of G, then $H' \subseteq G'$ (since commutators in H are commutators in G), so H' is abelian. Thus H is metabelian. Now suppose that $N \lhd G$. Then in $\frac{G}{N}$, each commutator $[Na, Nb] = N[a, b] \in NG'$. Hence $\left(\frac{G}{N}\right)' \subseteq \frac{NG'}{N} \cong \frac{G'}{N\cap G'}$. Since $\frac{G'}{N\cap G'}$ is abelian (being a factor group of the abelian group G'), it follows that $\left(\frac{G}{N}\right)'$ is abelian, that is $\frac{G}{N}$ is metabelian.

21. We are given subgroups H and K where $|H| = pq$, $|K| = q^2$, $p \neq q$ primes. Since $H \cap K \subseteq K$ and $|K| = q^2$, we have $|H \cap K| = 1$, q or q^2 by Lagrange's theorem. Similarly $H \cap K \subseteq H$ and $|H| = pq$ shows that $|H \cap K| = 1, p, q$ or pq. Hence $|H \cap K| = 1$ or q, and it remains to show $|H \cap K| \neq 1$. But $|H \cap K| = 1$ gives (by Theorem 4) $|HK| = |H||K| = pq^3$. Since $HK \subseteq G$, this contradicts $|G| < pq^3$.

22. (a) Let $a \neq b$ be elements of order 2. Since G is abelian, $H = \{1, a, b, ab\}$ is closed, and so is a subgroup. Then $|H| = 4$ divides $|G|$ by Lagrange's theorem.

23. The unity of M is $\{1\} = 1$. So if X is a unit in M, let $XY = 1$. This means that $xy = 1$ for all $x \in X$, $y \in Y$. So, given $y \in Y$, $x = y^{-1}$ for all $x \in X$. Thus X is a singleton, say $X = \{g\} = g \in G$. Conversely, each $g \in G$ is a unit in M because $gg^{-1} = 1$.

24. (a) We must show $\tilde{G}A = A\tilde{G}$. If $\sigma \in A = \text{aut } G$ and $\tau_a \in \tilde{G}$ ($\tau_a(g) = ag$ for all $g \in G$), then $\sigma\tau_a(g) = \sigma(ag) = \sigma(a) \cdot \sigma(g) = \tau_{\sigma(a)}\sigma(g)$. Thus $\sigma\tau_a = \tau_{\sigma(a)}\sigma$, so $A\tilde{G} \subseteq \tilde{G}A$. If we take $b = \sigma(a)$, this is $\sigma\tau_{\sigma^{-1}(b)} = \tau_b\sigma$, so $\tilde{G}A \subseteq A\tilde{G}$.

(c) Let $\tau_a \in \tilde{G}$ and $\lambda \in \tilde{G}A$, say $\lambda = \tau_b \sigma$. Then (using $\tau_a \sigma = \sigma \tau_{\sigma(a)}$ in (a)):

$$\lambda^{-1} u_a \lambda = \sigma^{-1} \tau_b^{-1} \tau_a \tau_b \sigma = \sigma^{-1} \tau_{b^{-1}ab} \sigma = \sigma^{-1}(\sigma \tau_{\sigma^{-1}(b^{-1}ab)}) = \tau_{\sigma^{-1}(b^{-1}ab)} \in G.$$

Thus $\tilde{G} \triangleleft \tilde{G}A$.

8.2   CAUCHY'S THEOREM

1. (a) $D_4 = \{1, a, a^2, a^3, b, ba, ba^2, ba^3\}$ where $o(a) = 4$, $o(b) = 2$ and $aba = b$. The classes are $\{1\}$, $\{a, a^3\}$, $\{a^2\}$, $\{b, ba^2\}$, $\{ba, ba^3\}$. Since normal subgroups are of orders $1, 2, 4, 8$, they are $\{1\}$, $\{1, a^2\}$, $\{1, a, a^2, a^3\}$, $\{1, b, a^2, ba^2\}$, $\{1, a^2, ba, ba^3\}$ and D_4.

3. If $a^m = g^{-1}ag$ with $g \in G$, then
$$g^{-2}ag^2 = g^{-1}a^m g = (g^{-1}ag)^m = (a^m)^m = a^{m^2}.$$

By induction we get $g^{-k}ag^k = a^{m^k}$ for all $k \geq 0$. Since G is finite, let $g^k = 1$, $k \geq 1$. Then $a = a^{m^k}$ whence $1 - m^k = qn$. This gives $1 = m^k + qn$, so $\gcd(m, n) = 1$.

5. Let $H = \text{class } a_1 \cup \ldots \cup \text{class } a_n$. Given $g \in G$ and $h \in H$, let $h \in \text{class } a_i$. Then $g^{-1}hg \in \text{class } a_i \subseteq H$, so $g^{-1}hg \in H$. Then $g^{-1}Hg \subseteq H$, as required.

7. Let $K = g^{-1}Hg$, so $H = gKg^{-1}$. We claim $N(K) = g^{-1}N(H)g$. Let $a \in N(K)$ so $a^{-1}Ka = K$. To show $a \in g^{-1}N(H)g$ it suffices to show $gag^{-1} \in N(H)$. But
$$(gag^{-1})^{-1}H(gag^{-1}) = ga^{-1}g^{-1}Hgag^{-1} = ga^{-1}Kag^{-1} = gKg^{-1} = H,$$

as required. Hence $N(K) \subseteq g^{-1}N(H)g$. Similarly $N(H) \subseteq gN(K)g^{-1}$ so $g^{-1}N(H)g \subseteq N(K)$.

9. If $|\text{class } a| = 2$ then $|G : N(a)| = 2$ by Theorem 2. Hence $N(a)$ is normal in G. Since $N(a) \neq G$ we are done if $N(a) \neq \{1\}$. But $N(a) = \{1\}$ implies $|G| = |G : N(a)| = 2$ so G is abelian and every conjugacy class is a singleton, a contradiction.

11. We have $H \subseteq N(H) \subseteq G$ and H has finite index m in G. So
$$m = |G : H| = |G : N(H)||N(H) : H|$$

by Exercise 31 §2.6. Thus $|G : N(H)|$ is finite, and H has $|G : N(H)|$ conjugates by Theorem 2.

13. Write $X = \{g^{-1}Hg \mid g \in G\}$ and $Y = \{N(H)g \mid g \in G\}$. Define $\varphi : X \to Y$ by $\varphi(g^{-1}Hg) = N(H)g$. Then:
$$g^{-1}Hg = a^{-1}Ha \iff (ag^{-1})^{-1}H(ag^{-1}) = H$$
$$\iff ag^{-1} \in N(H) \iff N(H)g = N(H)a.$$

Thus φ is well defined and one-to-one; it is clearly onto.

15. Let $\alpha = \gamma_1 \gamma_2 \ldots \gamma_r$ in S_n where the γ_i are disjoint cycles. Given $\sigma \in S_n$,
$$\sigma^{-1}\alpha\sigma = (\sigma^{-1}\gamma_1\sigma)(\sigma^{-1}\gamma_2\sigma)\cdots(\sigma^{-1}\gamma_n\sigma)$$

and this is the factorization of $\sigma^{-1}\alpha\sigma$ into disjoint cycles by Lemma 3 §2.8. Since γ_i and $\sigma^{-1}\gamma_i\sigma$ have the same length, $\sigma^{-1}\alpha\sigma$ has the same cycle structure as σ. Conversely let α and β have the same cycle structure, say

$$\alpha = \gamma_1 \gamma_2 \cdots \gamma_r \quad \text{and} \quad \beta = \delta_1 \delta_2 \cdots \delta_r$$

where the γ_i (the δ_i) are disjoint cycles, and γ_i and δ_i have equal length for each i. If $\gamma_i = (k_{i1}, k_{i2}, \ldots, k_{is})$ and $\delta_i = (l_{i1} l_{i2} \ldots l_{is})$ for all i define $\sigma \in S_n$ by

$$\sigma k_{ij} = l_{ij} \quad \text{for all } i,j \qquad \text{and} \qquad \sigma k = k \quad \text{if} \quad k \neq k_{ij} \quad \text{for all } i,j.$$

Then σ is a permutation and $\sigma^{-1} \gamma_i \sigma = \delta_i$ by Lemma 3 §2.8. Hence $\sigma^{-1} \alpha \sigma = \beta$, that is α and β are conjugate.

17. (a) S_4 consists of ε, three permutations of type $(a\,b)(c\,d)$, eight 3-cycles, six 2-cycles and six 4-cycles. These are the conjugacy classes by Exercise 15. So, by Theorem 1, each normal subgroup $H \lhd S_4$ is a union of these classes. Since $|H|$ divides $24 = |S_4|$, and since $\varepsilon \in H$, the only possibilities are $H = \varepsilon$, $H = K$, $H = A_4$ and $H = S_4$.

19. It suffices to show that G is abelian (then there are $|G|$ conjugacy classes). So assume G is not abelian. Then there must be a nonsingleton conjugacy class, say class a. Thus there is only one singleton class, $\{1\}$. This means $|G| = 1 + |\text{class } a| = 1 + |G : N(a)|$. Write $m = |G : N(a)|$ so that $|G| = md$, $d \geq 2$. But then $md = 1 + m$ gives $1 = m(d-1)$ so $m = 1$, that is $|\text{class } a| = 1$, contrary to assumption.

21. If G/K and K are p-groups, let $g \in G$. Then $(gK)^{p^n} = K$ in G/K for some n, so $g^{p^n} \in K$. Since K is itself a p-group, $(g^{p^n})^{p^m} = 1$ for some m; that is $g^{p^{n+m}} = 1$. Thus $o(g)$ is a power of p, so G is a p-group. Conversely, subgroups of p-groups are clearly p-groups, as are images because if $a^{p^n} = 1$ then $(Ka)^{p^n} = K$.

23. We have $|G| = p^n$ for some $n \geq 1$ so $g^{p^n} = 1$ for all $g \in G$. It follows that $[g_i]^{p^n} = [g_i^{p^n}] = 1$ in G^w for all $[g_i]$ in G^w, so G^w is a p-group. If $1 \neq g$ and $1 \neq h$ then $[g, 1, 1, \ldots), [1, g, 1, 1, \ldots), \ldots$ are all distinct in G^w. So $|G^w| = \infty$.

25. By Theorem 1, H is a union of G-conjugacy classes; suppose there are m of these which are singletons. The remaining classes in H have order a multiple of p by Theorem 3 (since G is a p-group) and p divides $|H|$ (since $H \neq \{1\}$). Hence $p|m$. Since $m > 0$ (because $\{1\} \subseteq H$) this means $m > 1$ and there exists $a \neq 1$, such that class $a = \{a\} \subseteq H$. But class $a = \{a\}$ means $a \in Z(G)$, so $a \in H \cap Z(G)$.

26. (a) Let $|G| = p^3$, G nonabelian. Write $Z = Z(G)$. Then $Z \neq \{1\}$ by Theorem 6 and $Z \neq G$ because G is nonabelian so $|Z| = p$ or p^2. If $|Z| = p^2$ then $|G/Z| = p$ so G/Z is cyclic and G is abelian by Theorem 2 §2.9, contrary to assumption. So $|Z| = p$. Then $|G/Z| = p^2$ so G/Z is abelian by Theorem 8. Hence $G' \subseteq Z$ so $G' = \{1\}$ or $G' = Z$. But $G' = \{1\}$ implies G is abelian. So $G' = Z$. Finally, if $K \lhd G$, $|K| = p$, then G/K is abelian so $G' \subseteq K$. Thus $|K| = p = |Z| = |G'|$ implies $K = G'$.

27. We proceed by induction on n where $|G| = p^n$. If $n = 1$ it is clear. In general $|H| = p^m$. If $H = \{1\}$ it is clear. If $H \neq \{1\}$ let $K = H \cap Z(G)$. Then $K \lhd G$ and $K \neq \{1\}$ by Exercise 25. Let $|K| = p^b$. We have $H/K \lhd G/K$ so, by induction, let $\frac{G}{K} = \frac{G_0}{K} \supset \frac{G_1}{K} \supset \cdots \supset \frac{G_t}{K} = \{K\}$ where $\frac{G_i}{K} \lhd \frac{G}{K}$ and $\left| \frac{G_i}{K} \Big/ \frac{G_{i+1}}{K} \right| = p$ for all i.

Thus $G_i \lhd G$ and $|G_i/G_{i+1}| = p$ for all i. Similarly

$$K = G_t \supset G_{t+1} \supset \cdots \supset G_b = \{1\}$$

where $G_i \lhd G$ for each i and $|G_i/G_{i+1}| = p$. These G_i do it.

29. Since $C \lhd G$, let $Z[G/C] = K/C$. Since $|G/C| > 1$, Theorem 6 shows $C \subset K$. But $K \lhd G$ so $K \nsubseteq H$ by Exercise 26 §2.8. If $k \in K$ then kC is in the center of G/C, so $h^{-1}k^{-1}hk \in C \lhd H$. Hence $k^{-1}Hk \subseteq H$, and similarly $kHk^{-1} \subseteq H$. Thus $k \in N(H)$ and we have shown $K \subseteq N(H)$.

31. The associativity is verified as follows:

$$[(x,y,z) \cdot (x_1,y_1,z_1)] \cdot (x_2,y_2,z_2)$$
$$= (x + x_1, y + y_1, z + z_1 - yx_1) \cdot (x_2,y_2,z_2)$$
$$= (x + x_1 + x_2, y + y_1 + y_2, (z + z_1 - yx_1) + z_2 - (y + y_1)x_2)$$
$$(x,y,z) \cdot [(x_1,y_1,z_1) \cdot (x_2,y_2,z_2)]$$
$$= (x,y,z) \cdot (x_1 + x_2, y_1 + y_2, z_1 + z_2 - y_1x_2)$$
$$= (x + x_1 + x_2, y + y_1 + y_2, z + (z_1 + z_2 - y_1x_2) - y(x_1 + x_2)).$$

These are equal. The unity is $(0,0,0)$ and $(x,y,z)^{-1} = (-x,-y,-z-xy)$ for all (x,y,z). The group is nonabelian because $(1,1,0) \cdot (1,0,0) = (2,1,-1)$ while $(1,0,0) \cdot (1,1,0) = (2,1,0)$. Finally $(x,y,z)^k = (kx, ky, kz - \binom{k}{2}xy)$ for all $k \geq 2$ by induction on k. If $k = p$, $(x,y,z)^p = (0,0,-\binom{p}{2}xy)$. Since $\binom{p}{2} = \frac{p(p-1)}{2}$ and p is odd, this is $(0,0,0)$ so $o((x,y,z)) = p$.

33. (a) Given $a \in G$, $|\text{class } a| = |G : N(a)|$ by Theorem 2. But

$$N(a) = \{g \in G \mid g^{-1}ag = a\} \supseteq Z(G)$$

so the fact that $|G : Z(G)|$ is finite implies $|G : N(a)|$ is finite.

(c) If $G = \langle X \rangle$ and $a \in G$, write $a = x_1^{k_1} x_2^{k_2} \cdots x_r^{k_r}$, $x_i \in X$, $k_i \in \mathbb{Z}$. We claim that $\bigcap_{i=1} N(x_i) \subseteq N(a)$. For if $g \in N(x_i)$ for all i then $gx_i = x_i g$ for all i, so $ga = ag$; that is $g \in N(a)$. But each $N(x_i)$ has finite index by Theorem 2, so $\bigcap_{i=1} N(x_i)$ has finite index by Poincaré's Theorem [Exercise 33 §2.6], so $N(a)$ has finite index. Thus $|\text{class } a|$ is finite.

(e) If $a,b \in G^*$ we have $N(a) \cap N(b) \subseteq N(ab)$. Since $N(a) \cap N(b)$ has finite index (Poincaré's theorem), so does $N(ab)$ by the following Lemma. Thus $ab \in G^*$.

(i) **Lemma.** If $K \subseteq H \subseteq G$ are groups and $|G : K|$ is finite, then $|G : H|$ is finite.
   Proof. We have $H = \cup\{hK \mid h \in H\}$, a finite union because $\{gK \mid g \in G\}$ is finite. If $H = h_1K \cup \cdots \cup h_nK$, a disjoint union, then

$$gH = gh_1K \cup \ldots \cup gh_nK.$$

Hence there are at most a finite number of cosets gH.

Next, $N(a^{-1}) = \{g \mid ga^{-1} = a^{-1}g\} = \{g \mid ga = ag\} = N(a)$, so $a^{-1} \in G^*$ too. Clearly $1 \in G^*$, so G^* is a subgroup. Finally, G^* is itself an FC-group.

Indeed, if $a \in G^*$

$$\text{class }_{G^*}(a) = \{g^{-1}ag \mid g \in G^*\} \subseteq \text{ class }_G(a).$$

Now let $\sigma : G \to G$ be an automorphism. If $a \in G^*$ then

$$\text{class }\sigma(a) = \{g^{-1}\sigma(a)g \mid g \in G\} = \{[\sigma(x)]^{-1} \cdot \sigma(a) \cdot \sigma(x) \mid x \in G\}$$
$$= \sigma(\text{class } a).$$

Since class a is finite, this shows class $\sigma(a)$ is finite, that is $\sigma(a) \in G^*$.

8.3 GROUP ACTIONS

1. (a) If $|G| = 20$, there is $a \in G$ with $o(a) = 5$ by Cauchy's theorem. Thus $|G : \langle a \rangle| = 4$ so there is a homomorphism $\theta : G \to S_4$ with $\ker \theta \subseteq \langle a \rangle$. Hence $|\langle a \rangle / \ker \ \theta|$ divides 24 so $\ker \ \theta \neq \{1\}$. Thus $\ker \ \theta = \langle a \rangle$ because $o(a)$ is prime, so $\langle a \rangle \triangleleft G$.

3. Assume $p \leq q$. By Cauchy's theorem, let $a \in G$, $o(a) = q$. Then $H = \langle a \rangle$ has index p so $H \triangleleft G$ by the Corollary to Theorem 1.

5. If $|H| = p$ then $|G : H| = m$ so let $\theta : G \to S_m$ be a homomorphism with $\ker \ \theta \subseteq H$. Now $\ker \ \theta = \{1\}$ is impossible since $|G| = pm$ does not divide $m!$. Since $|H| = p$, the only other possibility is $\ker \ \theta = H$, so $H \triangleleft G$ as asserted.

6. (a) If $|A_n : H| = p$ let $\theta : A_n \to S_p$ be a homomorphism with $\ker \theta \subseteq H$. Then θ is one-to-one because A_n is simple, so $\frac{1}{2}n!$ divides $p!$, say $q \cdot \frac{1}{2}n! = p!$. But $\frac{1}{2}n! = |A_n| = p|H,|$ so $p|\frac{1}{2}n!$ and hence $p \leq n$. By hypothesis $p < n$. Hence $\frac{q}{2} < 1$ so $q = 1$, $n! = p!$, a contradiction since $n \geq 5$.

7. If $U \subseteq V$ are subgroups then $g^{-1}Ug \subseteq g^{-1}Vg$ for all $g \in G$, so core $U \subseteq$ core V. Hence $\text{core}(H \cap K) \subseteq \text{core } H \cap \text{score } K$. If $x \in \text{core } H \cap \text{ core } K$, let $g \in G$. Thus $g^{-1}(H \cap K)g = (g^{-1}Hg) \cap (g^{-1}Kg)$ and so $x \in g^{-1}(H \cap K)g$. Thus $x \in \text{core }(H \cap K)$.

9. This is Exercise 26 §2.8.

10. (a) $H_0 \subseteq H$ because $H = 1_G(H)$. If $\tau \in \text{aut } G$ then $\tau^{-1}\sigma \in \text{aut } G$ for all $\sigma \in \text{aut } G$, so $H_0 \subseteq \tau^{-1}\sigma(H)$. Thus $\tau(H_0) \subseteq \sigma(H)$ for all σ, so $\tau(H_0) \subseteq H_0$. Similarly $\tau^{-1}(H_0) \subseteq H_0$, whence $\tau(H_0) = H_0$. Thus H_0 is characteristic in G.

11. (1) $\Rightarrow$ (2). If X is a nontrivial finite G-set, let $\theta : G \to S_X$ be the homomorphism in Theorem 2. Then $\theta(G) \neq \{1_X\}$ because the action is not trivial, so $\ker \ \theta \neq G$. Clearly $|G : \ker \ \theta| = \left|\frac{G}{\ker \ \theta}\right| = |\theta(G)|$ is finite.

13. We have $0 \cdot z = e^{i0}z = z$ and $b \cdot (a \cdot z) = e^{ib}(e^{ia}z) = e^{i(a+b)}z = (a+b) \cdot z$, so it is indeed an action. If $z = re^{i\theta}$ then $a \cdot z = re^{i(\theta+a)}$, so the action by a is to rotate z about the origin counterclockwise through a radians. If $z \in \mathbb{C}$, the orbit $G \cdot z = \{a \cdot z \mid a \in G\}$ is the circle, center at 0, radius $|z|$. Given $z \in \mathbb{C}$, the stabilizer is $S(z) = \{a \mid a \cdot z = z\} = \{a \mid ze^{ia} = z\}$. Hence $S(0) = G$; and if $z \neq 0$, then $S(z) = \{a \mid e^{ia} = 1\} = \{2\pi k \mid k \in \mathbb{Z}\} = \langle 2\pi \rangle$.

15. If $\sigma = (k_1 k_2 \cdots)(m_1 m_2 \cdots)(n_1 n_2 \cdots) \cdots$, the orbits are $G \cdot k_1 = \{k_1, k_2, \ldots\}$, $G \cdot m_1 = \{m_1, m_2, \ldots\}$, $G \cdot n_1 = \{n_1, n_2, \ldots\}$ and so on. Clearly $G \cdot k = \{k\}$ if and only if σ fixes k.

17. (a) We have $x \equiv x$ because $x = 1 \cdot x$; if $x \equiv y$ then $y = a \cdot x$ for some $a \in G$, so $x = a^{-1} \cdot y$, $y \equiv x$; if $x \equiv y$ and $y \equiv z$ then $y = a \cdot x$, $z = b \cdot y$, so $z = b \cdot (a \cdot x) = (ba) \cdot x$, that is $x \equiv z$.

19. Given G and $X = \{H \mid H$ is a subgroup of $G\}$, define $a \cdot H = aHa^{-1}$. Since aHa^{-1} is again in X, this is an action. The fixer here is $\{a \in G \mid a \cdot H = H$ for all $H\} = \{a \in G \mid aHa^{-1} = H$ for all $H\}$. This contains $Z(G)$. If $G = Q$ is the quaternion group, then every subgroup is normal, so $F = Q$. However $Z(G) = \{1, -1\}$.

21. Let $X = \{xH \mid x \in G\}$ and let $g \cdot xH = gxH$. Then $H \in X$ and
$$S(H) = \{g \mid gH = H\} = H.$$

23. (a) If $a, b \in S(x)$ then $(ab) \cdot x = a \cdot (b \cdot x) = a \cdot x = x$ and,
$$a^{-1} \cdot x = a^{-1} \cdot (a \cdot x) = (a^{-1} \cdot a) \cdot x = 1 \cdot x = x,$$
so $ab, a^{-1} \in S(G)$. Since $1 \cdot x = x$, $1 \in S(x)$ and we are done.

(c) Assume $bS(y)b^{-1} = S(x)$. Define $\sigma : G \cdot x \to G \cdot y$ by $\sigma(g \cdot x) = bg \cdot y$. Then
$$g \cdot x = h \cdot x \Leftrightarrow h^{-1}g \in S(x) \Leftrightarrow b^{-1}(h^{-1}g)b \in S(y)$$
$$\Leftrightarrow bg \cdot y = bh \cdot y.$$

Thus σ is well defined and one-to-one; it is clearly onto. Note that if G is finite this can be proved as follows:
$$|G \cdot x| = |G : S(x)| = |G : S(y)| = |G \cdot y|.$$

24. (a) The fixer is $F = \cap_{x \in X} S(x)$. If $y \in X$ then $G \cdot x = X = G \cdot y$ by hypothesis, so $S(x)$ and $S(y)$ are conjugate subgroups by the preceding exercise, say $S(y) = bS(x)b^{-1}$. Then $K \subseteq S(x)$ gives
$$K = bKb^{-1} \subseteq bS(x)b^{-1} = S(y).$$
Thus $K \subseteq \cap_{y \in X} S(y) = F$.

25. (a) This action is well defined: If $a \cdot x = b \cdot x$ then
$$(ha) \cdot x = h \cdot (a \cdot x) = h \cdot (b \cdot x) = (hb) \cdot x.$$
It is clearly an action.

27. Let G act on $X = \{H \mid H \subseteq G$ is a subgroup$\}$ by conjugation: $a \cdot H = aHa^{-1}$. Then $X_f = \{H \mid a \cdot H = H$ for all $a\}$ is the set of normal subgroups of G. Hence the nonnormal subgroups are partitioned into nonsingleton conjugacy classes, and each of these has $|G : N(H)|$ elements for some H. Since G is a p-group, this is a multiple of p, and the result follows.

29. If $|G : H| = p$ then $|G : \sigma^{-1}(H)| = p$ for all $\sigma \in$ aut G. Hence $K \subseteq \sigma^{-1}(H)$ so $\sigma(K) \subseteq H$. It follows that $\sigma(K) \subseteq K$, so K is characteristic; in particular normal. Now if G is a p-group, $H_i \neq N(H_i)$ for each i by Theorem 5, so since $|G : H_i| = p$ is prime, $H_i \triangleleft G$ by Exercise 6 §8.2 and $|G/H_i| = p$. Let $A = G/H_1 \times \cdots \times G/H_m$ and define $\theta : G \to A$ by $\theta(g) = (gH_1, \ldots, gH_m)$. This is a homomorphism with $\ker \theta = K$, so $G/K \cong \theta(G) \subseteq A$. The result follows.

31. We have $1 \cdot (h, k) = (h, k)$, and
$$b \cdot [a \cdot (h, k)] = b \cdot (ha^{-1}, ak) = (ha^{-1}b^{-1}, bak) = (ba) \cdot (h, k).$$

Thus it is an action. Let $X = H \times K$ so $|X| = |H|\,|K|$. Let $A = H \cap K$ and define $\lambda : A \to A \cdot (h,k)$ by $\lambda(a) = a \cdot (h,k) = (ha^{-1}, ak)$. This is clearly a bijection, so every orbit has $|A|$ elements. Finally, define

$$\mu : HK \to \{A \cdot (h,k) \mid (h,k) \in H \times K\}$$

by $(hk)\mu = A \cdot (h,k)$. Then

$$hk = h_1 k_1 \Rightarrow h_1^{-1}h = k_1 k^{-1} = a \in A \Rightarrow a \cdot (h,k)$$
$$= (ha^{-1}, ak) = (h_1, k_1) \Rightarrow A \cdot (h,k) = A \cdot (h_1, k_1),$$

so μ is well-defined. Conversely, $a \cdot (h,k) = (h_1, k_1)$ implies $hk = h_1 k_1$, so μ is a bijection.

32. (a) $(1,1) \cdot x = 1x1^{-1} = x$, $(h_1, k_1)((h,k) \cdot x)) = h_1(hxk^{-1})k_1^{-1}$
    $= (h_1 h, k_1 k) \cdot x = [(h_1 k_1) \cdot (h,k)] \cdot x$.
    The orbit is $(H \times K) \cdot x = \{(h,k) \cdot x \mid h \in H, k \in K\} = \{hxk^{-1} \mid h \in H, k \in K\} = HxK$—a *double coset*.

    (c) The size of the double coset is

    $$|HxK| = |HxK : S(x)| = \frac{|H \times K|}{|S(x)|} = \frac{|H|\,|K|}{|x^{-1}Hx \cap K|}$$

    by (b). Frobenius' theorem follows.

33. (a) Any two cosets aH and bH are in the same orbit because $bH = (ba^{-1}) \cdot (aH)$.

34. (a) This action is well-defined because $[x] = [x_1]$ gives $\varphi(x) = \varphi(x_1)$ so, since φ is a G-morphism,

    $$\varphi(a \cdot x) = a \cdot \varphi(x) = a \cdot \varphi(x_1) = \varphi(a \cdot x_1).$$

    Thus $[a \cdot x] = [a \cdot x_1]$ so the action is well-defined. Now $1 \cdot [x] = [1 \cdot x] = [x]$ and $a \cdot (b \cdot [x]) = a \cdot [b \cdot x] = [a \cdot (b \cdot x)] = [(ab) \cdot x] = (ab) \cdot [x]$. Hence $\frac{G}{\varphi}$ is a G-set.

8.4　THE SYLOW THEOREMS

1. Since $|S_4| = 2^3 \cdot 3$, the Sylow 3-subgroups are all cyclic of order 3, and thus have the form $P = \langle \gamma \rangle$, $\gamma = (i \ j \ k)$. Now $\sigma(1\ 2\ 3)\sigma^{-1} = (\sigma(1) \ \sigma(2) \ \sigma(3))$ for all $\sigma \in S_4$ (Lemma 3 §2.8) so let $\sigma = \begin{pmatrix} 1 & 2 & 3 & 4 \\ i & j & k & x \end{pmatrix}$ where $\{1, 2, 3, 4\} = \{i, j, k, x\}$. Then $\sigma(123)\sigma^{-1} = (i \ j \ k)$ so γ and $(1\ 2\ 3)$ are conjugate. Hence P and $\langle(1\ 2\ 3)\rangle$ are conjugate.

3. P is a Sylow p-subgroup of $N(P)$, being a p-subgroup of maximal order. It is unique because it is normal in $N(P)$.

5. Let $|G| = 1001 = 7 \cdot 11 \cdot 13$. We have $n_7 = 1, 11, 13, 143$ and $n_7 \equiv 1 \pmod 7$, so $n_7 = 1$. Similarly $n_{11} = 1, 7, 13, 91$ and $n_{11} \equiv 1 \pmod{11}$, so $n_{11} = 1$; and $n_{13} = 1, 7, 11, 77$ and $n_{13} \equiv 1 \pmod{13}$ so $n_{13} = 1$. Thus let $H \lhd G$, $K \lhd G$, $L \lhd G$ have order $|H| = 7$, $|K| = 11$ and $|L| = 13$. Then $H \cap K = \{1\}$ so $HK \cong H \times K \cong C_{77}$. Now $HK \cap L = \{1\}$ so $(HK)L \cong HK \times L \cong C_{1001}$. Thus $G = HKL \cong C_{1001}$ is unique up to isomorphism.

7. (a) If $|G| = 40 = 2^3 \cdot 5$, then $n_5 = 1, 2, 4, 8$ and $n_5 \equiv 1 \pmod 5$. Thus $n_5 = 1$ so the Sylow 5-subgroup is normal.

   (c) If $|G| = 48 = 2^4 \cdot 3$ and P is a Sylow 2-subgroup, then $|G : P| = 3$ so let $\theta : G \to S_3$ be a homomorphism. Clearly $\ker \theta \neq \{1\}$.

8. If $|G| = 520 = 2^3 \cdot 5 \cdot 13$, then $n_{13} = 1, 2, 4, 8, 18, 20, 40, 80$ and $n_{13} \equiv 1 \pmod{13}$, so $n_{13} = 1, 40$. Similarly $n_5 = 1, 2, 4, 8, 13, 26, 52, 104$, and $n_5 \equiv 1 \pmod 5$ so $n_5 = 1, 26$. If either $n_{13} = 1$ or $n_5 = 1$ we are done. Otherwise there are $40 \cdot 12 = 480$ elements of order 13 and $26 \cdot 4 = 104$ elements of order 5, giving $480 + 104 = 584$ in all, a contradiction.

9. (a) If $|G| = 70 = 2 \cdot 5 \cdot 7$ we have $n_5 = 1$ and $n_7 = 1$, so let $P \lhd G$ and $Q \lhd G$ satisfy $|P| = 5$ and $|Q| = 7$. Thus since
$$P \cap Q = \{1\}, \ PQ \cong P \times Q \cong C_5 \times C_7 \cong C_{35}.$$
   Thus $|G : PQ| = 2$ so $PQ \lhd G$.

   (c) If $|G| = 30 = 2 \cdot 3 \cdot 5$, then $n_3 = 1, 10$ and $n_5 = 1, 6$. Let $|P| = 3$ and $|Q| = 5$. If P is not normal, there are 10 Sylow 3-subgroups in G, and hence $10 \cdot 2 = 20$ elements of order 3. Similarly if Q is not normal there are $6 \cdot 4 = 24$ elements of order 5. So $P \lhd G$ or $Q \lhd G$; either way $PQ = H$ is a subgroup of G of index 2 (so $H \lhd G$). Since $|H| = 15$, $P \lhd H$ and $Q \lhd H$ by Sylow's third theorem, so $H = PQ \cong P \times Q \cong C_{15}$.

10. (a) If $|G| = 385 = 5 \cdot 7 \cdot 11$ then $n_{11} = 1$ and $n_7 = 1$, so let $P \lhd G$, $Q \lhd G$ satisfy $|P| = 11$ and $|Q| = 7$. Then
$$PQ \cong P \times Q \cong C_{11} \times C_7 \cong C_{77},$$
   and so $|G : PQ| = 5$. Since both $P \lhd G$ and $Q \lhd G$ we have $PQ \lhd G$.

11. (a) If $|G| = 105 = 3 \cdot 5 \cdot 7$, then $n_7 = 1, 15$ and $n_5 = 1, 21$. Let $|P| = 7$ and $|Q| = 5$. If neither P nor Q is normal in G, then G has $15 \cdot 6 = 90$ elements of order 7 and $21 \cdot 4 = 84$ elements of order 5, a contradiction. So $P \lhd G$ or $Q \lhd G$, whence PQ is a subgroup and $|PQ| = |P||Q| = 35$ because $P \cap Q = \{1\}$. Since $|G : PQ| = 3$, let $\theta : G \to S_3$ have $\ker \theta \subseteq PQ$. Clearly $|\ker \theta| \neq 1, 5, 7$ so $\ker \theta = PQ$ and $PQ \lhd G$. Finally $|PQ| = 35 = 7 \cdot 5$ means $P \lhd PQ$ and $Q \lhd PQ$, so $PQ \cong P \times Q \cong C_{35}$.

13. Let P be a Sylow p-subgroup of G. Then $|P| = p^n$, and $|G : P| = m$ because $p > m$. Thus, by Theorem 1 §8.3 there is a homomorphism $\theta : G \to S_m$ with $\ker \theta \subseteq P$. If $|\ker \theta| = p^k$ then $p^{n-k} = |G/\ker \theta|$ divides $m!$. Since $p > m$ this means that $k = n$, whence $P = \ker \theta \lhd G$.

15. $\alpha(P)$ is clearly a p-subgroup of G so $\alpha(P) \subseteq a^{-1}Pa$ for some $a \in G$ by Theorem 2. But $a^{-1}Pa = P$ here because $P \lhd G$.

17. For convenience, write $N = N(P)$. Let $gP \in N/P$ have order p^k, $g \in N$. We must show that $g \in P$. We have $g^{p^k} \in P$ so $o(g)$ is a power of p. Thus $\langle g \rangle$ is a p-subgroup of N. Since P is a Sylow p-subgroup of N, Theorem 2 gives $\langle g \rangle \subseteq a^{-1}Pa$ for some $a \in N$. But $P \lhd N$ so $a^{-1}Pa = P$. Thus $g \in P$, and $gP = P$.

118   8. p-Groups and the Sylow Theorems

18. (a) We have $P \subseteq N(P) \subseteq H$. If $a \in N(H)$ then $a^{-1}Pa \subseteq a^{-1}Ha = H$, so P and $a^{-1}Pa$ are both Sylow p-subgroups of H. By Sylow's second theorem, $h^{-1}(a^{-1}Pa)h = P$ for some $h \in H$. Thus $ah \in N(P) \subseteq H$, whence $a \in H$.

19. If Q is also a Sylow p-subgroup of G, then $Q = a^{-1}Pa$ by Sylow's second theorem. If $g \in N(Q)$ then $Q = g^{-1}Qg$; that is $a^{-1}Pa = g^{-1}a^{-1}Pag$. This implies that $aga^{-1} \in N(P) = P$, whence $g \in a^{-1}Pa = Q$.

21. K is clearly a p-subgroup. If $a \in G$ and P is a Sylow p-subgroup, then aPa^{-1} is also a Sylow p-subgroup, so $K \subseteq aPa^{-1}$. Hence $a^{-1}Ka \subseteq P$; since P was an arbitrary Sylow p-subgroup, $a^{-1}Ka \subseteq K$. Thus $K \lhd G$. Now let H be any normal p-subgroup of G. If H is a normal p-subgroup of G, then $H \subseteq aPa^{-1}$ for some $a \in G$. Thus $H = a^{-1}Ha \subseteq P$; whence $H \subseteq K$.

23. If $k|n$, let $n = kd$ and $a_1 = a^d$. Then $o(a_1) = k$ and $a_1ba_1 = a^dba^d = b$. Finally, let $k = 2m_1$, $2(m_1d) = kd = n = 2m$, so $m_1d = m$. Thus $b^2 = a^m = a_1^{m_1}$, and so $\langle a_1, b \rangle \cong Q_k$.

25. If p and q are distinct primes, any group of one of the following orders is not simple: p^n (Theorem 8 §8.2), pq (Example 5), p^2q (Exercise 14) and p^2q^2 (Example 9). The only remaining orders (apart from primes) in the range 2–59 are as follows, with the reason that such a group G is not simple (We also include $|G| = 36$ — see Theorem 4).

$|G|$	Reason
$24 = 2^3 \cdot 3$	$\theta : G \to S_3$ is not one-to-one
$30 = 2 \cdot 3 \cdot 5$	Exercise 9(c)
$40 = 2^3 \cdot 5$	Exercise 7(a)
$42 = 2 \cdot 3 \cdot 7$	$n_7 = 1$
$48 = 2^4 \cdot 3$	$\theta : G \to S_3$ is not one-to-one
$54 = 2 \cdot 3^3$	$n_3 = 1$
$56 = 2^3 \cdot 7$	Example 7
$36 = 2^2 \cdot 3^2$	$\theta : G \to S_4$ is not one-to-one

8.5 SEMIDIRECT PRODUCTS

1. (a). Write $\sigma = (1\ 2) \in S_n$ and $H = \langle \sigma \rangle$. Then $A_n \subseteq A_nH \subseteq S_n$ so, since $S_n/A_n \cong C_2$, either $A_nH = S_n$ or $A_nH = A_n$. Since $\sigma \notin A_n$ we have $S_n = A_nH$. Similarly, $A_n \cap H \neq \{\varepsilon\}$ means $A_n \cap H = H$ (because H is simple), again contradicting $h \notin A_n$. Hence $A_n \cap H = \{\varepsilon\}$ and the result follows from Theorem 2.

3. This is an instance of Theorem 3 (3), where $p = 3$ and $q = 13$. We have $q \equiv 1$ (mod p) so we look for m such that $1 \leq m \leq 12$ and $m^3 \equiv 1$ (mod 13). If $m = 1$ then $G \cong C_{13} \times C_3 \cong C_{55}$. The first solution with $m > 1$ is $m = 3$, whence $G \langle a, b \rangle$ where $|a| = 13$, $|b| = 3$ and $ab = ba^3$.

5. In G, Sylow-3 gives $n_3 = 1, 10$ and $n_5 = 1, 6$; if neither is 1 then G has $10 \cdot 2 = 20$ elements of order 3, and $6 \cdot 4 = 24$ elements of order 5, a contradiction as $|G| = 30$. So if P and Q are Sylow 3- and 5-subgroups, then $K = PQ$ is a subgroup of order $3 \cdot 5 = 15$ (as $P \cap Q = \{1\}$). Hence K has index 2, so $K \lhd G$. Moreover both $P \lhd K$ and $Q \lhd K$ by Sylow-3 applied to K, whence

$K \cong P \times Q \cong C_3 \times C_5 \cong C_{15}$, say $K = \langle a \rangle$ where $|a| = 15$. Let H be any Sylow 2-subgroup, say $H = \langle b \rangle$ where $|b| = 2$.

Let $b^{-1}ab = a^m$, so $b^{-k}ab^k = a^{m^k}$ for $k \geq 1$. Since $b^2 = 1$ this gives $a^{m^2} = 1$, that is $m^2 \equiv 1 \pmod{15}$. The solutions are $m = \pm 1$ and $m = \pm 4$.

Case 1. $m = 1$. Then $ab = ba$ so G is abelian, and we get $G \cong C_{15} \times C_3 \cong C_{30}$.

Case 2. $m = -1$. Then $ab = ba^{-1}$, so $aba = b$ and $G \cong D_{15}$.

Case 3. $m = 4$. Then $b^{-1}ab = a^4$. Take $a_1 = a^3$ so $|a_1| = 5$. Then $b^{-1}a_1b = (b^{-1}ab)^3 = a^{12} = a_1^{-1}$. It follows that $U = \langle a_1, b \rangle \cong D_5$. Define $V = \langle a^5 \rangle \cong C_3$. Then $U \cap V = \{1\}$ because $V \not\subseteq U$. It follows that $UV = G$. Finally, one verifies that $a^5b = ba^5$, so U and V commute elementwise. This gives $G \cong U \times V \cong D_5 \times C_3$.

Case 4. $m = -4$. Here $b^{-1}a^5b = (b^{-1}ab)^5 = a^{-20} = (a^5)^{-1}$. Hence $U = \langle a^5, b \rangle \cong D_3$. Take $V = \langle a^3 \rangle \cong C_5$, so that $U \cap V = \{1\}$. Hence $UV = G$. Moreover, since $b^{-1}a^3b = (b^{-1}ab)^3 = a^{-12} = a^3$, U and V commute elementwise, so $G \cong U \times V \cong D_3 \times C_5$.

Finally, observe that no two of the groups C_{30}, D_{15}, $D_5 \times C_3$ and $D_3 \times C_5$ are isomorphic: C_{30} is the only abelian one; D_{15} and $D_3 \times C_5$ have no element of order 6, while $D_5 \times C_3$ has 10; D_{15} has 15 elements of order 2, while $D_3 \times C_5$ has only 3.

8.6 AN APPLICATION TO COMBINATORICS

1. Let H act on G by $h \cdot x = hx$ for $x \in G$, $h \in H$. Then the orbits $H \cdot x = Hx$ are right cosets. Given

$$h \in H, \ F(h) = \{x \in G \mid hx = h\} = \begin{cases} \emptyset & \text{if } h \neq 1 \\ G & \text{if } h = 1 \end{cases}.$$

   Thus the Cauchy-Frobenius lemma gives the number of cosets as $\frac{1}{|H|}(|G|) = |G : H|$.

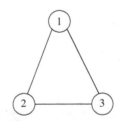

3. (a) If the vertices are labeled as shown, the group $G \subseteq S_3$ of motions is $G = \{\varepsilon, (2\ 3)\}$. Hence $|F(\varepsilon)| = q^3$ and $|F((2\ 3))| = q^2$, so the number of orbits is $\frac{1}{2}(q^3 + q^2)$ by Theorem 2.

4. (a) By Example 3 §2.7, the group of motions of the tetrahedron is A_4. Now $|A_4| = 12$ and A_4 consists of ε, eight 3-cycles, and $(1\ 2)(3\ 4)$, $(1\ 3)(2\ 4)$ and $(1\ 4)(2\ 3)$. Hence $|F(\sigma)| = q^2$ for all $\sigma \in A_4$ except $\sigma = \varepsilon$. Hence the number of colorings is $\frac{1}{12}(q^4 + 11q^2)$ by Theorem 2.

5. (a) Label the top and bottom as 1, 2, and the sides 3, 4, 5, 6 as shown. The group of motions is

$$G = \{\varepsilon, (3\ 4\ 5\ 6), (3\ 5)(4\ 6), (3\ 6\ 5\ 4),$$
$$(1\ 2)(3\ 5), (1\ 2)(4\ 6),$$
$$(1\ 2)(3\ 4)(5\ 6), (1\ 2)(3\ 6)(4\ 5)\}.$$

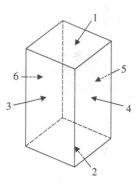

Here cyc $\varepsilon = 6$, cyc$(3\ 4\ 5\ 6) = 3$, cyc$(3\ 5)$ $(4\ 6) = 4$ and cyc $(1\ 2)(3\ 4)(5\ 6) = 3$. Hence Theorem 2 gives $\frac{1}{8}[q^6 + 2q^3 + 3q^4 + 2q^3]$ $= \frac{1}{8}q^3(q^3 + 3q + 4) = \frac{1}{8}q^3(q + 1)(q^2 - q + 4)$ as the number of colorings.

6. (a) The tetrahedron has 4 vertices so the number is $\frac{1}{12}q^2(q^2 + 11)$ as in Exercise 4.

7. Label the faces 1–6 as shown. Then a typical permutation σ in each category in the hint is

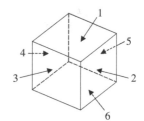

Category	σ	Number	$|F(\sigma)|$
1	ε	1	q^6
2	$(2\ 3\ 4\ 5)$	6	q^3
	$(2\ 4)(3\ 5)$	3	q^4
3	$(1\ 4)(2\ 6)(3\ 5)$	6	q^3
4	$(1\ 3\ 4)(2\ 6\ 5)$	8	q^2

Hence the number of orbits is

$$\frac{1}{24}[q^6 + 3q^4 + 12q^3 + 8q^2] = \frac{1}{24}q^2(q + 1)(q^3 - q^2 + 4q + 8).$$

8. (a) Number the sections 1-6 in order. Then $G = \langle \sigma \rangle$ where $\sigma = (1\ 2\ 3\ 4\ 5\ 6)$. Hence

$$G = \varepsilon, (1\ 2\ 3\ 4\ 5\ 6), (1\ 3\ 5)(2\ 4\ 6),$$
$$(1\ 4)(1\ 5)(3\ 6), (1\ 5\ 3)(2\ 6\ 4),$$
$$(1\ 6\ 5\ 4\ 3\ 2)\}.$$

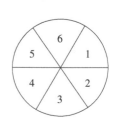

By Theorem 2, the number of orbits is

$$\frac{1}{6}[q^6 + 2q + 2q^2 + q^3] = \frac{1}{6}q(q^5 + q^2 + 2q + 2) = \frac{1}{6}q(q + 1)(q^4 - q^3 + q^2 + 2).$$

9. If $n = 2m$ then $\left\lfloor \frac{n+1}{2} \right\rfloor = \left\lfloor m + \frac{1}{2} \right\rfloor = m$.

| 1 | 2 | 3 | $\cdots$ | m | $m+1$ | $\cdots$ | n |

Number the strips $1, 2, \ldots, n$. Then $G = \{\varepsilon, (1\ n)(2\ n-1)\cdots(m\ m+1)\}$ so the number of orbits is $\frac{1}{2}(q^n + q^m)$, as required. If $n = 2m - 1$ then

$$\left\lfloor \frac{n+1}{2} \right\rfloor = \left\lfloor m - \frac{1}{2} \right\rfloor = m - 1.$$

Now $G = \{\varepsilon, (1\ n)(2\ n-1)\cdots(m-1\ m+1)\}$ so the number is $\frac{1}{2}\left\lfloor q^n + q^{m-1} \right\rfloor$, again as required.

11. If the vertices and edges are labeled as shown, the group of (vertex) motions is $G = \{\varepsilon, (1\ 3), (2\ 4), (1\ 3)(2\ 4)\}$. Each $\sigma \in G$ induces an edge permutation σ_e in S_5 as follows:

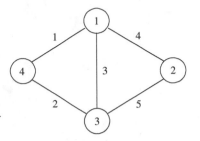

σ	ε	$(1\ 3)$	$(2\ 4)$	$(1\ 3)(2\ 4)$
σ_e	ε	$(1\ 2)(4\ 5)$	$(1\ 4)(2\ 5)$	$(1\ 5)(2\ 4)$

As in the preceding exercise, colorings are pairs (λ, μ) when $\lambda : \{1, 2, 3, 4\} \to C_v$ and $\mu : \{1, 2, 3, 4, 5\} \to C_e$. Hence

$$|F(\varepsilon)| = q^4 r^5 \quad |F((1\ 3))| = |F((2\ 4))| = q^3 r^3 \quad |F((1\ 3)(2\ 4))| = q^2 r^3.$$

Thus the number of orbits is

$$\tfrac{1}{4}(q^4 r^5 + 2q^3 r^3 + q^2 r^3) = \tfrac{1}{4}q^2 r^3 (q^2 r^2 + 2q + 1).$$

12 (a) If $X = \{(a, b) \in G \times G \mid ab = ba\}$ then $p(G) = \dfrac{|X|}{|G \times G|} = \dfrac{|X|}{|G|^2}$. Now

$$X = \{(a, b) \mid b^{-1}ab = a\} = \{(a, b) \mid b \in N(a)\}.$$

Hence $|X| = \sum_{a \in G} |N(a)| = |G|k(G)$ by the Corollary to the Cauchy-Frobenius lemma, and the result follows.

Chapter 9

Series of Subgroups

9.1 THE JORDAN-HÖLDER THEOREM

1. (a) If $C_8 = \langle g \rangle$, $o(g) = 8$, then $o(g^2) = 4$ and $o(g^4) = 2$. It follows that $C_8 \supset \langle g^2 \rangle \supset \langle g^4 \rangle \supset \{1\}$ is a composition series. Hence length $C_8 = 3$ and the factors are C_2, C_2, C_2.

(c) We have $D_4 = \{1, a, a^2, a^3, b, ba, ba^2, ba^3\}$ where $o(a) = 4$, $o(b) = 2$, $aba = b$. Hence $D_4 \supset \langle a \rangle \supset \langle a^2 \rangle \supset \{1\}$ is a composition series so length $D_4 = 3$ and the factors are C_2, C_2, C_2.

(e) $Q = \{\pm 1, \pm i, \pm j, \pm k\}$ and $o(i) = 4$. If $M = \langle i \rangle$, $K = \langle -1 \rangle$, then $Q \supset M \supset K \supset \{1\}$ is a composition series. Thus length $Q = 3$ and the factors are C_2, C_2, C_2.

3. (a) If M is a maximal normal subgroup of C_{24} then C_{24}/M is simple and abelian so $|C_{24}/M| = p$ is a prime. But p divides $|C_{24}| = 24 = 2^3 3$, so $p = 2$ or $p = 3$. Hence $|M| = 12$ or 8. Let H_d denote the unique subgroup of C_{24} of order d where $d \mid 24$. Then H_{12} and H_8 are the maximal normal subgroups, so any composition series $C_{24} \supset G_1 \supset \cdots \supset \{1\}$ must have $G_1 = H_{12}$ or H_8. In the same way, the maximal normal subgroups of H_{12} are H_4 and H_6; of H_8 is H_4; of H_6 are H_3 and H_2, and of H_4 is H_2. Each composition series must contain a maximal subgroup, so the various composition series are as shown. There are 4 in all.

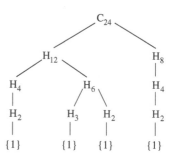

Student Solution Manual to Accompany Introduction to Abstract Algebra, Fourth Edition. W. Keith Nicholson.
© 2012 John Wiley & Sons, Inc. Published 2012 by John Wiley & Sons, Inc.

5. Write $G = C_4 \times C_2$. If M is a maximal subgroup then $|M| = 4$ so M is cyclic or M is the Klein group. If $C_4 = \langle a \rangle$, $o(a) = 4$, and $C_2 = \langle b \rangle$, $o(b) = 2$, the elements of G of order 4 are $(a,1), (a,b), (a^3,1)$ and (a^3,b), so the cyclic maximal subgroups are $M_1 = \langle (a,1) \rangle = \langle (a^3,1) \rangle$ and $M_2 = \langle (a,b) \rangle = \langle (a^3,b) \rangle$.

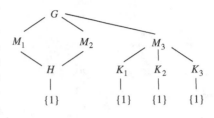

These have a unique subgroup $H = \langle (a^2,1) \rangle$ of order 2 leading to composition series $G \supset M_1 \supset H \supset \{1\}$ and $G \supset M_2 \supset H \supset \{1\}$. On the other hand, the only elements of order 2 in G are $(a^2,1), (a^2,b)$ and $(1,b)$, so $M_3 = \{(1,1),(a^2,1),(1,b),(a^2,b)\}$ is the unique maximal subgroup isomorphic to the Klein group. This has three subgroups of order 2: $K_1 = \langle (a^2,1) \rangle$, $K_2 = \langle (1,b) \rangle$ and $K_3 = \langle (a^2,b) \rangle$. This leads to three composition series $G \supset M_3 \supset K_1 \supset \{1\}$, $G \supset M_3 \supset K_2 \supset \{1\}$ and $G \supset M_3 \supset K_3 \supset \{1\}$. There are thus five composition series.

7. Write $D_{16} = \{1, a, \ldots, a^{15}, b, ba, \ldots, ba^{15}\}$ where $o(a) = 16$, $o(b) = 2$ and $aba = b$. Then $Z(D_{16}) = \{1, a^8\}$ by Exercise 26 §2.6. Write $Z = Z(D_{16})$. If $H = \langle a \rangle$, $K = \langle a^2 \rangle$ and $L = \langle a^4 \rangle$, then

$$D_{16} \supset H \supset K \supset L \supset Z \supset \{1\}$$

is a composition series containing Z. If

$$H = \{1, b\}, \quad I = \{1, a^8, b, ba^8\}, \quad J = \{1, a^4, a^8, a^{12}, b, ba^4, ba^8, ba^{12}\} \text{ and}$$
$$K = \{1, a^2, a^4, a^6, a^8, a^{10}, a^{12}, a^{14}, b, ba^2, ba^4, ba^6, ba^8, ba^{10}, ba^{12}, ba^{14}\},$$

then $G \supset K \supset J \supset I \supset H \supset \{1\}$ is a composition series not containing Z.

8. (a) Let $n = p_1 p_2 \cdots p_m$ where the p_i are distinct primes. Then C_n has length $1 + 1 + \cdots + 1 = m$ by Example 8.

9. Define $G_0, G_1, G_2, \ldots$ by $G_0 = G$,

$$G_1 = \{1\} \times K_1 \times K_2 \times \cdots \times K_r,$$
$$G_2 = \{1\} \times \{1\} \times K_2 \times \cdots \times K_r, \ldots.$$

Then $G_{i+1} \triangleleft G_i$ for each i and $G = G_0 \supset G_1 \supset \cdots \supset G_r = \{1\}$. Since

$$\frac{G_i}{G_{i+1}} = \frac{\{1\} \times \cdots \times K_i \times K_{i+1} \times \cdots \times K_r}{\{1\} \times \cdots \times \{1\} \times K_{i+1} \times \cdots \times K_r} \cong \frac{K_i}{\{1\}} \cong K_i$$

for each i, we are done.

11. Induct on n. If $n = 1$ then $G = G_0 \supset G_1 = \{1\}$ so $G \cong G_0/G_1$ is finite. In general, G_1 is finite by induction, and $G/G_1 = G_0/G_1$ is finite by hypothesis. Thus G consists of $|G/G_1|$ cosets, each with $|G_1|$ elements. Hence G is finite. Now $|G| = |G_0/G_1| \cdot |G_1|$, and the formula follows by induction.

13. Since G has a composition series it follows that $G_1 \triangleleft G$, $G_2 \triangleleft G_1 \ldots$ all have composition series by Theorem 2. Hence G_i/G_{i+1} has a composition series (again by Theorem 2):

$$\frac{G_i}{G_{i+1}} \supset \frac{G_{i1}}{G_{i+1}} \supset \frac{G_{i2}}{G_{i+1}} \supset \cdots \supset \frac{G_{ir}}{G_{i+1}} = \{G_{i+1}\}.$$

Hence we obtain a subnormal series in G_i :

$$G_i \supset G_{i1} \supset G_{i2} \supset \cdots \supset G_{ir} = G_{i+1}.$$

Moreover $\dfrac{G_{ik}}{G_{i(k+1)}} \cong \dfrac{G_{ik}/G_{i+1}}{G_{i(k+1)}/G_{i+1}}$ is simple for all i and k, so piecing these together gives a composition series for G.

14. (a) By Exercise 13, if $H \triangleleft G$ and $K \triangleleft G$ choose composition series for G refining

$$G \supset HK \supset K \supset H \cap K \supset \{1\} \qquad (*)$$

$$G \supset HK \supset H \supset H \cap K \supset \{1\}. \qquad (**)$$

Now the factors in $(*)$ between HK and K are the same as those in $(**)$ between H and $H \cap K$ because $\frac{HK}{K} \cong \frac{H}{H \cap K}$. If $H \cong K$ the factors in $(**)$ between H and $H \cap K$ are the same as those in $(*)$ between K and $H \cap K$ (they are the factors of $K \cong H$ omitting those of $H \cap K$). Hence the factors in $(*)$ between HK and K are the same as those between K and $H \cap K$. By hypothesis, this implies $H = K$.

15. (a) If $M \subseteq C_n$ is maximal normal, then C_n/M has order a prime q (being simple and abelian) and, since q divides $|C_n| = n$, q is one of the p_i. Thus $|M| = \frac{n}{p_i}$ for some $i = 1, 2, \ldots, r$. Since C_n is cyclic, it has exactly one subgroup of order $\frac{n}{p_i}$ by Theorem 9 §2.4.

17. See for instance: Rose, John S., *A Course on Group Theory*, Cambridge University Press, 1978, pp. 122–125.

9.2 SOLVABLE GROUPS

1. No. The group S_4 is solvable (Example 4) but $Z(S_4) = \{\varepsilon\}$.

3. No. S_4 is solvable (Example 4) but $S_4' = A_4$ is not abelian. Indeed $S_4' \subseteq A_4$ because S_4/A_4 is abelian. Thus $S_4' = A_4$, $\{\varepsilon\}$ or $K = \{\varepsilon, \ (1\,2)(3\,4), (1\,3)(2\,4), (1\,4)(2\,3)\}$. But $S_4/\{\varepsilon\}$ and S_4/K are not abelian (see Exercise 30 §2.9).

5. If $G = A_5$ then G is not solvable being nonabelian and simple. Since $|A_5| = 60 = 2^2 \cdot 3 \cdot 5$ the Sylow subgroups have orders 4, 3 or 5, and so are abelian.

7. G need not be solvable. If $G = A_5 \times C_2$ then $K = \{\varepsilon\} \times C_2$ is an abelian normal subgroup which is maximal because $G/K \cong A_5$ is simple. But G is not solvable because A_5 is not solvable.

8. (1) This is because $\alpha[a, b] = [\alpha(a), \alpha(b)]$ and the fact that G' consists of products of commutators.

   (3) Every commutator from H is a commutator from G.

9. By Exercise 14 §8.4, let $K \triangleleft G$ where $K \neq \{1\}, G$. If $|G| = p^2 q$ let $K \triangleleft G$, $K \neq \{1\}$, $K \neq G$. Then $|K| = p, q, p^2$ or pq, so $|G/K| = pq, p^2, q$ or p. Thus both K and G/K are either abelian or of order pq, and hence are both solvable by Example 5. So G is solvable by Theorem 4.

11. (a) View F as an additive group and define $\theta : G \to F \oplus F$ by

$$\theta \begin{bmatrix} 1 & a & b \\ 0 & 1 & c \\ 0 & 0 & 1 \end{bmatrix} = (a, c).$$

Then θ is an onto homomorphism and $\ker \theta = \left\{ \begin{bmatrix} 1 & 0 & b \\ 0 & 1 & 0 \\ 0 & 0 & 1 \end{bmatrix} \middle| b \in F \right\} \cong F$ via

$\begin{bmatrix} 1 & 0 & b \\ 0 & 1 & 0 \\ 0 & 0 & 1 \end{bmatrix} \mapsto b$. Hence both $\ker \theta$ and $G/\ker \theta$ are abelian (hence solvable),
so G is solvable.

13. If odd order groups are solvable then the only odd order simple groups are abelian by Corollary 2 of Theorem 3. Hence the nonabelian simple finite groups have even order. Conversely, let $|G|$ be odd. If G is abelian it is solvable. Otherwise it is not simple by hypothesis, so let $K \triangleleft G$, $K \neq \{1\}$, $K \neq G$. Then both K and G/K have odd order less than $|G|$, so they are both solvable by induction. Thus G is solvable by Theorem 4.

15. If G is solvable and $G = G_0 \supset G_1 \supset \cdots \supset G_p = \{1\}$ is a composition series, each simple factor is abelian and hence finite. Hence $|G| = \left| \frac{G_0}{G_1} \right| \cdot \left| \frac{G_1}{G_2} \right| \cdots \left| \frac{G_{n-1}}{G_n} \right|$ is finite (see Exercise 11 §9.1). The converse holds because *every* finite group has a composition series.

17. If $G/(H \cap K)$ is solvable, so are its images G/H and G/K (by Theorem 7 §8.1). Hence $G/H \times G/K$ is solvable by Theorem 4. The converse is because $G/(H \cap K)$ is isomorphic to a subgroup of $(G/H) \times (G/K)$ via

$$g(H \cap K) \mapsto (gH, gK).$$

19. HK is a subgroup because $K \triangleleft G$, and $\frac{HK}{K}$ is solvable because $\frac{HK}{K} \cong \frac{H}{H \cap K}$ is solvable by Theorem 3 because H is solvable. Since K is solvable too, HK is solvable by Theorem 4.

20. (a) If $G = \{1\}$, it is solvable. If $G \neq \{1\}$ then $Z_1 = Z(G) \neq \{1\}$ by hypothesis (with $K = \{1\}$). If $Z_1 \neq G$ let $\frac{Z_2}{Z_1} = Z\left(\frac{G}{Z_1}\right)$. Thus $Z_2 \supset Z_1$ and $Z_2 \triangleleft G$.
    If $Z_2 \neq G$ let $\frac{Z_3}{Z_2} = Z\left(\frac{G}{Z_2}\right)$. So $Z_3 \triangleleft G$, $Z_3 \supset Z_2 \supset Z_1 \supset \{1\}$. Since G is finite, $Z_m = G$ for some m, so $G = Z_m \supset \cdots \supset Z_1 \supset \{1\}$ is a solvable series for G.

21. (a) It suffices to show that $G' \neq G$, since then $G/G' \neq \{1\}$ is abelian. But if $G' = G$ then $G^{(2)} = (G')' = G' = G$, and it follows by induction that $G^{(k)} = G \neq 1$ for each $k \geq 0$, and this cannot happen in a solvable group by Theorem 2.

22. (2) $\Rightarrow$ (1). Assume (2). Then there exists $G_1 \triangleleft G$ such that $\frac{G}{G_1}$ is abelian and $G_1 \subset G$. If $G_1 = \{1\}$ we are done. If not let $G_2 \triangleleft G_1$, $G_2 \subset G_1$, $\frac{G_1}{G_2}$ abelian. Thus $G \supset G_1 \supset G_2$. This process continues and, since G is finite, $G_n = \{1\}$ at some stage. Then $G \supset G_1 \supset G_2 \supset \cdots \supset G_n = \{1\}$ is a solvable series.

23. Write $R = R(G)$ and $R_1 = R(H)$.

(a) Write $\{K \triangleleft G \mid G/K \text{ solvable}\} = \{K_1, K_2, \ldots, K_m\}$. This set is nonempty as it contains G. Then $R = \bigcap_1^m K_i$ is normal and G/R is solvable by Exercise 18. If $K \triangleleft G$ and G/K solvable, then $R \subseteq K$ by definition.

(c) Define $K = \{k \in G \mid \alpha(k) \in R_1\}$. We must show $R \subseteq K$. We have $G \xrightarrow{\alpha} H \xrightarrow{\varphi} H/R_1$ where φ is the coset map, and

$$\ker(\varphi\alpha) = \{g \mid \alpha(g) \in \ker \varphi\} = \{g \mid \alpha(g) \in R_1\} = K.$$

This shows that $K \triangleleft G$ and $G/K \cong \varphi\alpha(G) \subseteq H/R_1$. Since H/R_1 is solvable, this shows that G/K is solvable. It follows that $R \subseteq K$ by definition.

24. Write $S = S(G)$.

(a) Write $\{K \triangleleft G \mid K \text{ is solvable}\} = \{K_1, K_2, \ldots, K_m\}$. This set is nonempty as it contains $\{1\}$. Now $K_1 \triangleleft G$ is solvable. Next $K_1 K_2 \triangleleft G$ and is solvable by Exercise 19. If $K_1 K_2 \cdots K_n$ is normal and solvable, so is $(K_1 K_2 \cdots K_n)K_{n+1}$. Hence S is normal and solvable by induction. If K is normal and solvable then $K = K_i$ for some i so $K \subseteq S$.

(c) $\alpha(S) \triangleleft H$ because α is onto, and it is solvable by Theorem 3. Hence $\alpha(S) \subseteq S(H)$.

(e) Write $S\left(\frac{G}{S}\right) = \frac{K}{S}$. Then $K \triangleleft G$ and K is solvable, so $K \subseteq S$ by (a). Hence $K = S$ and we are done.

25. (a) If $G = G_0 \supset G_1 \supset \cdots \supset G_n = \{1\}$ is a solvable series, then G_i/G_{i+1} is abelian so, by taking a subgroup of prime index repeatedly we get

$$\frac{G_i}{G_{i+1}} = \frac{G_{i0}}{G_{i+1}} \supset \frac{G_{i1}}{G_{i+1}} \supset \cdots \supset \frac{G_{ik}}{G_{i+1}} = \{G_{i+1}\}$$

where $\frac{G_{ij}}{G_{i\,j+1}} \cong \frac{G_{ij}/G_{i+1}}{G_{i\,j+1}/G_{i+1}}$ is of prime order. Piece these series together.

(c) If $G = G_0 \supset G_1 \supset \cdots \supset G_n = \{1\}$ has G_i/G_{i+1} cyclic for all i, then Lemma 1 §9.1 shows that if $K \triangleleft G$ then K and G/K have such a series because subgroups and images of cyclic groups are cyclic.

(e) (iii) $\Rightarrow$ (i). G is solvable and finitely generated because $G \triangleleft G$. Now let $G = G_0 \supset G_1 \supset \cdots \supset G_n = \{1\}$ be a solvable series where $G_i \triangleleft G$ for all i (for example the derived series). Then each G_i is finitely generated by (iii) so G_i/G_{i+1} is finitely generated and abelian. It is thus a direct product of cyclic groups by Theorem 3 §7.2 and so is polycyclic. Hence, piecing together series, we get a polycyclic series for G.

27. (a) Write $V = \mathcal{V}(G)$. Then $V \triangleleft G$ because the intersection of normal subgroups is normal. Note that the intersection is not empty because $G \triangleleft G$ and G/G is in $\mathcal{V}$. If $V = K_1 \cap K_2 \cap \cdots \cap K_n$ then G/V embeds in $\frac{G}{K_1} \times \cdots \times \frac{G}{K_n}$ (as in Exercise 18) and $\frac{G}{K_1} \times \cdots \times \frac{G}{K_n}$ is in $\mathcal{V}$ by induction because $\mathcal{V}$ is closed under taking direct products. Hence G/V is in $\mathcal{V}$, being isomorphic to a subgroup of a group in $\mathcal{V}$.

(c) If $K \triangleleft G$ and G/K is in $\mathcal{V}$, we must show $\mathcal{V}(H) \triangleleft K$. Now $\frac{H}{H \cap K} \cong \frac{HK}{K} \subseteq \frac{G}{K}$ so $\frac{H}{H \cap K}$ is in $\mathcal{V}$ because $\frac{G}{K}$ is in $\mathcal{V}$. Then $\mathcal{V}(H) \subseteq H \cap K \subseteq K$.

28. (a) If $k = 0$ it is clear. Write $V_k = \mathcal{V}_k(G)$. If $A(V_k) \subseteq V_k(H)$ then, using the preceding exercise, $\alpha(V_{k+1}) = \alpha(\mathcal{V}(V_k)) \subseteq \mathcal{V}(V_k(H)) = V_{k+1}(H)$.

(c) If G is $\mathcal{V}$-solvable let $G = G_0 \supseteq G_1 \supseteq \cdots \supseteq G_n = \{1\}$ be a $\mathcal{V}$-series. Write $V_k = \mathcal{V}_k(G)$, and show $V_k \subseteq G_k$ by induction on k. It is clear if $k = 0$. If $V_k \subseteq G_k$ then $V_{k+1} = \mathcal{V}(V_k) \subseteq \mathcal{V}(G_k)$ so it suffices to show that $\mathcal{V}(G_k) \subseteq G_{k+1}$. But $G_k/G_{k+1} \in \mathcal{V}$ so this follows from the definition of $\mathcal{V}(G)$. The converse is clear because $G \supset \mathcal{V}_1(G) \supset \mathcal{V}_2(G) \supset \cdots \supset \mathcal{V}_n(G) = \{1\}$.

(e) If G is $\mathcal{V}$-solvable and $\mathcal{V}(H) = H$ for some H, then $\mathcal{V}_k(H) = H \neq \{1\}$ for all k, contradicting (d). Conversely $\mathcal{V}_{k+1}(G) \subset \mathcal{V}_k(G)$ for all k by hypothesis, so $G \supset \mathcal{V}_1(G) \supset \mathcal{V}_2(G) \supset \cdots$. Since G is finite, it reaches $\{1\}$.

9.3 NILPOTENT GROUPS

1. (a) $Z(A_4) = \{\varepsilon\}$ and $Z(A_n) = \{\varepsilon\}$ if $n \geq 5$ because A_n is simple and non-abelian. Proceed as in Example 1.

2. (1) $[H, K] = [K, H]$ because $[h, k]^{-1} = [k, h]$, so inverses of products of commutators are again products of commutators.

(3) If $H \lhd G$ and $K \lhd G$ then $a^{-1}[h, k]a = [a^{-1}ha, a^{-1}ka] \in [H, K]$ for all $h \in H$, $k \in K$. Hence if $c_1, c_2, \ldots, c_k$ are commutators, then

$$a^{-1}(c_1 c_2 \ldots c_k)a = (a^{-1}c_1 a)(a^{-1}c_2 a) \cdots (a^{-1}c_k a)$$

is a product of commutators.

3. $\alpha([h, k]) = [\alpha(h), \alpha(k)]$ for all $h \in H$, $k \in K$, so $\alpha([H, K]) \subseteq [\alpha(H), \alpha(K)]$. Since each element of $[\alpha(H), \alpha(K)]$ is a product of commutators of the form $[\alpha(h), \alpha(k)] = \alpha[h, k]$, this is equality. The rest is clear if $\alpha : G \to G$ is any inner automorphism (any automorphism).

5. One verifies that $[H, H] = H'$ for any group H, so $[H', H'] = (H')' = [H, H]'$. Now proceed by induction on k. Clearly $[G^{(0)}, G^{(0)}] = [G, G] = G' = G^{(1)}$. In general, if $[G^{(k)}, G^{(k)}] = (G^{(k+1)})$ then

$$[G^{(k+1)}, G^{(k+1)}] = [(G^{(k)})', (G^{(k)})']'] = [G^{(k)}, G^{(k)}]' = (G^{(k+1)})' = G^{(k+2)}.$$

6. (a) It suffices to show $\Gamma_i(G \times H) \subseteq \Gamma_i(G) \times \Gamma_i(H)$ holds for all $i \geq 0$ (then use induction on the number n of groups). Do this by induction on i. If $i = 0$ then $\Gamma_0(G \times H) = G \times H = \Gamma_0(G) \times \Gamma_0(H)$. If it holds for some i, then

$$\Gamma_{i+1}(G \times H) = [\Gamma_i(G \times H), G \times H] \subseteq [\Gamma_i(G) \times \Gamma_i(H), G \times H]$$

by induction. So it suffices to show that, if $A \subseteq G$ and $B \subseteq H$, then

$$[A \times B, G \times H] \subseteq [A, G] \times [B, H].$$

But $[(a, b), (g, h)] = ([a, g], [b, h])$ so this is clear.

7. If $i = 0$ we have $G_n = \{1\} = Z_0(G)$. Write $Z_i = Z_i(G)$ for each i, and assume inductively that $G_{n-i} \subseteq Z_i$. Let $a \in G_{n-i-1}$. We have $\frac{G_{n-i-1}}{G_{n-i}} \subseteq Z\left[\frac{G}{G_{n-i}}\right]$ so aG_{n-i} commutes with gG_{n-i} for all $g \in G$; that is $[a, g] \in G_{n-i} \subseteq Z_i$. Hence aZ_i commutes with gZ_i, so aZ_i is in $Z\left[\frac{G}{Z_i}\right] = \frac{Z_{i+1}}{Z_i}$. Thus $a \in Z_{i+1}$ and we have shown $G_{n-i-1} \subseteq Z_{i+1}$.

9. If $n = 2^k$ then $|D_n| = 2^{k+1}$ so D_n is nilpotent by Example 3. Conversely, let $n = 2^k m$, $m > 1$ odd. Then $\langle a^{2^k}, b \rangle \cong D_m$ so D_m is nilpotent by Theorem 1. But in this case $\{1, b\}$ is a Sylow 2-subgroup, and it is not normal because $a^{-1}\{1, b\}a = \{1, ba^2\}$ and $ba^2 \neq b$, contradicting Theorem 4.

11. A group G is nilpotent of class 2 if and only if $Z_2(G) = G$ and $Z_1(G) \neq G$. Since $Z_1(G) = Z(G)$, we have $Z(\frac{G}{Z(G)}) = \frac{Z_2(G)}{Z(G)}$. Hence

$$Z_2(G) = G \Leftrightarrow Z(\tfrac{G}{Z(G)}) = \tfrac{G}{Z(G)} \Leftrightarrow \tfrac{G}{Z(G)} \text{ is abelian} \Leftrightarrow G' \subseteq Z(G).$$

Also, G is not abelian if and only if $Z_1(G) \neq G$. Hence G is nilpotent of class 2 if and only if G is nonabelian and $G' \subseteq Z(G)$, as required.

13. $K \cap Z(G) \neq \{1\}$ by Theorem 7. Thus $K \cap Z(G) = K$ by the condition on K, that is $K \subseteq Z(G)$. But then every subgroup of K is normal in G, so $|K|$ is prime.

15. If $|G| = p_1^{n_1} \cdots p_r^{n_r}$, let $m = p_1^{m_1}, \ldots, p_r^{m_r}$. Thus $G \cong P_1 \times P_2 \times \cdots \times P_r$ where $|P_i| = p_i^{n_i}$ for each i (Theorem 3). By Theorem 8 §8.2 let $Q_i \lhd P_i$, $|Q_i| = p_i^{m_i}$. Then $Q_1 \times \cdots \times Q_r \lhd P_1 \times \cdots \times P_r$ and $|Q_1 \times \cdots \times Q_r| = m$.
Conversely, the hypothesis implies that every Sylow subgroup of G is normal in G, so G is nilpotent by Theorem 4.

17. If $M \subseteq G$ is maximal, we show that $M \lhd G$ and invoke Theorem 4. If $K \subseteq M$ then M/K is maximal in G/K, so $M/K \lhd G/K$ by hypothesis, whence $M \lhd G$. If $K \nsubseteq M$ then $KM = G$ because M is maximal. If $g \in G$, say $g = km$, $m \in M$, $k \in K$, then $gMg^{-1} = kmMm^{-1}k^{-1} = kMk^{-1} = M$ (because $k \in Z(G)$). So $M \lhd G$ after all.

18. (a) H is itself nilpotent by Theorem 1. Hence $Z(H) \neq \{1\}$ by Theorem 7.

19. Write $\Gamma_i = \Gamma_i(G)$. Then $\Gamma_1 = G'$ so G/Γ_1 is cyclic. Consider the group $\frac{G}{\Gamma_2}$. The subgroup $\frac{\Gamma_1}{\Gamma_2} \subseteq Z(\frac{G}{\Gamma_2})$ by definition, and $\frac{G/\Gamma_2}{\Gamma_1/\Gamma_2} \cong \frac{G}{\Gamma_1}$ is cyclic. By Theorem 2 §2.9, $\frac{G}{\Gamma_2}$ is abelian. This means $[g, h] \in \Gamma_2$ for all g, h in G, whence $\Gamma_1 \subseteq \Gamma_2$. Hence $\Gamma_1 = \Gamma_2$. Now consider $\frac{G}{\Gamma_3}$; $\frac{\Gamma_1}{\Gamma_3} = \frac{\Gamma_2}{\Gamma_3}$ is central and the factor is $\frac{G}{\Gamma_1}$, so $\frac{G}{\Gamma_3}$ is abelian and $\Gamma_3 = \Gamma_1$. Continuing: $\Gamma_1 = \Gamma_2 = \cdots = \Gamma_n = \{1\}$. Thus $G' = \Gamma_1 = \{1\}$ and G is abelian.

21. Let $D_n = \langle a, b \rangle$ with the usual presentation. Note first that $\bar{H}$ in the Hint is closed so it is a subgroup (D_n is finite). Since $|\bar{H}| = 2|H|$, it has index m.

(a) Let $H = \langle a^2 \rangle$ so $\bar{H} = \{1, a^2, b, ba^2\}$. Both are maximal (index 3) so $\Phi \subseteq H \cap \bar{H} = \{1, a^2\}$. If $\Phi = \{1\}$ then $D_4' = \{1\}$ by (3) of Theorem 5 because D_4 is nilpotent ($|D_4| = 8$). But $D_4' \neq \{1\}$ because D_4 is not abelian. So $\Phi = \{1, a^2\}$.

(c) If $H = \langle a^p \rangle$ then $\bar{H} = H \cup Hb$ is a subgroup of index p and so is maximal. Similarly, if $K = \langle a^q \rangle$ then $\bar{K}$ has index q. Hence $\Phi \subseteq \bar{H} \cap \bar{K} = \{1, b\}$. But $\langle a \rangle$ is also maximal (index 2) and so $\Phi \subseteq \langle a \rangle$. It follows that $\Phi = \{1\}$.

23. Write $Z = Z(G)$. We have $Z = G'$ and $|Z| = p$ by Exercise 26 §8.2. Since G is nilpotent (Example 3), $G' \subseteq \Phi$ by Theorem 5. But $\Phi \neq G$, so the only other possibility is $|\Phi| = p^2$. But then Φ is itself maximal, and so is the unique

maximal subgroup of G. In particular every subgroup $H \neq G$ is contained in Φ (it is contained in *some* maximal subgroup). If $a \notin \Phi$, then $\langle a \rangle \not\subseteq \Phi$ so $\langle a \rangle = G$. This means G is abelian contrary to hypothesis so $|\Phi| = p$, whence $\Phi = Z = G'$.

24. (1) $\Rightarrow$ (2). We show that $G' \subseteq \Phi$, that is $G' \subseteq M$ for every maximal subgroup M of G. But $M \triangleleft G$ by Theorem 3, so G/M has order p for a prime p by maximality. Hence G/M is abelian, whence $G' \subseteq M$.

25. Let $\alpha : G \to G/K$ be the coset map. Then Theorem 5 gives
$$\alpha(\Phi(G)) \subseteq \Phi[\alpha(G)] = \{K\}$$
so $\Phi(G) \subseteq \ker \alpha = K$.

26. (a) Write $\Phi(G) = \Phi$ and $\Phi\left(\frac{G}{K}\right) = \frac{F}{K}$ where $F \triangleleft G$. Then we must show that $F = \Phi$. If M is maximal in G then $K \subseteq M$ (because $K \subseteq \Phi \subseteq M$) and $\frac{M}{K}$ is maximal in $\frac{G}{K}$. Hence $\frac{F}{K} \subseteq \frac{M}{K}$ whence $F \subseteq M$. It follows that $F \subseteq \Phi$. Conversely, if $\alpha : G \to G/K$ is the coset map, then $\alpha(\Phi) \subseteq \Phi(G/K) = \frac{F}{K}$ by Theorem 5. Hence $x \in \Phi$ implies $xK \in \frac{F}{K}$; so $x \in F$. Thus $\Phi \subseteq F$ and so $\Phi = F$ as required.

27. If $(x,y) \in \Phi(G \times H)$ let $M \subseteq G$ be maximal. Then $M \times H \subseteq G \times H$ is maximal, so $(x,y) \in M \times H$. Thus $x \in M$, whence $x \in \Phi(G)$. Similarly $y \in \Phi(H)$, so $\Phi(G \times H) \subseteq \Phi(G) \times \Phi(H)$. Now define $\sigma : G \to G \times H$ by $\sigma(g) = (g,1)$. Then $\Phi(G) \times \{1\} = \sigma(\Phi(G)) \subseteq \Phi(G \times H)$ by Exercise 26. Similarly we obtain $\{1\} \times \Phi(H) \subseteq \Phi(G \times H)$, and we're done.

29. (a) Write $Z = Z(G)$. Suppose $Z \not\subseteq M$. Then $M \subset MZ \subseteq G$ so $MZ = G$ because M is maximal in G. If $m \in M$ and $a \in G$, let $a = m_1 z$, $m_1 \in M$, and $z \in Z$. Then $a^{-1}ma = z^{-1}m_1^{-1}mm_1z = m^{-1}mm_1 \in M$; that is $M \triangleleft G$. Hence $|G/M|$ is a prime so G/M is abelian. But then $G' \subseteq M$.

31. Since G is a p-group, $M \subseteq G$ maximal implies $N(M) \neq M$, whence $M \triangleleft G$. Thus $G/M \cong C_p$, and so $G' \subseteq M$ and $g^p \in M$ for all $g \in G$. Then $\langle G' \cup X \rangle \subseteq \Phi(G)$ where $X = \{g^p \mid g \in G\}$. For the converse, write $K = \langle G' \cup X \rangle$. Then $G' \subseteq K$ so $K \triangleleft G$, and G/K is an elementary abelian p-group because $(gK)^p = g^pK = K$ for all g. It follows that $\Phi(G/K) = 1$, and hence that $\Phi(G) \subseteq K$ by Exercise 26(b).

Chapter 10

Galois Theory

10.1 GALOIS GROUPS AND SEPARABILITY

1. First $\varepsilon \in \mathrm{gal}(E:F)$ because $\varepsilon(a) = a$ for all $a \in F$. If σ, $\tau \in \mathrm{gal}(E:F)$ then $\sigma(a) = a$ for all $a \in F$, so $a = \sigma^{-1}(a)$ for all a; hence (since σ^{-1} is an automorphism) $\sigma^{-1} \in \mathrm{gal}(E:F)$. Finally $\sigma\tau(a) = \sigma[\tau(a)] = \sigma(a) = a$, so $\sigma\tau \in \mathrm{gal}(E:F)$.

3. Let $\sigma(u_i) = \tau(u_i)$ for all i where σ, $\tau \in \mathrm{gal}(E:F)$. If $v \in E$, write $v = \sum_{i=1}^{n} a_i u_i$, $a_i \in F$. Then

$$\sigma(v) = \sum_{i=1}^{n} \sigma(a_i) \cdot \sigma(u_i) = \sum_{i=1}^{n} a_i \cdot \tau(u_i) = \sum_{i=1}^{n} \tau(a_i) \cdot \tau(u_i) = \tau(v).$$

As $v \in E$ was arbitrary, this shows that $\sigma = \tau$.

5. Let $\sigma : E \to E$ be an automorphism. Then if $n \in \mathbb{Z}$, $\sigma(n) = n \cdot \sigma(1) = n$. Thus $\sigma\left(\frac{n}{m}\right) = \sigma(nm^{-1}) = \sigma(n) \cdot [\sigma(m)]^{-1} = nm^{-1} = \frac{n}{m}$ for all $\frac{n}{m} \in \mathbb{Q}$. Hence $\sigma \in \mathrm{gal}(E:\mathbb{Q})$.

7. Put $u = e^{2\pi i/6}$. Then u satisfies $x^6 - 1 = (x^2 - 1)(x^2 + x + 1)(x^2 - x + 1)$, and the only roots in $\mathbb{C}$ are the sixth roots of unity $1, u, u^2, u^3, u^4, u^5$. But $u^3 = -1$ so u satisfies $x^3 + 1 = (x + 1)(x^2 - x + 1)$. Hence u is a root of $m = x^2 - x + 1$. Since m is irreducible over $\mathbb{Q}$, it is the minimal polynomial of u. The other root of m is u^5 (because $(u^5)^3 = (u^3)^5 = (-1)^5 = -1$), so Theorem 1 gives $\mathrm{gal}(E:\mathbb{Q}) = \{\varepsilon, \sigma\}$ where $\sigma(u) = u^5$. Note that $\sigma^2 = \varepsilon$ because $\sigma^2(u) = \sigma[\sigma(u)] = \sigma(u^5) = [\sigma(u)]^5 = u^{25} = u$. Hence $\mathrm{gal}(E:\mathbb{Q}) \cong C_2$.

9. The minimal polynomial of i is $x^2 + 1$, and that of $\sqrt{3}$ is $x^2 - 3$. By Lemma 2, there exists a $\mathbb{Q}$-isomorphism $\sigma_0 : \mathbb{Q}(i) \to \mathbb{Q}(i)$ with $\sigma_0(i) = -i$. This extends to an automorphism σ of $E = \mathbb{Q}(i)(\sqrt{3}) = \mathbb{Q}(-i)(\sqrt{3})$ satisfying $\sigma(\sqrt{3}) = \sqrt{3}$. Thus $\sigma \in \mathrm{gal}(E:\mathbb{Q})$, $\sigma(i) = -i$ and $\sigma(\sqrt{3}) = \sqrt{3}$. Similarly, there exists $\tau \in \mathrm{gal}(E:\mathbb{Q})$ with $\tau(i) = i$ and $\tau(\sqrt{3}) = -\sqrt{3}$. Note that $\sigma(i) \in \{i, -i\}$

Student Solution Manual to Accompany Introduction to Abstract Algebra, Fourth Edition. W. Keith Nicholson.

and $\tau(\sqrt{3}) \in \{\sqrt{3}, -\sqrt{3}\}$ so $|\text{gal}(E : \mathbb{Q})| \leq 2 \cdot 2 = 4$ by Theorem 2. Now observe that

$$\sigma\tau(i) = -i = \tau\sigma(i) \qquad \text{and} \qquad \sigma\tau(\sqrt{3}) = -\sqrt{3} = \tau\sigma(\sqrt{3}).$$

Hence $\tau\sigma = \sigma\tau$ by Theorem 2. Since $o(\sigma) = 2 = o(\tau)$, it follows that $\{\varepsilon, \sigma, \tau, \sigma\tau\} \cong C_2 \times C_2$. But $\{\varepsilon, \sigma, \tau, \sigma\tau\} \subseteq \text{gal}(E : \mathbb{Q})$ and $|\text{gal}(E : \mathbb{Q})| \leq 4$, so $\text{gal}(E : \mathbb{Q}) = \langle \sigma, \tau \rangle \cong C_2 \times C_2$.

10. (a) Write $u = \sqrt[4]{2}$. Then u satisfies $x^4 - 2$, irreducible by the Eisenstein criterion. The roots of $x^4 - 2$ in $\mathbb{C}$ are $u, -u, iu$ and $-iu$, and the only ones in $E = \mathbb{Q}(u)$ are u and $-u$. Thus Theorem 1 gives $|\text{gal}(E : \mathbb{Q})| = 2$ so $\text{gal}(E : \mathbb{Q}) = \{\varepsilon, \sigma\} \cong C_2$ where $\sigma(u) = -u$.

11. Choose $u \in E$, $u \notin F$. Thus $[F(u) : F] \neq 1$ so, since $[E : F] = 2$ is prime, $E = F(u)$ by Theorem 5 §6.2 (the multiplication theorem). If m is the minimal polynomial of u over F, then $\deg m = 2$. Thus since one root u lies in E, so does the other (their sum is the negative of the coefficient of x in m). Now the result is clear by Theorem 1.

13. Write $u = \sqrt[4]{2}$. Now $x^4 - 2$ is irreducible (Eisenstein) so $x^4 - 2$ and $x^2 + 1$ are the minimal polynomials of u and i. The roots are $\{u, -u, iu, -iu\}$ and $\{i, -i\}$, and all are in E so $|G| \leq 8$ by Theorem 1 where $G = \text{gal}(E : \mathbb{Q})$. By Lemma 1 there exists $\sigma_0 : \mathbb{Q}(u) \to \mathbb{Q}(iu)$ satisfying $\sigma_0(u) = iu$. Extend it to an automorphism σ of $E = \mathbb{Q}(u)(i) = \mathbb{Q}(iu)(i)$ where $\sigma(i) = i$. Thus $\sigma(u) = iu$ and $\sigma(i) = i$, and so $\sigma^2(u) = -u$, $\sigma^3(u) = -iu$, $\sigma^4(u) = u$. Thus $o(\sigma) = 4$. Next let $\tau_0 : \mathbb{Q}(i) \to \mathbb{Q}(i)$ have $\tau_0(i) = -i$, extend τ_0 to an automorphism τ of $E = \mathbb{Q}(i)(u) = \mathbb{Q}(-i)(u)$ where $\tau(u) = u$. Then $\tau^2 = \varepsilon$ so $o(\tau) = 2$. Finally

$$\sigma\tau\sigma(u) = \sigma\tau(iu) = \sigma(-iu) = u = \tau(u),$$
$$\sigma\tau\sigma(i) = \tau\sigma(i) = \sigma(-i) = -i = \tau(i).$$

Thus $\sigma\tau\sigma = \tau$, whence $\langle \sigma, \tau \rangle \cong D_4$. Thus $|\langle \sigma, \tau \rangle| = 8$ so, since $\langle \sigma, \tau \rangle \subseteq G$ and $|G| = 8$, $G = \langle \sigma, \tau \rangle \cong D_4$.

15. (a) If $v = \sqrt{3}$ and $w = \sqrt{5}$, then $\frac{v_i - v}{w - w_j}$ is 0 or $\frac{-\sqrt{3}}{\sqrt{5}}$. So $a = 1$ is none of these, whence $E = \mathbb{Q}(v + 1w) = \mathbb{Q}(\sqrt{3} + \sqrt{5})$ by the proof of Theorem 6.

16. (a) If $v = \sqrt{p}$ and $w = \sqrt{q}$, then $\frac{v_i - v}{w - w_j}$ is 0 or $\frac{-\sqrt{p}}{\sqrt{q}}$. So $a = 1$ is neither of these, whence $E = \mathbb{Q}(v + 1w) = \mathbb{Q}(\sqrt{p} + \sqrt{q})$ by the proof of Theorem 6.

17. If $u < v$ then $v - u = w^2$ for some $w \in \mathbb{R}$. If $\sigma \neq \varepsilon$ in $\text{gal}(\mathbb{R} : \mathbb{Q})$, then $\sigma(v) - \sigma(u) = [\sigma(w)]^2$ and $\sigma(u) < \sigma(v)$. Now let $u \in \mathbb{R}$. If $u \neq \sigma(u)$ let $u < \sigma(u)$. If $u < a < \sigma(u)$, $a \in \mathbb{Q}$, then $\sigma(u) < \sigma(a) = a$, a contradiction. A similar argument dimates $\sigma(u) < u$.

19. We proceed by induction on n. If $n = 1$ then $E = F(u_1) = \{f(u_1) \mid f \in F[x]\}$. Hence $\sigma(f(u_1)) = f(\sigma(u_1)) = f(\tau(u_1)) = \tau(f(u_1))$ for all f, as required. In general, write $K = F(u_1, u_2, \ldots, u_{n-1})$ so that $E = K(u_n)$. By induction, $\sigma = \tau$ on K, so $\sigma, \tau \in \text{gal}(K : F)$. Since $\sigma(u_n) = \tau(u_n)$ the result follows from the case $n = 1$.

21. Let $\sigma \in \text{gal}(F(t) : F)$. If $\sigma(t) = \frac{f(t)}{g(t)}$, f, g relatively prime in $F[t]$, then $\sigma\left(\frac{p(t)}{q(t)}\right) = \frac{\sigma[p(t)]}{\sigma[q(t)]} = \frac{p(\sigma(t))}{q(\sigma(t))}$, so if $\lambda = \frac{p}{q}$, $\sigma(\lambda(t)) = \lambda(\sigma(t))$. Thus σ is determined

completely by f and g. Clearly $f \neq 0$, $g \neq 0$. Now suppose $\sigma^{-1}(t) = \frac{f_1}{g_1}$. Then $x = \sigma\left(\frac{f_1}{g_1}\right) = \frac{f_1(f/g)}{g_1(f/g)}$. Suppose

$$f_1(t) = a_0 + a_1 t + \cdots + a_n t^n, \quad g_1(t) = b_0 + b_1 t + \cdots + b_m t^m.$$

Then

$$t\left(b_0 + b_1 \cdot \frac{f}{g} + \cdots + b_m \cdot \frac{f^m}{g^m}\right) = \left(a_0 + a_1 \cdot \frac{f}{g} + \cdots + a_n \cdot \frac{f^n}{g^n}\right).$$

Suppose $n > m$. Then

$$tg^{n-m}(b_0 g^m + b_1 fg^{m-1} + \cdots + b_n f^m) = a_0 g^n + a_1 fg^{n-1} + \cdots + a_n f^n.$$

Since $a_n f^n$ is the only term not involving g, it follows that $g | a_n f^n$. Hence $g | a_n$ since f, g are relatively prime, so deg $g = 0$. Similarly $n < m$ implies that $g | b_m t f^m$ so, again, deg $g \leq 1$. Thus either deg $g \leq 1$ or $n = m$. But if $n = m$ we have $t(b_0 g^n + b_1 fg^{n-1} + \cdots + b_n f^n) = a_0 g^n + \cdots + a_n f^n$. This yields $g | (a_n - b_n t) f^n$ and deg $g \leq 1$ in all cases. Now observe that $\sigma(\frac{1}{t}) = \frac{g}{f}$ so $\frac{1}{t} = \sigma^{-1}\left(\frac{g}{f}\right) = \frac{f_1(g/f)}{g_1(g/f)}$. Hence $g_1\left(\frac{g}{f}\right) = tf_1\left(\frac{g}{f}\right)$ and the same type of argument implies deg $f \leq 1$. Hence we have proved that $\sigma(t) = \frac{at+b}{ct+d}$ so that $\sigma = \sigma_M$ where $M = \begin{bmatrix} a & b \\ c & d \end{bmatrix}$. Now observe that $\sigma_M[\sigma_N(\lambda(x))] = \sigma_{NM}\lambda(x)$ holds for all $\lambda(t) \in F(t)$. In particular if $\sigma \in \mathrm{gal}(F(t) : F)$ and $\sigma = \sigma_M$, and if $\sigma^{-1} = \sigma_N$, we have $t = \sigma_M \sigma_N(t) = \sigma_{NM}(t) = \left(\frac{a''t+b''}{c''t+d''}\right)$ where $NM = \begin{bmatrix} a'' & b'' \\ c'' & d'' \end{bmatrix}$. It follows that $c''x^2 + d''x = a''x + b''$ so $c'' = b'' = 0$, $a'' = d'' \neq 0$ so $NM = \begin{bmatrix} a'' & 0 \\ 0 & a'' \end{bmatrix}$ and so $M^{-1} = \frac{1}{a''}N$. Thus the map $M \mapsto \sigma_M$ from $GL_2(F) \to \mathrm{gal}\,(F(t) : F)$ is an onto group homomorphism. Moreover the above shows that the kernel is Z.

22. (a) (3) $\Rightarrow$ (1). Let $E \supseteq F$ be a field and suppose f has a repeated root u in E. Then $f(u) = 0 = f'(u)$ by Theorem 3 §6.4. But (3) implies $1 = fg + f'h$ in $F[x]$ with $g, h \in F[x]$, and this is valid in $E[x]$. But then, taking $x = u$ gives $1 = 0$ in E, a contradiction.

23. Here $f' = nx^{n-1} - 1$. Write $d = \gcd(f, f')$; we must show $d = 1$ by the preceding exercise. Let $E \supseteq F$ be a splitting field for f. If $d \neq 1$ then deg $d > 1$ so d has a root u in E. Thus $f(u) = 0 = f'(u)$ so $u^n = u$ and $nu^{n-1} = 1$. Hence $(n-1)u = 0$, a contradiction if char $F = 0$. But if char $F = p$ it implies $p | (n-1)$, contrary to hypothesis.

25. If $E \supseteq F$ and $f \in F[x]$ is separable over F, let q be any irreducible factor of f in $E[x]$. We must show that q has no repeated root in E. Now $f = p_1 p_2 \cdots p_r$ in $F[x]$ where the p_i are irreducible and (by hypothesis) separable. But $f = p_1 p_2 \cdots p_r$ in $E[x]$ so $q | p_i$ for some i. Hence any repeated root of q in some splitting field of f would be a repeated root of p_i, contrary to assumption. So q is separable.

27. Suppose $f = x^p - a$ is not a power of a linear polynomial in $F[x]$; we must show it is irreducible. If u is a root of f in an extension $E \supseteq F$, then $u^p = a$ so $f = x^p - u^p = (x - u)^p$ in $E[x]$ because the characteristic is p. Then $u \notin F$

and $F(u)$ is a splitting field of f over F. Let q be an irreducible factor of f in $F[x]$. Then $q|(x-u)^p$ in $E[x]$ so $q = (x-u)^t$. Then $t > 1$ because $u \notin F$ so $q' = 0$ by Lemma 5. Hence $q = g(x^p)$ by Theorem 4. Thus, the factorization of f into irreducibles in $F[x]$ takes the form

$$f = x^p - a = g_1(x^p)g_2(x^p) \cdots g_r(x^p).$$

Since def $f = p$, it follows that $f = g_i(x^p) = q_i$ for some i, so f is irreducible.

28. (a) (3) $\Rightarrow$ (1). If $E \supseteq F$ is algebraic and $u \in E$, the minimal polynomial of u is separable by (3), and so $E \supseteq F$ is a separable extension.

    (c) Let $E \supseteq F$ be algebraic, F perfect. If $K \supseteq E$ is algebraic, then $K \supseteq F$ is algebraic by Corollary 1, Theorem 6 §6.2, hence separable by hypothesis.

29. (a) If F is perfect, and $a \in F$, let E be the splitting field of $f = x^p - a$. If $u \in E$ is a root of f then $u^p = a$, so $f = x^p - a = x^p - u^p = (x-u)^p$. Let q be an irreducible factor of f in $F[x]$. Then $q = (x-u)^m$. But $E \supseteq F$ is finite, hence separable by hypothesis. Thus q has distinct roots, whence $m = 1$ and $q = x - u$. But then $u \in F$ and $u^p = a$, as required.

    Conversely, let q be irreducible in $F[x]$. If q is not separable then $q' = 0$ so $q = f(x^p)$ for some $f \in F[x]$ by Theorem 4, say $f = \sum_{i=0}^n a_i(x^p)^i$. By hypothesis, let $a_i = b_i^p$, $b_i \in F$. Then

    $$q = b_0^p + b_1^p x^p + b_2^p x^{2p} + \cdots + b_n^p x^{np} = (b_0 + b_1 x + \cdots + b_n x^n)^p.$$

    This contradicts the irreducibility of q.

30. (a) Let q be the minimal polynomial of u over F. If $K = F(u^p)$ let $m \in K[x]$ be the minimal polynomial of u over K. Then $q \in K[x]$ and $q(u) = 0$, so $m|q$. But q has distinct roots by hypothesis, so m has distinct roots. On the other hand, $x^p - u^p \in K[x]$ and $x^p - u^p = (x-u)^p$ in $E[x]$. Hence $m|(x-u)^p$ so $m = (x-u)^r$. Since m has distinct roots, $r = 1$ and so $u \in K$.

    (c) Extend $\{w_1, \ldots, w_k\}$ to a basis $\{w_1, \ldots, w_k, \ldots, w_n\}$ of E over F. Suppose $w \in E = F(E^p)$, say $w = \sum_{i=1}^m a_i u_i^p$. Write $u_i = \sum_{j=1}^n b_{ij} w_j$, so $u_i^p = \sum_{j=1}^n b_{ij}^p w_j^p$, whence $w = \sum_{j=1}^n \left(\sum_{i=1}^m a_i b_{ij}^p \right) w_j^p$. Thus $\{w_1^p, \ldots, w_k^p, \ldots, w_n^p\}$ spans E and so is F-independent because dim $E = n$ (Theorem 7 §6.1). But then $\{w_1^p, \ldots, w_k^p\}$ is F-independent.

31. If $E \supseteq F$ is separable, the result is Exercise 26. If $E \supseteq K$ and $K \supseteq F$ are separable, we may assume char $F = p$ by the Corollary to Theorem 4. By the preceding exercise, $E = K(E^p)$ and $K = F(K^p)$. If $u \in E$ then $u = \sum_{i=1}^m v_i u_i^p$ where $v_i \in K$ and $u_i \in E$. But $v_i = \sum_{j=1}^n a_{ij} v_{ij}^p$ where $a_{ij} \in F$ and $v_{ij} \in K$ for all i, j. Thus $u = \sum_{i=1}^m \sum_{j=1}^n a_{ij}(u_i v_{ij})^p \in F(E^p)$. Thus $E \supseteq F$ is separable by the preceding exercise.

32. (a) Let p and q be the minimal polynomials of u over F and K respectively. Then $p \in K[x]$ and $p(u) = 0$, so $q|p$. Since p has distinct roots in some splitting field $L \supseteq K$, q is separable over K.

    (c) Let $u, v \in S$. Then $F(u) \supseteq F$ is separable and, since v is separable over $F(u)$ by (a), $F(u, v) = F(u)(v) \supseteq F(u)$ is also separable. Hence $F(u, v) \supseteq F$ is separable by the preceding exercise. Since $u \pm v$ and uv are in $F(u, v)$, they are separable over F, so S is a field. Clearly $S \supseteq F$ since the minimal polynomial of $a \in F$ is $x - a$, and $S \supseteq F$ is separable by the definition of S. If $E \supseteq K \supseteq F$ and $K \supseteq F$ is separable, then each $u \in K$ is separable over F; that is $u \in K$. Hence $K \subseteq S$.

10.2 THE MAIN THEOREM OF GALOIS THEORY

1. (a) $E = \mathbb{Q}(u)$ is the splitting field of $x^5 - 1$ over $\mathbb{Q}$, and $E \supseteq \mathbb{Q}$ is separable as char $\mathbb{Q} = 0$. By Example 4 §10.1, $\mathrm{gal}(E : \mathbb{Q}) = \langle \sigma \rangle \cong C_4$ where $\sigma(u) = u^2$ (because $\mathbb{Z}_5^* = \langle 2 \rangle$). The lattices are:

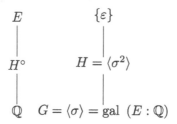

Thus H° is the only intermediate field, $H^\circ \supseteq \mathbb{Q}$ is Galois (as $H \triangleleft G$), and $[H^\circ : \mathbb{Q}][H^\circ : G^\circ] = |G : H| = 2$. We have $\sigma^2(u) = u^4 = u^{-1}$ so $\sigma^2(u + u^4) = (u + u^4)$. Hence $\mathbb{Q}(u + u^4) \subseteq H^\circ$ so, since $u + u^4 \notin \mathbb{Q}$, $H^\circ = \mathbb{Q}(u + u^4)$. Of course $E = \mathbb{Q}(u)$.

(c) $E = \mathbb{Q}(i, \sqrt{3})$ splits $(x^2 + 1)(x^2 - 3)$ over $\mathbb{Q}$, so it is a Galois extension. Clearly $x^2 + 1$ is the minimal polynomial of i, and has roots $\{i, -i\}$ in E; and $x^2 - 3$ is the minimal polynomial of $\sqrt{3}$, and has roots $\{\sqrt{3}, -\sqrt{3}\}$. It follows by Theorem 1 that $|G| \leq 4$ where $G = \mathrm{gal}(E : Q)$. Construct $\sigma_0 : \mathbb{Q}(\sqrt{3}) \to \mathbb{Q}(\sqrt{3})$ such that $\sigma_0(\sqrt{3}) = \sqrt{3}$; then extend σ_0 to $\sigma : \mathbb{Q}(\sqrt{3})(i) \to \mathbb{Q}(\sqrt{3})(i)$ where $\sigma(i) = -i$. Hence $\sigma \in G$ and $\sigma(\sqrt{3}) = \sqrt{3}$. Similarly, construct $\tau_0 : \mathbb{Q}(i) \to \mathbb{Q}(i)$ such that $\tau_0(i) = i$; then extend τ_0 to $\tau : \mathbb{Q}(i)(\sqrt{3}) \to \mathbb{Q}(i)(\sqrt{3})$ where $\tau(\sqrt{3}) = -\sqrt{3}$. Hence $\tau \in G$ and $\tau(i) = i$. It follows easily that $\sigma^2 = \varepsilon$ and $\tau^2 = \varepsilon$, so $|\sigma| = 2 = |\tau|$. Now consider $\langle \sigma, \tau \rangle \subseteq G$. Since $|\langle \sigma, \tau \rangle| \geq 3$ and $|G| \leq 4$, it follows that $G = \langle \sigma, \tau \rangle \cong C_2 \times C_2$. If $H_0 = \langle \sigma \rangle$, $H_1 = \langle \tau \rangle$ and $H_2 = \langle \sigma \tau \rangle$, the lattices are

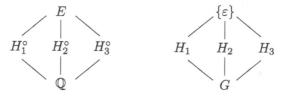

Here each $H_i^\circ \supseteq \mathbb{Q}$ is Galois (because $H_i \triangleleft G$) and $[H_i^\circ : \mathbb{Q}] = 2$. Hence $\sigma(\sqrt{3}) = \sqrt{3}$ means $\mathbb{Q}(\sqrt{3}) \subseteq H_1^\circ$; then $H_1^\circ = \mathbb{Q}(\sqrt{3})$ because $\sqrt{3} \notin \mathbb{Q}$. Similarly $H_2^\circ = \mathbb{Q}(i)$. Finally, $\sigma\tau(i\sqrt{3}) = \sigma(-i\sqrt{3}) = -(-i\sqrt{3}) = i\sqrt{3}$, so $H_3^\circ = \mathbb{Q}(i\sqrt{3})$.

(e) $E = \mathbb{Q}(\sqrt[4]{2}, i) = \mathbb{Q}(u, i)$, $u = \sqrt[4]{2}$; so $E = \mathbb{Q}(u + i)$ by Theorem 6 §10.1. Then E is the splitting field of $x^4 - 2 = (x - u)(x + u)(x - iu)(x + iu)$ so

$E \supseteq \mathbb{Q}$ is Galois. By Exercise 13 §10.1, $G = \mathrm{gal}(E : \mathbb{Q}) = \langle \sigma, \tau \rangle \cong D_4$ where $o(\sigma) = 4$, $o(\tau) = 2$, $\sigma\tau\sigma = \tau$ and $\begin{array}{ll} \sigma(u) = iu & \tau(u) = u \\ \sigma(i) = i & \tau(i) = -i \end{array}$. The lattice diagrams are:

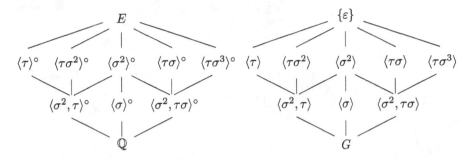

Here $\langle \sigma^2 \rangle, \langle \sigma^2, \tau \rangle, \langle \sigma \rangle$ and $\langle \sigma^2, \tau\sigma \rangle$ are normal in G (because $\langle \sigma^2 \rangle = Z(G)$), so $\langle \sigma^2 \rangle^\circ \supseteq \mathbb{Q}$, $\langle \sigma^2, \tau \rangle^\circ \supseteq \mathbb{Q}$, $\langle \sigma \rangle^\circ \supseteq \mathbb{Q}$ and $\langle \sigma^2, \tau\sigma \rangle^\circ \supseteq \mathbb{Q}$ are Galois.

(1) $\langle \sigma^2 \rangle^\circ = \mathbb{Q}(u^2, i) = \mathbb{Q}(\sqrt{2}, i)$: Hence $\sigma^2(i) = i$; $\sigma(u^2) = -u^2$ so
$$\sigma^2(u) = \sigma(-u^2) = u^2.$$
Thus $\mathbb{Q}(u^2, i) \subseteq \langle \sigma^2 \rangle^\circ$. Also $[\langle \sigma^2 \rangle^\circ : \mathbb{Q}] = |G : \langle \sigma^2 \rangle| = 4$, and so we are done because $[\mathbb{Q}(u^2, i) : \mathbb{Q}] = 4$. Now $\mathbb{Q}(u^2, i) = \mathbb{Q}(u^2 + i)$ by Theorem 6 §10.1.

(2) $\langle \sigma \rangle^\circ = \mathbb{Q}(i)$: We have $\mathbb{Q}(i) \subseteq \langle \sigma \rangle^\circ$; and $[\langle \sigma \rangle^\circ : \mathbb{Q}] = |G : \langle \sigma \rangle| = 2$.

(3) $\langle \sigma^2, \tau \rangle^\circ = \mathbb{Q}(u^2) = \mathbb{Q}(\sqrt{2})$: We have $\tau(u^2) = u^2$ and $\sigma^2(u^2) = u^2$ (in (1)) so $\mathbb{Q}(u^2) \subseteq \langle \sigma^2, \tau \rangle^\circ$. But $[\langle \sigma^2, \tau \rangle^\circ : \mathbb{Q}] = |G : \langle \sigma^2, \tau \rangle| = 2$ and $u^2 = \sqrt{2} \notin \mathbb{Q}$.

(4) $\langle \sigma^2, \tau\sigma \rangle^\circ = \mathbb{Q}(iu^2)$: Check $\mathbb{Q}(iu^2) \subseteq \langle \sigma^2, \tau\sigma \rangle^\circ$ and $[\langle \sigma^2, \tau\sigma^2 \rangle^\circ : \mathbb{Q}] = 2$.

(5) $\langle \tau \rangle^\circ = \mathbb{Q}(u)$: Clearly $\mathbb{Q}(u) \subseteq \langle \tau \rangle^\circ$ and $[\mathbb{Q}(u) : \mathbb{Q}] = 4$ as $x^4 - 2$ is irreducible.

(6) $\langle \tau\sigma \rangle^\circ = \mathbb{Q}(u - iu)$: We have $\mathbb{Q}(u - iu) \subseteq \langle \tau\sigma \rangle^\circ$ as $\tau\sigma(iu) = -u$ and $\tau\sigma(u) = -iu$. This is equality as $x^4 + 8$ is the minimum polynomial of $u - iu = \sqrt[4]{2}(1 - i)$.

(7) $\langle \tau\sigma^2 \rangle^\circ = \mathbb{Q}(iu)$: Compute $\mathbb{Q}(iu) \subseteq \langle \tau\sigma^2 \rangle^\circ$; $iu = i\sqrt[4]{2}$ has minimum polynomial $x^4 - 2$.

(8) $\langle \tau\sigma^3 \rangle^\circ = \mathbb{Q}(u + iu)$: We have $\mathbb{Q}(u + iu) \subseteq \langle \tau\sigma^3 \rangle^\circ$ because $\tau\sigma^3(u) = iu$ and $\tau\sigma^3(iu) = u$. The minimum polynomial of $u + iu = \sqrt[4]{2}(1 + i)$ is $x^4 + 8$.

2. (a) If $G = \mathrm{gal}(E : F)$ then $|G| = p^2$ implies G is abelian, so $G \cong C_{p^2}$ or $G \cong C_p \times C_p$ by Theorem 7, §8.2.

If $G \cong C_{p^2} = \langle \sigma \rangle$, the lattices are:

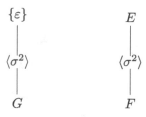

Here $\langle \sigma^2 \rangle^\circ \supseteq F$ is Galois because $\langle \sigma^2 \rangle \lhd G$, and $[\langle \sigma^2 \rangle^\circ : F] = 2$.

If $G \cong C_p \times C_p = \langle \sigma, \tau \rangle$ the lattices are:

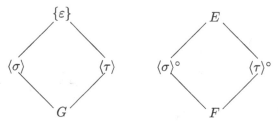

Each of $\langle \sigma \rangle^\circ \supseteq F$ and $\langle \tau \rangle^\circ \supseteq F$ are Galois because $[\langle \sigma \rangle^\circ : F] = 2$ and $[\langle \tau \rangle^\circ : F]$.

3. By Example 6 §10.1, $G = \mathrm{gal}(E : \mathbb{Z}_p) = \langle \sigma \rangle \cong C_n$ where $\sigma : E \to F$ is the Frobenius automorphism given by $\sigma(u) = u^p$. By the Dedekind-Artin theorem $[E : G^\circ] = |G| = n$. But $[E : \mathbb{Z}_p] = n$ so, since $\mathbb{Z}_p \subseteq G^\circ$, $G^\circ = \mathbb{Z}_p$. Thus $E \supseteq \mathbb{Z}_p$ is a Galois extension by Lemma 4.

Now the (inverted) lattice of subgroups of $G = \langle \sigma \rangle$, when $o(\sigma) = 12$, is shown below at the right. Write $E_k = \langle \sigma^k \rangle^\circ = \{u \in E \mid \sigma^k(u) = u\}$ for each divisor k of n. Then the subfield lattice is shown below on the left.

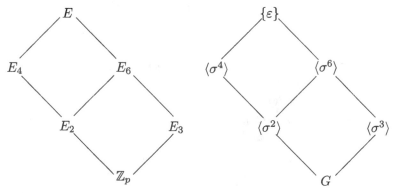

Note that
$$[E_k : \mathbb{Z}_p] = [\langle \sigma^k \rangle^\circ : G^\circ] = |G : \langle \sigma^k \rangle| = k$$
so $|E_k| = p^k$, and $E_k \cong GF(p^k)$.

4. (a) $H = \{\sigma_1^{k_1} \sigma_2^{k_2} \cdots \sigma_n^{k_n} \mid n \geq 1, \sigma_i \in X, k_i \in \mathbb{Z}\}$. Thus $u \in H$ is fixed by all $\tau \in H$ if and only if $\sigma(u) = u$ for all $\sigma \in X$.

5. (a) Let $r = r(t) = \frac{f(t)}{g(t)} \in E$. Then

$$r \in E_G \Leftrightarrow \sigma(r) = r \Leftrightarrow r(t) = r(-t) \Leftrightarrow f(t)g(-t) = f(-t)g(t).$$

If char $F = 2$, this always holds and $K = E_G = E$. Thus $t \in K$ so $m = x - t$.

If char $F \neq 2$, write $h(t) = f(t)g(-t)$. Then $h(t) = h(-t)$ so $h(t) = k(t^2)$ for some polynomial k (because char $F \neq 2$). Thus $f(t)g(-t) = k(t^2)$. Similarly and $g(t)g(-t) = l(t^2)$ for some polynomial l. Thus $r(t) = \frac{k(t^2)}{g(t)g(-t)} = \frac{k(t^2)}{l(t^2)}$, from which $K = F(t^2)$. It follows that $m = x^2 - t^2$. Note that this shows $[E : K] = 2 = |G|$ in this case, as the Dedekind-Artin theorem guarantees.

7. K' is a subgroup of the abelian group G, so it is normal. Then the main theorem applies.

9. (a) If $H \to H^\circ$ is onto, and K is an intermediate field, then $K = H^\circ$ for a subgroup H, so $K^{\circ\prime} = H^{\circ\prime\circ} = H' = K$. Thus K is closed. Conversely, if all intermediate fields K are closed then $K = K'^\circ$ is the image of K', and the map is onto.

11. Write $G = \text{gal}(E : F) = \{\sigma \in \text{aut } E \mid \sigma \text{ fixes } F\}$. Since $K \supseteq F$ we have

$$\text{gal}(E : K) = \{\sigma \in \text{aut } E \mid \sigma \text{ fixes } K\} = \{\sigma \in G \mid \sigma \text{ fixes } K\} = K'.$$

Thus:

$$\begin{aligned}
E \supseteq K \text{ is Galois} &\Leftrightarrow K \text{ is the set of elements of } E \text{ fixed by } \text{gal}(E : K) \\
&\Leftrightarrow K = \{u \in E \mid \sigma(u) = u \text{ for all } \sigma \in K'\} \\
&\Leftrightarrow K = K'^\circ \\
&\Leftrightarrow K \text{ is closed in } E \supseteq F.
\end{aligned}$$

13. If $[E : K] = 6$ then $6 = [E : K] = |K' : E'| = |K' : \{1\}| = |K'|$, so K' is a subgroup of A_4 of order 6. There is no such subgroup (Exercise 34 §2.6).

15. (a) By its definition, $K \vee L$ is the smallest intermediate field containing both K and L. Since the Galois connection $K \to K'$ is order reversing, $(K \vee L)'$ is the largest subgroup of $G = \text{gal}(E : F)$ contained in both K' and L'; that is $(K \vee L)' = K' \cap L'$.

17. If $K = \sigma(K_1)$ let $\lambda \in K_1'$. If $v \in K$ write $v = \sigma(u)$, $u \in K_1$. Then

$$\sigma\lambda\sigma^{-1}(v) = \sigma\lambda(u) = \sigma(u) = v$$

so $\sigma K'\sigma^{-1} \subseteq K'$. On the other hand, if $\mu \in K'$ then $\mu(v) = v$ for all $v \in K$, so $\mu[\sigma(u)] = \sigma(u)$ for all $u \in K_1$. Hence $\sigma^{-1}\mu\sigma \in K_1'$, so $\mu \in \sigma K_1'\sigma^{-1}$. This proves $\sigma K_1'\sigma^{-1} = K'$.

Conversely, assume $\sigma^{-1}K_1'\sigma = K$, that is $K_1' = \sigma K'\sigma^{-1}$. If $u \in K_1$ then $\mu(u) = u$ for all $\mu \in K_1'$, so $\sigma^{-1}\lambda\sigma(u) = u$ for all $\lambda \in K'$, that is $\lambda[\sigma(u)] = \sigma(u)$ for all $\lambda \in K'$, that is $\sigma(u) \in K'^\circ$. Since $E \supseteq F$ is Galois, $K'^\circ = K$, so this shows $\sigma(u) \in K$, $u \in \sigma^{-1}(K)$. Hence $K_1 \subseteq \sigma^{-1}(K)$, so $\sigma(K_1) \subseteq K$. Similarly $K \subseteq \sigma(K_1)$.

19. (a) Let $E = F(u_1, u_2, \ldots, u_m)$ where $X = \{u_1, u_2, \ldots, u_m\}$ is the set of distinct roots of f in E. If $\sigma \in G$ then $\sigma(u_i) \in X$ for all i so $\sigma : X \to X$ is one-to-one (hence onto). Thus σ induces a permutation of these roots:

$$\sigma(u_i) = u_{\bar{\sigma}(i)}, \quad \bar{\sigma} \in S_m.$$

Now the map $\sigma \mapsto \bar{\sigma}$ is a group homomorphism $G \to S_m$ because

$$u_{\overline{\sigma\tau}(i)} = \sigma\tau(u_i) = \sigma(u_{\overline{\tau}(i)}) = u_{\bar{\sigma}\bar{\tau}(i)}$$

for all σ, τ in G. Moreover $\bar{\sigma} = \varepsilon$ means $\sigma(u_i) = u_i$ for all i, so $\sigma = \varepsilon$ in G by Theorem 3 §10.1. Thus $\sigma \mapsto \bar{\sigma}$ is an embedding.

20. (a) If $N(u) = v$ then, given

$$\tau \in G, \; \tau(v) = \tau\left(\prod_{\sigma \in G} \sigma(u)\right) = \prod_{\sigma \in G} \tau\sigma(u).$$

But $\tau\sigma$ traces all of G as σ does, so $\tau(v) = v$. This is true for all $\tau \in G$, so $v \in G^\circ = F$. The proof that $T(u)$ is in F is analogous.

(c) If $K \supseteq F$ is Galois and $H = \mathrm{gal}(K : F)$, then $|H| = n$ by Theorem 3, Corollary 3. Now p splits in K and has distinct roots $u_1, u_2, \ldots, u_m$ (Theorem 3). Hence $p = (x - u_1)(x - u_2)\cdots(x - u_n)$. For each i there exists $\sigma_i \in H$ such that $\sigma_i(u) = u_i$, so $p = \prod_{\sigma \in G}(x - \sigma(u))$. Thus the coefficients of x^{n-1} and 1 are, respectively

$$a_{n-1} = \sum_{\sigma \in H}[-\sigma(u)] = -T_{K/F}(u), \quad \text{and}$$

$$a_0 = \prod_{\sigma \in H}[-\sigma(u)] = (-1)^n N_{K/F}(u).$$

21. If $f = v_0 + v_1 x + \cdots + v_m x^m$ define $f^\tau = \tau(v_0) + \tau(v_1)x + \cdots + \tau(v_n)x^m$ for $\tau \in G$. Thus $f^\tau = \prod_{\sigma \in G}[x - \tau\sigma(u)] = f$ because $\tau\sigma$ runs through G as σ does. It follows that $\tau(v_j) = v_j$ for all $\tau \in G$, whence $v_j \in G^\circ = F$ for each j. Thus $f \in F[x]$. Since $f(u) = 0$, $p|f$ where p is the minimal polynomial of u over F. Write $f = p^m g$ where p and g are relatively prime in $F[x]$. If $g \neq 1$, let q be an irreducible factor of g. Since $q|f$, $0 = q[\sigma(u)] = \sigma[q(u)]$ for some $\sigma \in G$, so $q(u) = 0$. Since $p(u) = 0$ this is a contradiction because $\gcd(p, q) = 1$ in $F[x]$. In fact, $1 = pr + qs$ gives $1 = 0$ when we substitute $x = u$.

10.3 INSOLVABILITY OF POLYNOMIALS

1. (a) $E = \mathbb{Q}(\sqrt{3}, \sqrt[3]{5}, \sqrt[5]{7})$ contains $\sqrt{3}(\sqrt[3]{5} - \sqrt[5]{7})$, and is radical over $\mathbb{Q}$ because

$$\mathbb{Q} \subseteq \mathbb{Q}(\sqrt{3})(\sqrt[3]{5}) \subseteq \mathbb{Q}[\sqrt{3}, \sqrt[3]{5}](\sqrt[5]{7}) = E$$

has the required properties.

2. (a) If $f = x^5 - 4x - 2$ then $f' = 5x^4 - 4$ is zero at $\pm a$, $\pm ai$ where $a = \sqrt[4]{\frac{4}{5}}$. We find $f = x(x^4 - 4) - 2$ so $f(a) = a(\frac{4}{5} - 4) - 2 = \frac{-16a}{5} - 2 < 0$ while $f(-a) = -a(\frac{4}{5} - 4) - 2 = \frac{16a}{5} - 2 > 0$. As in Example 1, this shows that f has three real roots and two (conjugate) nonreal roots. Since f is irreducible by the Eisenstein criterion, its Galois group is S_5 as in Example 1.

3. Take $p = x^7 - 14x + 2$. Then $p' = 7(x^6 - 2)$. If $a = \sqrt[6]{2} \simeq 1.22$, then $p' = 0$ if $n = \pm a, aw, aw^2, aw^4, aw^5$ where $w = e^{2\pi i/6}$. Also $p = x(x^6 - 14) + 2$ so

$$p(a) = a(2 - 12) + 2 = 2 - 10a < 0$$

and

$$p(-a) = -a(2 - 12) + 2 = 2 + 10a > 0.$$

Thus p has three distinct real roots and the rest complex (conjugate pairs). If $E \supseteq \mathbb{Q}$ is the splitting field, view $G = \text{gal}[E : \mathbb{Q}]$ as a subgroup of S_X where $X \subseteq \mathbb{C}$ is the set of roots. If we identify G with as a subgroup of S_X where X is the set of p distinct roots of f, then conjugation gives a transposition in $\mathbb{C}$; if u is a real root, then $[\mathbb{Q}(u) : \mathbb{Q}] = 7$ because p is the minimal polynomial of u over $\mathbb{Q}$ (it is irreducible by Eisenstein). Since $[\mathbb{Q}(u) : \mathbb{Q}] = 7$ divides $|G| = [E : \mathbb{Q}]$, G has an element of order 7 by Cauchy's theorem (Theorem 4 §8.2). Since the 7-cycles are the only elements in S_7 of order 7, the proof in Example 1 goes through.

5. Let X denote the set of roots of p in the splitting field $E \supseteq F$ where $p \in F[x]$. Then $G = \text{gal}(E : F)$ is isomorphic to a subgroup of S_X. Since $|X| \leq 4$, G embeds in S_4. Now S_4 is solvable ($S_4 \supseteq A_4 \supseteq K \supseteq \{\varepsilon\}$ has abelian factors) so every subgroup is solvable by Theorem 3 §9.2.

7. Since $f' = 3(x^2 - 1)$, $f(1) = -1$ and $f(-1) = 3$, f has three real roots. Here $b = -3$ and $c = 1$ so, in the cubic formula, p^3 and q^3 are roots of $x^2 + x + 1$ which satisfy $p^3 + q^3 = -1$ and $pq = 1$. The roots are w and w^2 ($w = e^{2\pi i/3}$) so take $p^3 = w$, $q^3 = w^2$. Thus

$$p = e^{\frac{2\pi}{9}i}, \qquad e^{\frac{8\pi}{9}i}, \qquad \text{or} \qquad e^{\frac{14\pi}{9}i}$$

$$q = e^{\frac{4\pi}{9}i}, \qquad e^{\frac{10\pi}{9}i}, \qquad \text{or} \qquad e^{\frac{16\pi}{9}i}.$$

We need $pq = 1$ so take $p = e^{2\pi i/9}$, $q = e^{16\pi i/9} = e^{-2\pi i/9} = \bar{p}$. The roots are

$$u_1 = p + q = p + \bar{p} = 2 \, \cos\left(\tfrac{2\pi}{9}\right)$$
$$u_2 = wp + w^2 q = e^{8\pi i/9} + e^{10\pi i/9} = 2 \, \cos\left(\tfrac{8\pi}{9}\right)$$
$$u_3 = w^2 p + wq = e^{14\pi i/9} + e^{4\pi i/9} = 2 \, \cos\left(\tfrac{4\pi}{9}\right).$$

If $E \subseteq \mathbb{C}$ is the splitting field, then $E \subseteq \mathbb{R}$ and, since f is irreducible (the roots are not in $\mathbb{Q}$) and separable

$$|\text{gal}[E : \mathbb{Q}]| = [E : \mathbb{Q}] = \deg f = 3.$$

Thus $\text{gal}[E : \mathbb{Q}] \cong C_3$.

8. (a) $\sigma(\Delta^2) = \Delta^2$ for all $\sigma \in G$ because σ permutes the roots u_i. Since $E \supseteq F$ is Galois (f is separable because the u_i are distinct) this means $\Delta^2 \in G^\circ = F$.

   (b) The transposition $\gamma = (u_i \; u_j)$ in G (regarding $G \subseteq S_X$) changes the sign of $(u_i - u_j)$ in Δ, interchanges $(u_i - u_k)$ and $(u_j - u_k)$, and fixes $(u_k - u_m)$. Hence $\gamma(\Delta) = -\Delta$. The even (odd) permutations are products of an even (odd) number of transpositions, so the result follows.

   (c) If $A = \{\sigma \in G \mid \sigma$ is an even permutation of $X\}$ then (b) gives

   $$F(\Delta)' = \text{gal}(E : F(\Delta)) = \{\sigma \in G \mid \sigma(\Delta) = \Delta\} \supseteq A. \qquad (*)$$

   Write $A_X = \{\sigma \in S_X \mid \sigma \text{ even}\}$ so that $A = G \cap A_X$. We have $|S_X : A_X| = 2$ so (Exercise 15 §2.8) either $G \subseteq A_X$ or $|G : G \cap A_X| = |G : A| = 2$. If

$G \subseteq A_X$ then $A = G = F(\Delta)'$. If $|G : A| = 2$ then (*) gives $F(\Delta)' = G$ or $F(\Delta)' = A$. But $F(\Delta)' = G$ implies $\sigma(\Delta) = \Delta$ for all $\sigma \in G$ so $G \subseteq A_X$ a contradiction. Thus $F(A)' = A$ in any case.

(d) Using (c): $\Delta \in F \iff F(\Delta) = F \iff A = F(\Delta)' = G \iff G \subseteq A_X$.

(e) If $f = x^2 + bx + c = (x - u_1)(x - u_2)$ then $u_1 + u_2 = -b$, $u_1 u_2 = c$, so $\Delta^2 = (u_1 - u_2)^2 = (u_1 + u_2)^2 - 4u_1 u_2 = b^2 - 4c$.

(f) If $f = x^3 + bx + c = (x - u)(x - v)(x - w)$ then

$$u + v + w = 0 \tag{1}$$
$$uv + uw + vw = b \tag{2}$$
$$uvw = -c \tag{3}$$

Now (1) and (2) give $b = uv + (u + v)(-u - v) = uv - (u + v)^2$; so $(u + v)^2 = uv - b$. Thus $(u - v)^2 = (u + v)^2 - 4uv = -b - 3uv$. Permuting u, v, w:

$$(u - v)^2 = -b - 3uv$$
$$(u - w)^2 = -b - 3uw$$
$$(v - w)^2 = -b - 3vw.$$

Hence:

$$
\begin{aligned}
-\Delta^2 &= (b + 3uv)(b + 3uw)(b + 3vw) \\
&= b^3 + 3(uv + uw + vw)b^2 + 9(u^2vw + uv^2w + uvw^2)b + 27(uvw)^2 \\
&= b^3 + 3(b)b^2 + 9(uvw)(u + v + w)b + 27(-c)^2 \\
&= 4b^3 - 27c^2.
\end{aligned}
$$

10.4 CYCLOTOMIC POLYNOMIALS AND WEDDERBURN'S THEOREM

1. (a) $\Phi_8(x) = \dfrac{x^8 - 1}{\Phi_1(x)\Phi_2(x)\Phi_4(x)} = \dfrac{x^8 - 1}{(x - 1)(x + 1)(x^2 + 1)} = \dfrac{x^8 - 1}{(x^4 - 1)} = x^4 + 1.$

(c) $\Phi_{12}(x) = \dfrac{x^{12} - 1}{(\Phi_1\Phi_2\Phi_3\Phi_6)\Phi_4} = \dfrac{x^{12} - 1}{(x^6 - 1)(x^2 + 1)} = \dfrac{x^6 + 1}{x^2 + 1} = x^4 - x^2 + 1.$

(e) $\Phi_{18}(x) = \dfrac{x^{18} - 1}{(\Phi_1\Phi_2\Phi_3\Phi_6)\Phi_9} = \dfrac{x^{18} - 1}{(x^6 - 1)\Phi_9} = \dfrac{x^{12} + x^6 + 1}{\Phi_9}.$

Now $\Phi_9(x) = \dfrac{x^9 - 1}{\Phi_1\Phi_3} = \dfrac{x^9 - 1}{x^3 - 1} = x^6 + x^3 + 1.$

Finally $\Phi_{18}(x) = \dfrac{x^{12} + x^6 + 1}{x^6 + x^3 + 1} = x^6 - x^3 + 1.$

3. Since n is odd, $d|2n$ if and only if either $d|n$ or $d = 2b$, $b|n$. Thus, by induction

$$x^{2n} - 1 = \prod_{d|2n} \Phi_d(x) = \left(\prod_{d|n} \Phi_d(x)\right)\left(\prod_{d|n} \Phi_{2d}(x)\right) = (x^n - 1)\prod_{d|n} \Phi_{2d}(x).$$

Now observe that $\Phi_2(x) = -\Phi_1(-x)$, and (by induction) $\Phi_{2d}(x) = \Phi_d(-x)$ if $d < n$. Hence

$$x^{2n} - 1 = (x^n - 1)\left[-\prod_{\substack{d|n \\ d \neq n}} \Phi_d(-x)\right] \cdot \Phi_{2n}(x) = \frac{-(x^n-1)\Phi_{2n}(x)}{\Phi_n(-x)}\left[\prod_{d|n}\Phi_d(-x)\right]$$

$$= \frac{-(x^n-1)\Phi_{2n}(x)((-x)^n - 1)}{\Phi_n(-x)} = \frac{-(x^n-1)(-1)(x^n+1)\Phi_{2n}(x)}{\Phi_n(-x)}.$$

Thus $\Phi_n(-x) = \Phi_{2n}(x)$, as required.

5. If $w_k = e^{2\pi i/k}$, these fields are $\mathbb{Q}(w_{mn})$ and $\mathbb{Q}(w_m, w_n)$ respectively. Now $w_{mn}^n = w_m$ and $w_{mn}^m = w_n$, so $\mathbb{Q}(w_m, w_n) \subseteq \mathbb{Q}(w_{mn})$. But $\gcd(m,n) = 1$, so write $1 = rm + sn$; $r, s \in \mathbb{Z}$. Then $\frac{1}{mn} = \frac{r}{n} + \frac{s}{m}$, so $w_{mn} = w_n^r \cdot w_m^s$. Thus $\mathbb{Q}(w_{mn}) \subseteq \mathbb{Q}(w_m, w_n)$.

6. (a) If $S \subseteq R$ is a finite subring of the division ring S, then char $S \neq 0$, say char $S = p$. Let $\mathbb{Z}_p \subseteq S$ and let $\dim_{\mathbb{Z}_p}(S) = n$. Then if $0 \neq s \in S$, $1, s, \dots, s^n$ are not linearly independent over $\mathbb{Z}_p$, so $r_0 + r_1 s + \cdots + r_n s^n = 0$, $r_i \in \mathbb{Z}_p$. We may assume $r_0 \neq 0$ (can cancel s) so

$$s[r_0^{-1}(-r_1 - r_2 s - \cdots - r_n s^{n-1})] = 1.$$

7. Write $\sigma(n) = \sum_{d|n} \mu(d)$. If $n = 1$, $\sigma(n) = \mu(1) = 1$. If $n = p_1^{n_1} p_2^{n_2} \cdots p_r^{n_r}$ then $\mu(d) = 0$ for any $d|n$ with $p_i^2|d$. If we write $m = p_1 p_2 \cdots p_r$, then $\sigma(n) = \sigma(m)$. If $d|m$, then $\mu(d) = 1$ if and only if d is the product of an even number (possibly 0) of the p_i, and $\mu(d) = -1$ otherwise. Since half the divisors d are in each category, $\sigma(m) = 0$.

8. (a) The sums are equal by replacing d by n/d throughout. Use the hint:

$$\sum_{d|n} \mu(d)\alpha\left(\tfrac{n}{d}\right) = \sum_{d|n} \mu(d)\left[\sum_{c|\frac{n}{d}} \beta(c)\right] = \sum_{cd|n} \mu(d)\beta(c)$$

$$= \sum_{c|n} \beta(c)\left[\sum_{d|\frac{n}{c}} \mu(d)\right] = \beta(n),$$

But $\sum_{d|n} \mu(d) = 1$ if and only if $n = c$ by the preceding exercise. We have $x^n - 1 = \prod_{d|n} \Phi_d(x)$. Fix x and take a formal logarithm:

$$\log(x^n - 1) = \sum_{d|n} \log(\Phi_d(x)).$$

Let $\sigma(n) = \log(x^n - 1)$ and $\beta(n) = \log(\Phi_n(x))$. By Möbius we get

$$\log(\Phi_n(x)) = \sum_{d|n} \mu\left(\frac{n}{d}\right) \log(x^d - 1) = \log[\Pi(x^d - 1)^{\mu(n/d)}].$$

The result follows.

If formal logarithms are distasteful, repeat Exercise 8(a) with Σ replaced by Π and all coefficients replaced by exponents.

9. Write $n/m = k$ so $n = km$. Thus $d|m$ if and only if $kd|n$. Now Exercise 8(b) gives

$$\Phi_m(x^k) = \prod_{d|m} [(x^k)^d - 1]^{\mu\left(\frac{m}{d}\right)} = \prod_{kd|n} [x^{kd} - 1]^{\mu\left(\frac{n}{kd}\right)} = \prod_{b|n} (x^b - 1)^{\mu(n/b)}$$

$$= \Phi_n(x).$$

Chapter *11*

Finiteness Conditions for Rings and Modules

11.1 WEDDERBURN'S THEOREM

1. If $_RM$ is simple, then $M = Rx$ for any $0 \neq x \in M$. Hence $\alpha : R \to M$ given by $\alpha(r) = rx$ is an onto R-linear map, so $M \cong R/L$ where $L = \ker(\alpha)$, and L is maximal because M is simple. Conversely, R/L is simple for every maximal left ideal L. Conversely, R/L is simple for every maximal left ideal L again by Theorem 6 §8.1.

3. Define $\sigma : Ra \to Rb$ by $\sigma(ra) = rb$. This is well defined because $ra = 0$ implies $rbR = r(aR) = 0$, so $rb = 0$. So σ is an onto homomorphism of left R-modules with $\sigma(a) = b$. Finally if $\sigma(ra) = 0$ then $rb = 0$ so $(ra)R = r(aR) = r(bR) = 0$, so $ra = 0$. Hence σ is one-to-one.

5. If $L_1 \subseteq L_2 \subseteq \cdots$ are left ideals of eRe, then $RL_1 \subseteq RL_2 \subseteq \cdots$ are left ideals of R so $RL_n = RL_{n+1} = \cdots$ for some n by hypothesis. If $i \geq n$, the fact that $L_i \subseteq eRe$ for each i, gives $L_i = eL_i \subseteq eRL_i = eRL_n = eReL_n \subseteq L_n$. Hence $L_n = L_{n+1} = \cdots$, as required.

7. If $_RM$ is finitely generated, then M is an image of R^n for some $n \geq 1$ by Theorem 5 §7.1 and its Corollaries. Since R^n is noetherian as a left R-module (Corollary 1 of Lemma 2), the same is true of M by Lemma 2.

8. (a) Observe that $\operatorname{ann}(X)$ is a left ideal for any $X \subseteq M$ (note that $\operatorname{ann}(\varnothing) = R$). Hence use the artinian condition to choose $\operatorname{ann}(X)$ minimal in

$$\mathcal{S} = \{\operatorname{ann}(X) \mid X \subseteq M \text{ and } X \text{ is finite}\}.$$

Then $\operatorname{ann}(M) \subseteq \operatorname{ann}(X)$ is clear; we prove equality. If $a \in \operatorname{ann}(X)$, suppose that $a \notin \operatorname{ann}(M)$, say $am \neq 0$ with $m \in M$. Put $Y = X \cup \{m\}$. Then

Student Solution Manual to Accompany Introduction to Abstract Algebra, Fourth Edition. W. Keith Nicholson. © 2012 John Wiley & Sons, Inc. Published 2012 by John Wiley & Sons, Inc.

ann$(Y) \subseteq$ ann(X), so ann$(Y) =$ ann(X) by minimality. But $a \in ann(Y)$ and so $am = 0$, a contradiction. Hence $a \in ann(M)$ and we have proved (a).

9. Let $X = \{\frac{m}{p^n} \mid m \in \mathbb{Z}$ and p does not divide $m\}$. If $\mathbb{Z} \subset Y \subset X$, where Y is a subgroup of X, choose $\frac{m}{p^n} \in Y$ with n maximal (n exists because $Y \neq X$, $X = \mathbb{Z}\frac{1}{p} \cup \mathbb{Z}\frac{1}{p^2} \cup \mathbb{Z}\frac{1}{p^3} \cup \cdots$, and $\mathbb{Z}\frac{1}{p} \subseteq \mathbb{Z}\frac{1}{p^2} \subseteq \mathbb{Z}\frac{1}{p^3} \subseteq \cdots$). Then $\frac{1}{p^n} \in Y$ by Theorem 4 §1.2 because m and p^n are relatively prime. Hence $\mathbb{Z}\frac{1}{p^n} \subseteq Y$; we claim that this is equality. To see this, let $y = \frac{m'}{p^k} \in Y$ where p does not divide m'. Then $k \leq n$ by the maximality of n, so $\frac{1}{p^k} = p^{n-k}\frac{1}{p^n} \in \mathbb{Z}\frac{1}{p^n}$, whence $y \in \mathbb{Z}\frac{1}{p^n}$.

11. (a). If $x \in M$ then $x - \pi(x) \in \ker \pi$, and it follows that $M = \pi(M) + \ker \pi$. If $y \in \pi(M) \cap \ker \pi$ write $y = \pi(z)$, $z \in M$. Since $y \in \ker \pi$, we have $0 = \pi(y) = \pi^2(z) = \pi(z) = y$, so $\pi(M) \cap \ker \pi = 0$. This shows that $M = \pi(M) \oplus \ker \pi$.
(c). We have $\ker \pi = (1 - \pi)(M)$ because $\pi(x) = 0$ if and only if $(1 - \pi)(x) = x$. And $\pi(M) = ker(1 - \pi)$ because $(1 - \pi)(x) = 0$ if and only if $x = \pi(x)$.

13. (a). Suppose that $K = K_1 \oplus K_2 \oplus \cdots$ is an infinite direct sum of nonzero submodules of M. Then $K \supset K_2 \oplus K_3 \oplus \cdots \supset K_3 \oplus K_4 \supset \cdots$ contradicts the DCC, and $K_1 \subset K_1 \oplus K_2 \subset K_1 \oplus K_2 \oplus K_3 \subset \cdots$ contradicts the ACC.

15. Here $R = \begin{bmatrix} F & F \\ 0 & F \end{bmatrix}$, and we consider $e = \begin{bmatrix} 0 & 0 \\ 0 & 1 \end{bmatrix}$. Then $e^2 = e$ and $Re = \begin{bmatrix} 0 & F \\ 0 & F \end{bmatrix}$ is not simple as $\begin{bmatrix} 0 & F \\ 0 & 0 \end{bmatrix}$ is a submodule. But $end(Re) \cong eRe = \begin{bmatrix} 0 & 0 \\ 0 & F \end{bmatrix} \cong F$ as rings.

17. (a). We have $\ker(\alpha) \subseteq \ker(\alpha^2) \subseteq \ker(\alpha^3) \subseteq \cdots$ so, since M is noetherian, there exists $n \geq 1$ such that $\ker(\alpha^n) = \ker(\alpha^{n+1}) = \cdots$. If $x \in \ker(\alpha)$ then (since α^n is onto) write $x = \alpha^n(y)$ for some $y \in M$. Then $0 = \alpha(x) = \alpha^{n+1}(y)$, so $y \in \ker(\alpha^{n+1}) = \ker(\alpha^n)$. Hence $x = \alpha^n(y) = 0$, proving that $\ker(\alpha) = 0$.

19. (a). The easy verification that θ is an F-linear homomorphism of additive groups is left to the reader. To show that θ is one-to-one, suppose that $\theta(r) = 0$; that is, $u_i r = 0$ for each i. If $1 = \sum_i a_i u_i$, then $r = 1 \cdot r = \sum_i a_i u_i r = 0$, so $\ker \theta = 0$ and θ is one-to-one. Finally, let $\theta(s) = [s_{ij}]$ so that $u_i s = \sum_{j=1}^n s_{ij} u_j$. Then:

$$u_i rs = \left(\sum_k r_{ik} u_k\right) s = \sum_k r_{ik} \left(\sum_j s_{kj} u_j\right) = \sum_j \left(\sum_k r_{ik} s_{kj}\right) u_j.$$

Thus $\theta(rs) = [\sum_k r_{ik} s_{kj}] = [r_{ij}][s_{ij}] = \theta(r) \cdot \theta(s)$, and the proof is complete.

(c). $a + bi + cj + dk \mapsto \begin{bmatrix} a & b & c & d \\ -b & a & d & -c \\ -c & -d & a & b \\ -d & c & -b & a \end{bmatrix}$

(e). $a + bu \mapsto \begin{bmatrix} a & b \\ b & a \end{bmatrix}$

11.2 THE WEDDERBURN-ARTIN THEOREM

1. The $\mathbb{Z}$-modules are the (additive) abelian groups, so the simple ones are the simple abelian groups. These are the prime cycles $\mathbb{Z}_p$ (in additive notation)

where p is a prime. Hence the semisimple $\mathbb{Z}$-modules are the the direct sums of copies of these $\mathbb{Z}_p$ for various primes p. The homogeneous semisimple $\mathbb{Z}$-modules are the direct sums of copies of $\mathbb{Z}_p$ for a fixed prime p.

2. (a). Since $_RR$ is complemented by Theorem 1, we have $R = L \oplus M$ for some left ideal M. Write $1 = e + f$ where $e \in L$ and $f \in M$. Hence $Re \subseteq L$, and we claim this is equality. If $x \in L$ then $x - xe = xf \in L \cap M = 0$ because $x - xe \in L$ and $xf \in M$. Hence $x = xe \in Re$, so $L \subseteq Re$, as required.

3. (a). Let $\alpha, \beta \in E$. Then $(\alpha + \beta) \cdot x = (\alpha + \beta)(x) = \alpha(x) + \beta(x) = \alpha \cdot x + \beta \cdot x$ so Axiom M2 holds (see Section 11.1). Similarly,

$$(\alpha\beta) \cdot x = (\alpha\beta)(x) = \alpha[\beta(x)] = \alpha \cdot (\beta \cdot x)$$

proves Axiom M3. The other axioms are routinely verified.

5. Let $N_1, N_2, \ldots, N_m$ be maximal submodules of M. Define

$$\alpha : M \to \frac{M}{N_1} \oplus \frac{M}{N_2} \oplus \cdots \oplus \frac{M}{N_m}$$

by $\alpha(x) = (x + N_1, x + N_2, \ldots, x + N_m)$ for all $x \in M$. Then α is R-linear and $\ker(\alpha) = \{x \in M \mid x + N_i = 0 \text{ for each } i\} = \cap_i N_i$. Since each $\frac{M}{N_i}$ is simple, $M/(\cap_i N_i) \cong \alpha(M)$ is semisimple by Corollary 1 of Theorem 2.

7. If M is a finitely generated semisimple module then (Lemma 4) M is a finite direct sum of simple submodules. By Theorem 2 $M = H_1 \oplus \cdots \oplus H_n$ where the H_i are the homogeneous components of M. By the preceding exercise, $\mathrm{end}(M) \cong \mathrm{end}(H_1) \times \cdots \times \mathrm{end}(H_n)$, so it suffices to show that $\mathrm{end}(H)$ is semisimple for any finitely generated, homogeneous, semisimple module H. But then H is a finite direct sum of isomorphic simple modules (using Lemma 4), and so $H \cong K^n$ for some simple module K. Now Lemma 3 §11.1 gives $\mathrm{end}(H) \cong M_n(\mathrm{end}\, K)$. Since $\mathrm{end}\, K$ is a division ring by Schur's lemma, we are done.

9. Let K be a simple left ideal of R. If R is a domain and then $K^2 \neq 0$ so Brauer's lemma shows that $K = Re$ where $e^2 = e$. But $e(1 - e) = 0$ so, again since R is a domain, $e = 0$ or $e = 1$. Since $K \neq 0$ we must have $e = 1$, whence $R = Re$ is simple as a left R-module. But then, given $0 \neq a \in R$, we have $Ra = R$, say $ba = 1$. Again, $b \neq 0$ so $Rb = R$, say $cb = 1$. Now compute:

$$a = 1a = (cb)a = c(ba) = c1 = c.$$

Hence $ab = cb = 1$ and we have proved that a is a unit with inverse b. Since $a \neq 0$ was arbitrary, this shows that R is a division ring.

11. If A is an ideal of R, we must show that every left ideal $\mathcal{L}$ of the ring R/A is semisimple as an R/A-module. But $\mathcal{L}$ is an R-submodule of the left R-module R/A because $(r + A)(s + A) = rs + A = r(s + A)$. Hence $\mathcal{L}$ is semisimple as a left R-module by Corollary 1 of Theorem 1 (since $_RR$ is semisimple). Since the R-action and the R/A-action on $\mathcal{L}$ are the same, it follows that $\mathcal{L}$ is semisimple as a left R/A-submodule. This is what we wanted.

13. (2) $\Rightarrow$ (1). Assume that eRe is a division ring, and that R is semiprime. Choose $0 \neq x \in Re$; we must show that $Rx = Re$, that is $e \in Rx$. Since R is semiprime we have $xRx \neq 0$, say $xax \neq 0$, $a \in R$. Since $x \in Re$, we have $xe = x$, so $xe(ax) \neq 0$. This means $eax \neq 0$ so $eaxe \neq 0$ (again since $x = xe$). As

eRe is a division ring, there exists $t \in eRe$ such that $t(eax) = e$. Thus $e \in Rx$, as required.

15. Let $_R R$ be semisimple. Then Lemma 4 implies that $R = K_1 \oplus K_2 \oplus \cdots \oplus K_n$ where the K_i are simple left ideals. Hence R is left noetherian by the Corollary to Lemma 2 §11.1. Since R_R is also semisimple by Theorem 5, the same argument shows that R is right noetherian. [Note that this also shows that R is left and right artinian.]

16. Let R be a semiprime ring.

    (a). If $LM = 0$ where L and M are left ideals, then
    $$(ML)^2 = MLML = M0L = 0.$$
    Since ML is a left ideal and R is semiprime, this gives $ML = 0$.

    (c). If $rA = 0$ then $(Ar)^2 = ArAr = A0r = 0$, Hence $Ar = 0$ because Ar is a left ideal and R is semiprime. The converse is similar.

17. Write $R = R_1 \times R_2 \times \cdots \times R_n$. By Exercise 4, every ideal A of has the form $A = A_1 \times A_2 \times \cdots \times A_n$ where each A_i an ideal of R_i. Hence $A^2 = 0$ if and only if $A_i^2 = 0$ for each i. The result follows.

19. If R is semiprime, let $\mathcal{A}$ be an ideal of $M_n(R)$. By Lemma 3 §3.3, $\mathcal{A}$ has the form $\mathcal{A} = M_n(A)$ for some ideal A of R. Hence if $\mathcal{A}^2 = 0$ then $A^2 = 0$, whence $A = 0$, so $\mathcal{A} = 0$. This shows that $M_n(R)$ is semiprime for any $n \geq 1$. Conversely, assume that $M_k(R)$ is semiprime for some k. If $A^2 = 0$ where A is an ideal of R, then $M_k(A)$ is an ideal of $M_k(R)$ and $M_k(A)^2 = 0$. Hence $M_k(A) = 0$, whence $A = 0$. This shows that R is semiprime.

20. (a). If $ab = 0$ in R then $(Ra)(Rb) = 0$ because R is commutative. Hence, if R is prime either $Ra = 0$ or $Rb = 0$, that is $a = 0$ or $b = 0$. Thus R is a domain. Conversely, if R is a domain and $AB = 0$ then, if $A \neq 0 \neq B$, choose $0 \neq a \in A$ and $0 \neq b \in B$. Then $ab \in AB = 0$, a contradiction. So either $A = 0$ or $B = 0$; that is R is a prime ring.

21. Write $P = P_1 \oplus P_2 \oplus \cdots \oplus P_n$, and view P as an internal direct sum. Then each P_i is isomorphic to a direct summand of a free module F_i; by Lemma 9 we may assume that $F_i = P_i \oplus Q_i$. Then $F = F_1 \oplus F_2 \oplus \cdots \oplus F_n$ is also free, and $F = (P_1 \oplus P_2 \oplus \cdots \oplus P_n) \oplus (Q_1 \oplus Q_2 \oplus \cdots \oplus Q_n)$. Hence $P_1 \oplus P_2 \oplus \cdots \oplus P_n$ is projective by Theorem 3.

22. (a). If $\alpha : M \to N$ is R-linear and $K \subseteq M$ is simple then either $\alpha(K) = 0$ or $\gamma(K) \cong K$ by Schur's lemma. Either way, $\alpha(K) \subseteq \operatorname{soc} N$. Since $\operatorname{soc} M$ is a sum of simple submodules of M, it follows that $\alpha(\operatorname{soc} M) \subseteq \operatorname{soc} N$.

    (c). If $M = N_1 \oplus N_1$ it is clear that $\operatorname{soc}(N_1) \oplus \operatorname{soc}(N_2) \subseteq \operatorname{soc} M$ because $\operatorname{soc}(N_i)$ is semisimple for each i. For the other inclusion, define the projection $\pi_1 : M \to N_1$ by $\pi_1(n_1 + n_2) = n_1$, and define $\pi_2 : M \to N_2$ analogously. If $K \subseteq M$ is any simple module then $\pi_i(K) \subseteq \operatorname{soc} N_i$ for each i [$\pi_i(K)$ is either 0 or isomorphic to K by Schur's lemma]. Hence if $k \in K$ then $k = \pi_1(k) + \pi_2(k) \in \operatorname{soc}(N_i) \oplus \operatorname{soc}(N_i)$. It follows that
    $$\operatorname{soc}(M) \subseteq \operatorname{soc}(N_1) \oplus \operatorname{soc}(N_2).$$

Appendices

APPENDIX A: COMPLEX NUMBERS

1. (a) $x = 3$ (c) $x = 0$, $x = 4i$

2. (a) $7 - 9i$ (c) $-\frac{6}{13} + \frac{4}{13}i$ (e) $-i$ (g) -4

3. (a) $z = 1 - 3i$ (c) $z = \pm\frac{1}{\sqrt{2}}(1 - i)$ (e) $z = 2 + 3i$

4. Write $z = a + bi$ and $w = c + di$.

   (a) $\operatorname{im}(iz) = \operatorname{im}(-b + ai) = a = \operatorname{re} z$.

   (c) $z + \bar{z} = 2a = 2\operatorname{re} z$.

   (e) $\operatorname{re}(z + w) = a + c = \operatorname{re} z + \operatorname{re} w$; $\operatorname{re}(tz) = ta = t\operatorname{re} z$.

5. (a) unit circle   (c) line $y = x$ (e) 0 and the positive real axis

7. $|w + z|^2 = (w + z)\overline{(w + z)} = (w + z)(\bar{w} + \bar{z}) = w\bar{w} + z\bar{z} + w\bar{z} + z\bar{w}$
   $$= |w|^2 + |z|^2 + w\bar{z} + z\bar{w}.$$

9. (a) If $z = a + bi$ and $w = c + di$, then $z - w = (a - c) + (b - d)i$, so
   $$|z - w|^2 = (a - c)^2 + (b - d)^2.$$

   Take positive square roots.

10. (a) $3\sqrt{2}\,e^{-\pi i/4}$ (c) $2e^{5\pi i/6}$ (e) $7e^{-\pi i/2}$

11. (a) -3 (c) $-\sqrt{2} + \sqrt{2}i$ (e) $-\frac{1}{\sqrt{2}} - \frac{1}{\sqrt{2}}i$

12. (a) $-2 - 2\sqrt{3}i$ (c) 16   (e) -64

13. (a) $\cos 2\theta + i\sin 2\theta = e^{2i\theta} = \left(e^{i\theta}\right)^2 = (\cos\theta + i\sin\theta)^2$
    $$= (\cos^2\theta - \sin^2\theta) + i(2\cos\theta\sin\theta)$$

14. (a) $\pm\frac{1}{\sqrt{2}}(1 + i)$, $\pm\frac{1}{\sqrt{2}}(1 - i)$

    (c) $3i$, $\frac{3}{2}(\sqrt{3} - i)$, $\frac{3}{2}(-\sqrt{3} - i)$

Student Solution Manual to Accompany Introduction to Abstract Algebra, Fourth Edition.
W. Keith Nicholson.
© 2012 John Wiley & Sons, Inc. Published 2012 by John Wiley & Sons, Inc.

15. (a) $z = r(\cos\theta + \sin\theta)$ so $\bar{z} = r(\cos\theta - \sin\theta)$. But $\sin(-\theta) = -\sin\theta$ and $\cos(-\theta) = \cos\theta$, so $re^{-i\theta} = r[\cos(-\theta) + i\sin(-\theta)] = \bar{z}$. If $z^{-1} = se^{i\phi}$, then $1 = zz^{-1} = rse^{i(\theta+\phi)}$. Thus $rs = 1$, so $s = \frac{1}{r}$, and $\theta + \phi = 0$ so (one choice for ϕ is) $\phi = -\theta$. Hence $z^{-1} = \frac{1}{r}e^{-i\theta} = \frac{1}{r}\bar{z}$.

16. (a) Let $w = e^{2\pi i/n}$ so that the kth root of unity is $w_k = w^k$ by DeMoivre's theorem. Now $(1 - w^n) = (1 - w)(1 + w + w^2 + \cdots + w^{n-1})$ by the Hint so,

$$(1 - w)(1 + w + w^2 + \cdots + w^{n-1}) = 0$$

because $w^n = 1$. Since $w \neq 1$ this implies that $1 + w + w^2 + \cdots + w^{n-1} = 0$, as required.

17. (a) Have $z_i = e^{i\theta_i}$ for angles θ_i. The angles between them all equal $\beta = \frac{2\pi}{5}$ (because they are equally spaced). Let $\theta_1 = \alpha$ as in the diagram. Then

$$z_1 = e^{i\alpha},\ z_2 = e^{i(\alpha+\beta)},\ z_3 = e^{i(\alpha+2\beta)},$$
$$z_4 = e^{i(\alpha+3\beta)},\ z_5 = e^{i(\alpha+4\beta)}.$$

If we write $z = e^{i\beta}$, then $z^5 = 1$. Now use the hint:

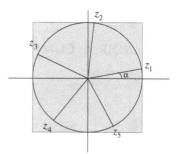

$$\begin{aligned} & z_1 + z_2 + z_3 + z_4 + z_5 \\ &= e^{i\alpha}(1 + z + z^2 + z^3 + z^4) \\ &= e^{i\alpha}\left(\frac{1 - z^5}{1 - z}\right) = 0. \end{aligned}$$

19. Let $f(x) = \sum_{i=1}^{n} a_i x^i$, $a_i \in \mathbb{R}$. If $z \in \mathbb{C}$ is a root of $f(x)$, then $f(z) = 0$. Hence, $0 = \bar{0} = \overline{f(z)} = \overline{\sum_{i=1}^{n} a_i z^i} = \sum_{i=1}^{n} \bar{a}_i \bar{z}^i = \sum_{i=1}^{n} a_i \bar{z}^i = f(\bar{z})$ because $\bar{a}_i = a_i$ for each i (being real).

21. Let $z = re^{i\theta}$. (a) If $t > 0$, $tz = tre^{i\theta}$ has the same angle θ as z. Hence tz is on the line through 0 and z, on the same side of 0 as z. (b) If $t = -s$, $s > 0$, then $tz = sre^{i(\theta+\pi)}$, and so is on the line through 0 and z, on the other side of 0 from z.

23. (a) If $a + b\sqrt{2} = a_1 + b_1\sqrt{2}$, then $a - a_1 = (b_1 - b)\sqrt{2}$. If $b_1 \neq b$, then $\sqrt{2} = \frac{a-a_1}{b_1-b}$ is rational, an impossibility. Hence, $b_1 = b$ whence $a = a_1$.

(c) $pq = (ac + 2bd) + (ad + bc)\sqrt{2}$, so:

$$\widetilde{pq} = (ac + 2bd) - (ad + bc)\sqrt{2} = (a - b\sqrt{2})(c - d\sqrt{2}) = \tilde{p}\tilde{q}.$$

(e) Using (c) and (d) above: $[pq] = (pq)\widetilde{(pq)} = pq\tilde{p}\tilde{q} = (p\tilde{p})(q\tilde{q}) = [p][q]$.

APPENDIX B: MATRIX ARITHMETIC

1. If A is invertible then $B = IB = (A^{-1}A)B = A^{-1}0 = 0$, contrary to assumption.

2. (a) By the definition of matrix multiplication column k of AB is exactly AB_k.

3. $AB = I_2$ but $BA = \begin{bmatrix} -9 & -10 & 5 \\ -6 & -5 & 3 \\ -30 & -30 & 16 \end{bmatrix}$.

5. $A = \begin{bmatrix} 1 & 0 \\ 0 & 1 \end{bmatrix}$ and $B = \begin{bmatrix} 1 & 0 \\ 0 & -1 \end{bmatrix}$ are both invertible (in fact $A^{-1} = A$ and $B^{-1} = B$), but $A + B = \begin{bmatrix} 2 & 0 \\ 0 & 0 \end{bmatrix}$ is not invertible by Theorem 3 because $\det(A + B) = 0$. For a simpler example, take $A = U$ and $B = -U$ for any invertible matrix U.

7. $(A + B)(A - B) = A^2 + AB - BA - B^2$, so $(A - B)(A + B) = A^2 - B^2$ if and only if $AB - BA = 0$.

9. It is routine that $A^3 = I$, so $AA^2 = I = A^2A$. This implies that $A^{-1} = A^2$.

11. (a) $AA^{-1} = I = A^{-1}A$ shows A is the inverse of A^{-1}, that is $(A^{-1})^{-1} = A$. Next, $(AB)(B^{-1}A^{-1}) = AIA^{-1} = AA^{-1} = I$, and similarly $(B^{-1}A^{-1})(AB) = I$. Hence AB is invertible and $(AB)^{-1} = B^{-1}A^{-1}$.

13. (a) If $AB = I$ then $\det A \det B = \det(AB) = \det I = 1$. Hence $\det A$ is a unit so, by Theorem 7, A is invertible. But then $A^{-1} = A^{-1}I = A^{-1}(AB) = B$, whence $BA = A^{-1}A = I$ as required.

15. (a) The only nonzero entry in $E_{ik}E_{mj}$ must come from entry k of row i of E_{ik} and column m of E_{mj}. The result follows.

   (c) For each i and j, $a_{ij}E_{ij}$ has a_{ij} in the (i, j)-entry and zeros elsewhere. So their sum is $[a_{ij}]$.

APPENDIX C: ZORN'S LEMMA

1. Let $K \subseteq M$ be modules, M finitely generated, say
$$M = Rx_1 + Rx_2 + \cdots + Rx_n.$$
If $\mathcal{S} = \{X \subseteq M \mid K \subseteq X \subset M\}$, we must show that $\mathcal{S}$ contains maximal members. Suppose $\{X_i \mid i \in I\}$ is a chain from $\mathcal{S}$, and put $U = \cup_{i \in I} X_i$. It is clear that U is a submodule and that $K \subseteq U$, and we claim that $U \neq M$. For if $U = M$ then each $x_i \in U$, and so each $x_i \in X_k$ for some k. Since the X_i form a chain, this means that $\{x_1, x_2, \ldots, x_n\} \subseteq X_m$ for some m. Since the x_i generate M this means that $M \subseteq X_m$, contradicting the fact that $X_m \in \mathcal{S}$. This shows that $U \in \mathcal{S}$, and so U is an upper bound for the X_i. Hence $\mathcal{S}$ has maximal members by Zorn's lemma, as required.

2. Let $K \subseteq M$ be modules.
   (a) Let $\mathcal{S} = \{X \subseteq M \mid X \text{ is a submodule and } K \cap X = 0\}$. Then $\mathcal{S}$ is nonempty because $0 \in \mathcal{S}$, so let $\{X_i \mid i \in I\}$ be a chain from $\mathcal{S}$ and put $U = \cup_{i \in I} X_i$. It is clear that U is a submodule, and $K \cap U = 0$ because $K \cap U \subseteq K \cap X_i = 0$ for each i. Hence U is an upper bound for the chain $\{X_i \mid i \in I\}$, so $\mathcal{S}$ contains maximal members by Zorn's lemma.

3. Let $\mathcal{S}$ be the set of all prime ideals of R, and partially order $\mathcal{S}$ downward: Let $P \leq Q$ mean $P \supseteq Q$. We must find a maximal element in $\mathcal{S}$. Let $\{P_i \mid i \in I\}$ be a chain from $\mathcal{S}$ and put $Q = \cap_{i \in I} P_i$. We claim that Q is an upper bound

on $\{P_i \mid i \in I\}$. Clearly $Q \subseteq P_i$ for each i, so it remains to show that Q is a prime ideal. If $rs \in Q$ where $r, s \in R$; we must show that either $r \in Q$ or $s \in Q$. Suppose on the contrary that $r \notin Q$ and $s \notin Q$. Then $r \notin P_i$ for some i, and $s \notin P_j$ for some j. Since the P_i form a chain, one of $P_i \subseteq P_j$ or $P_j \subseteq P_i$ must hold; assume $P_i \subseteq P_j$. Then $r \notin P_j$ and $s \notin P_j$ but $rs \in P_j$ (because $rs \in Q \subseteq P_j$). This contradicts the fact that P_j is a prime ideal. Since we also obtain a contradiction if $P_j \subseteq P_i$, this proves that Q is a prime ideal.

Printed in the USA
CPSIA information can be obtained
at www.ICGtesting.com
CBHW081927151024
15715CB00004B/8